Topics in Biodiversity and Conservation

Volume 16

More information about this series at http://www.springer.com/series/7488

Russell Paterson • Nelson Lima
Editors

Bioprospecting

Success, Potential and Constraints

Springer

Editors
Russell Paterson
Centre of Biological Engineering
University of Minho
Braga, Portugal

Nelson Lima
Centre of Biological Engineering
University of Minho
Braga, Portugal

ISSN 1875-1288 ISSN 1875-1296 (electronic)
Topics in Biodiversity and Conservation
ISBN 978-3-319-47933-0 ISBN 978-3-319-47935-4 (eBook)
DOI 10.1007/978-3-319-47935-4

Library of Congress Control Number: 2016961311

Printed on acid-free paper

This Springer imprint is published by Springer Nature
The registered company is Springer International Publishing AG
The registered company address is: Gewerbestrasse 11, 6330 Cham, Switzerland

Foreword

Living organisms have been tested for their potential as food and medicines from the earliest days of humankind. However, it was really only after the spectacular demonstrations in the 1930s and 1940s of the efficacy of penicillin, first named in 1929 from what we now know to have been *Penicillium rubens* (Houbraken et al. 2011), that industry started to look for novel exploitable products in earnest. A gold rush started, and bacteria and fungi in particular were isolated in huge numbers from soils and other substrata worldwide. The penicillins were soon followed by other wonder drugs such as the cephalosporins, erythromycin, nystatin and streptomycin. Major pharmaceutical companies across the world had massive screening programmes through into the early 1990s, some employing staff overseas to isolate strains and others requesting staff to collect soil samples whilst on vacation around the world. Tens of thousands of strains were being screened annually by major companies for targeted activities, often with highly automated systems. Databases of known natural products were also developed which meant that active compounds found could be quickly checked to see if they were novel. Further spectacular discoveries that resulted from bacteria and fungi included the immunosuppressant cyclosporins that make transplant surgery safer than ever before, antifungal azole drugs, strobilurin fungicides, antifungal pneumocandins, anticancer taxol and cholesterol-lowering statins. Plants have given us the pyrethroid insecticides, anticancer paclitaxel and Alzheimer's-mediating galantamine, amongst others (Atanasov et al. 2015). Living organisms, especially insects and fungi able to attack insect pests, were sought in countries where they were native and transported for the biocontrol of the pests and diseases of crops in glasshouses and in the fields; *Metarhizium acridum,* for example, is the active component of the biopesticides such as Green Muscle used for the control of desert locust. Interest has also extended to animal products, with antiplatelet drug tirofiban from venom being of particular note. Industry has also been active in the search for exploitable enzymes and organisms that might be of value in the bioremediation of oil- and metal-contaminated soils.

In searching for organisms with potentially exploitable attributes, attention has also been directed to medicinal uses of plants and macrofungi which have traditionally been used by different tribal groups but where the actual efficacy and nature of the active ingredients were unknown. Companies were further aware that many described organisms had never been screened for any useful attributes and that only a small proportion of the estimated biodiversity on the planet had been characterized and named, probably less than 5 % in the case of the fungi.

Against this background, there was much excitement, with many small venture-capital companies becoming established to look for new wonder drugs, often with particular target organisms in mind. This search for novel and exploitable products started to be referred to as biodiversity prospecting, or bioprospecting, in the early 1990s (Reid et al. 1993) following the heightened global interest in the conservation of biodiversity. The Convention on Biological Diversity (UNEP 1992) was coming into force, and amongst the objectives cited in Article 1 was "the fair and equitable sharing of benefits arising out of the utilization of genetic resources". This was to have unforeseen consequences for the bioprospecting companies. The expectation of many less developed countries was high, as they felt they should have a right to a proportion of the profits made from organisms taken from their territories. A move to restrict access to national biological resources soon started to be developed in several tropical countries. Matters were inflamed by attempts by more developed countries to take out patents on not only new discoveries but traditional practices, such as applications of the neem tree and the use of ayahuasca and turmeric. Major multinational pharmaceutical companies started to be charged with "biopiracy", a term coined by Canadian Pat Mooney (2000). Anxious not to tarnish their names and risking reduced sales, the response of the largest pharmacological and drug discovery companies was to stop all work on natural products. Staff were made redundant, and in some cases, substantial collections of living material built up over decades for screening purposes were destroyed. Instead, companies increasingly turned to combinatorial chemistry, constructing novel compounds and screening them for desired activities.

The potential problems and complexity for bioprospecting, those involved in plant breeding and bodies holding ex situ collections were recognized already in the 1990s (Kirsop and Hawksworth 1994; Lesser 1998) and have become increasingly complex as countries started to formulate national policies and regulations on access to genetic resources (i.e. living organisms) and benefit sharing. Various authors have struggled with this issue, some making constructive proposals (e.g. Sampath 2005), not least the parties to the Convention on Biological Diversity. The matter has come to a head in the last 2 years with the adoption by the parties to the Convention of the Nagoya Protocol on "Access to Genetic Resources and the Fair and Equitable Sharing of Benefits Arising from Their Utilization" (Convention on Biological Diversity 2011) which came into force in October 2014. This has led to great concern in the bioprospecting community as, whilst well intentioned, it will inevitably hinder drug discovery from natural products because of the bureaucracy needing to be put in place for each sample or group of samples (Verkeley 2015) and furthermore restrict the traditional exchange of cultures or seeds for research purposes.

This is a most unfortunate situation to have arisen at a time when antibiotic resistance to both bacteria and certain fungi is emerging as a major global health problem.

There are now few antimicrobials on the way to becoming available (Butler and Cooper 2012), and we may soon have no drugs available to attack strains of bacteria and fungi that have evolved resistance to widely used. This is now a major concern of health authorities worldwide, and the UK's Chief Medical Officer has called for antimicrobial resistance to be placed on the "national risk register and taken seriously by politicians at an international level, including the G8 and World Health Organisation" (Davies 2013).

In some countries, the search for new natural products is compounded by plant health regulations that require living organisms being brought into a country to first be named, and then to have a risk assessment made, something that is impossible for ones that are new to science and as yet have no names, i.e. perhaps those most likely to generate compounds of interest. As pointed out elsewhere (Hawksworth and Dentinger 2013), such restrictions are misplaced when transport of propagules on goods, vehicles and people is uncontrolled. The risk of an organism escaping from a sterile laboratory environment where it is being handled according to standard microbiological safety protocols for unknowns in a safety cabinet is negligible in comparison.

As described in this volume, there are enormous opportunities for the discovery of novel organisms, and the screening of those can provide new and hitherto unsuspected candidates for further testing. Further, despite the constraints mentioned above, examples are given here of some encouraging flagship projects in which organizations have gone to considerable trouble to make arrangements complying with national regulations with industrial partners and scientists to facilitate bioprospecting. Sadly, the need to conclude formal agreements before any living material can be moved to a second country inevitably deters serendipitous discovery, when no arrangements were made in advance of a visit as no collecting was anticipated; a scientist on holiday not expecting to find anything of interest, for example, will now often be unable to bring that material out of the country legally for further investigation.

The potential of what is "out there" is indisputably enormous. Indeed, there is every reason to expect more wonder drugs to be found, but neither what organisms they will be from nor what their activity will be can be forecast. The contributions here show something of the potential, but also what can still be done under the new legal constraints. They should serve as encouragement to others working in the field and guidelines to examples of arrangements that can be made. For the moment, the editors are to be congratulated on bringing such a diverse collection of studies together, at a time when investigators are having to re-evaluate how they operate and nations are drawing up their regulations under the Nagoya Protocol. In time, however, I anticipate that as the urgency to discover more antimicrobials and ways of limiting crop losses through pests and disease rises up political agendas, we may find regulation easing to the long-term benefit of not just bioprospecting companies but the people of all nations.

Royal Botanic Gardens, Kew, UK
5 September 2016

David L. Hawksworth

References

Atanasov AG, Waltenberger B, Pferschy-Wenzig E-M, Linder T, Wawrosch C, Uhrin P, Temml V, Wang L, Schwaiger S, Heiss EH, Rollinger JM, Schuster D, Breuss JM, Bochkov V, Mihovilovic MD, Kopp B, Bauer R, Dirsch VM, Stumppner H (2015) Discovery and resupply of pharmacologically active plant-derived natural products: a review. Biotech Adv 33:1528–1614

Butler MS, Cooper MA (2012) Screening strategies to develop new antibiotics. Curr Drug Targets 13: 373–387

Convention on Biological Diversity (2011) Nagoya protocol on access to genetic resources and the fair and equitable sharing of benefits arising from their utilization to the convention on biological diversity. UNEP, Montreal

Davies S (2013) Annual Report of the Chief Medical Officer 2011. Vol. 2. Department of Health, London

Hawksworth DL, Dentinger BTM (2013) Antibiotics: relax UK import rule on fungi. Nature 496:169

Houbraken J, Frisvad JC, Samson RA (2011) Fleming's penicillin producing strain is not *Penicillium chrysogenum* but *P. rubens*. IMA Fungus 2:87–95

Kirsop B, Hawksworth DL (eds) (1994) The biodiversity of microorganisms and the role of microbial resource centres. World Federation for Culture Collections, Braunschweig

Lesser W (1998) Sustainable use of genetic resources under the convention on biological diversity: exploring access and benefit sharing issues. CAB International, Wallingford

Mooney P (2000) Why we call it biopiracy. In: Svarstad H, Dhillion S (eds) Bioprospecting: from biodiversity in the South to medicines in the North. Spartacus Forlag, Oslo, pp 37

Reid WA, Laird SA, Meyer CA, Gámez R, Sittenfeld A, Janzen DH, Gollin MA, Juma C (eds) (1993) Biodiversity prospecting: using genetic resources for sustainable development. World Resources Institute, Washington DC

Sampath PG (2005) Regulating Bioprospecting: institutions for doing research, access, and benefit sharing. United Nations University Press, Tokyo

UNEP (1992) Convention on biological diversity. United Nations Environment Programme, Nairobi

Verkeley G (2015) How will the nagoya protocol affect our daily work? IMA Fungus 6:3–5

Contents

Chapter 1
Bioprospecting: An Industrial Perspective

Robert P. Borris

Abstract The discovery of promising new biologically active molecules from natural products was a mainstay of the pharmaceutical industry for many years. While interest in this field has fluctuated over the years, it entered a boom period following the Earth Summit and Rio Convention in the early 1990s lasting into the twenty-first century. This boom period was followed by the industry's almost complete exit from this field of endeavor. This chapter presents a discussion of the changes in the capabilities of natural products research and factors contributing to the demise of natural products in the drug discovery programs in Big Pharma.

1.1 Introduction

When first approached about writing this chapter I had very mixed emotions. Over 20 years ago, I had the privilege of contributing a lecture entitled, "Natural Products Research: Perspectives from a Major Pharmaceutical Company," at a conference on *Intellectual Property Rights, Naturally-derived Bioactive Compounds and Resource Conservation* held in San Jose, Costa Rica (during October 1994). That paper was eventually published (Borris 1996), and has often been cited by later authors. Much has happened since that paper was written. Most, if not all, of the major American pharmaceutical corporations (including my former employer, Merck) have exited the field of natural products research, at least as in house ventures, not even maintaining a small core of scientists experienced in the field to resuscitate their programs at a later date. The technologies that enable this kind of research continue to evolve at an ever accelerating rate. It could be an exciting time to be a young natural products researcher! It should be an exciting time to be a young natural products researcher! Is it?

Of necessity, this chapter will not be a review of the role industry has played in biodiversity prospecting. Little information has been published on screening strategies, hit rates, successes and failures to make such a review meaningful. It is, as the

R.P. Borris (✉)
School of Pharmaceutical Science and Technology, Tianjin University,
Bldg 24/A206, 92 Weijin Road, Nankai District, Tianjin 300072, China
e-mail: rborris@tju.edu.cn

R. Paterson, N. Lima (eds.), *Bioprospecting*, Topics in Biodiversity and Conservation 16, DOI 10.1007/978-3-319-47935-4_1

title suggests, one perspective of the industrial experience from a person who has spent an entire career in this field. Any opinions expressed are my own and do not necessarily express the opinion of past or present employers.

This chapter will compare the current state of natural products research with the state of the art at the height of the biodiversity boom in the mid-1990s. At that time, it was still relatively unusual for an academic laboratory to be involved in the patenting of inventions or other forms of protecting their intellectual property. This has changed dramatically. Some granting agencies now require a plan for the protection of intellectual property and many labs routinely apply for patents on their promising discoveries. Many faculty members and their institutions have formed startup companies to develop these inventions. In certain ways this trend is making academic labs take on a more industrial perspective. The advancement of science may still be the primary driving force of academic research, but the allure of an eventual (big) payday is also an incentive.

One often ignored fact must be mentioned. In the late twentieth century, and perhaps still today, industrial researchers were expected to contribute to the understanding of the basic science of their disciplines, not just the applied research leading to product discovery. Much energy went into the discovery of molecules that helped to elucidate the nature of biological targets under investigation, driving forward our understanding of the etiology of disease.

It is difficult to overstate the continuing importance of natural products in the drug discovery process. In the most recent of their reviews on the subject (Newman and Cragg 2012), David Newman and Gordon Cragg have shown that over the 30 year period from 1981 to 2010, 33.9 % of all small molecule new chemical entities (NCEs) approved by the U.S. Food and Drug Adninistration (FDA) were either natural products or compounds derived from natural products. In contrast to the 363 natural products or natural product derivatives, to the best of my knowledge only one approved compound, sorafenib (Nexavar), has been produced by *de novo* combinatorial chemistry since the advent of that technique (Wilhelm et al. 2008). The interested reader is referred to the excellent series of Newman and Cragg reviews on natural product drug discovery (Newman and Cragg 2007, 2012; Cragg et al. 1997; Newman et al. 2003), the detail of which is clearly beyond the scope of this chapter.

1.2 Pharmaceutical Research and Development

It is a given that the *raison d'etre* for research at a pharmaceutical company is to discover and develop new drugs that will address medical needs and make a profit for the shareholders. Even at their highest point, natural product research programs were only a small part of the drug discovery effort. As one of the earliest stages of the discovery process, natural product discovery was often the source of the original hit in a new bioassay or therapeutic area. Compounds discovered from natural sources had the potential to become actual products, as did lovastatin (Mevacor), but more frequently served as leads for medicinal chemistry programs or tools for assay refinement. While a lead compound had to demonstrate reasonable potency and some specificity, it was expected that the medicinal chemist would modify the structure to build in better potency, selectivity, and appropriate ADME (*A*dsorption, *D*istribution, *M*etabolism and *E*xcretion) characteristics, while reducing any toxicities which may have been observed. This is an iterative process. Compounds were first optimized for the *in vitro* assay in which they were discovered. Oftentimes, compounds optimized for an *in vitro* system require further optimization when evaluated in animals and still more optimization before introduction into the clinic. It is not difficult to see why the native natural product is unlikely to become a marketed product.

1.3 Screening in the Boom Years

Finding a lead compound is the necessary second step in the drug discovery process. Preceding the evaluation of the first samples is the discovery of a biological target and development of an assay to measure the interaction of the drug/lead with that target. Assays have evolved significantly over the last 50–60 years. Phenotypic screening using animals was the prevalent methodology in the 1950s and into the 1970s. These assays were not terribly sensitive, often subjective and difficult to interpret. The interaction of a chemical with an intact macro-organism can be a complicated process. Numerous metabolic processes may be affected in both positive and negative ways. Administration of a crude plant extract containing hundreds, if not thousands, of compounds dramatically increases the complexity of the problem. That said, observation of a positive result in a phenotypic screen had much significance. A substance could only show activity in such an assay if it was biologically available via the route of administration employed and acted at one or more steps in a metabolic pathway that was relevant to the disease or metabolic process under investigation. On the down side, phenotypic screening assays were unable to elucidate exactly how a drug/lead elicited its activity, i.e., which steps in the metabolic pathway or pathways were being affected.

Humanity's understanding of biochemistry and metabolism has evolved dramatically over the last 60 years. The details of many metabolic pathways have been elucidated allowing us to understand, or at least hypothesize, how some derangement in a particular step may result in an observed disease. As this mechanistic knowledge of metabolism developed, so too did a dissatisfaction with the level of detail offered by the results of phenotypic screening. Science, and especially industry, was no longer satisfied by the knowledge that a sample could evoke a desired biological response. We needed to know why and how that happened. To answer these questions, science in general, and industry in particular, turned away from phenotypic screening in favor of the newly developed biochemical mechanism of action screens. Thus was born industry's "one drug – one target" paradigm for drug discovery.

Much can be said for mechanism of action screening. When an enzyme is isolated from an intact cell, its activity can be accurately measured using standard biochemical techniques. Inhibition of that activity can then be measured quantitatively. Similarly, the interaction of a receptor and its natural ligand can also be accurately measured, and inhibition of that interaction can also be quantitated. It was a straightforward task to miniaturize these biochemical tests so that they required less sample and less reagent (meaning less money). Format standardization using the now ubiquitous 96-well, 384-well and 1536-well microplates allowed a proliferation of laboratory equipment specifically for these assays, leading inevitably to automation. While any good technician can easily perform pipetting tasks in 96-well plates, I would challenge anyone with even the slightest hint of astigmatism to try pipetting to or from a 384-well plate. Miniaturization and automation effectively removed throughput as a barrier to discovery. While the miniaturization and automation of biochemical assays facilitated natural products research in the industrial environment, it must be recognized that these same technologies also enabled the development of combinatorial chemistry and chemical library screening as potentially viable areas of endeavor.

Of course, every silver lining has its gray cloud. Mechanism of action screening was fast, relatively inexpensive and quantitative, but there was a cost for this information. These screening techniques could no longer tell us whether an activity was actually biologically relevant, or whether a sample had any bioavailability or any observable toxicity. These parameters would all need to be addressed elsewhere in the discovery and development process.

Once an assay, biochemical or biological, has been developed it can be used to evaluate a range of samples, natural product extracts, synthetic libraries, combinatorial chemistry libraries, corporate compound collections, looking for one or more initial hits. These first hits must then be refined to pure compounds active in the assay which can be used to test the hypothesis that the target is truly relevant to the disease state being studied. If relevance is established, then broad screening for new leads spanning chemical space will commence. If not, the assay must be refined or discarded. In Pharma, there is a constant turnover of assays used in screening.

For natural products, it is possible to hypothesize a generic process for creating extract samples for screening. While the actual nature of each kind of sample, i.e., plants, microorganisms, insects, marine invertebrate, reptile/amphibian, etc., may be quite different, the initial goal is always to separate the interesting small molecules from the rest of the biomass. (NOTE, for the remainder of this chapter the terms lead and drug will be used to refer only to small molecules. This is a bias of the author.). Over the years, there have been many discussions touting the merits of various extraction schemes over all others, but in the final analysis the only real criteria for acceptability are that the extraction scheme produces samples that are reasonably representative of the chemical diversity of the material being extracted and compatible with the assays in which they will be tested. Large screening programs tend to establish an extraction protocol for plant samples, a separate protocol for microbial fermentation broths, another separate protocol for solid-phase fermentations, and so on. Each protocol is optimized for the kind of sample on which it will be used. This level of standardization generates samples that can easily be compared to all other similar samples in the extract collection. Some organizations choose to retain bulk samples of all extracts in a library, ready to be re-evaluated as new assays come on line. Others do not. This choice is often based on the cost per sample and the perceived level of difficulty or uncertainty in obtaining a new supply if/when needed. Simply stated, there is no single "correct" way of processing samples or making extracts for screening.

1.4 Pharmaceutical Bioprospecting

Collecting samples legally and in a cost effective manner was and is, of course, of great importance. Multinational pharmaceutical corporations could not afford the fallout that could ensue from making collections without the necessary permits. The availability of permits, or lack of same, was often a criterion for selecting locations for collection activities. Countries having interesting floras but lacking a government framework for making permits were generally considered off limits, as were countries that actively discouraged or prohibited collections.

As research has progressed, it has become evident that terrestrial plants produce a fairly well proscribed suite of secondary metabolites, fungi produce a different (but overlapping) suite, marine invertebrates another different suite, prokaryotes a still different suite and so on. In order to capture the breadth of the chemical space occupied by secondary metabolites, it is necessary to sample all of these groups. Such a broad based approach is expensive and requires a long term commitment from corporate management. In practice, most pharma programs focused on one or two groups of organisms and perhaps dabbled in others.

The 1990s saw a near perfect storm for a resurgence in interest in natural products as a source of new drug leads. The 1992 Earth Summit and Rio Convention on Biological Diversity focused attention on the loss of habitat and loss of species that is still ongoing on the planet. The potential impact of these losses on drug discovery was not lost on the senior scientists running research at that time for the major pharmaceutical companies. Fueled by an earlier generation of billion dollar blockbuster products derived from natural products, e.g., Mevacor and the Avermectins, at Merck, decisions were made to broaden and accelerate the screening of natural products across the full range of disease targets. Using Merck as an example, ongoing collaborations with the New York Botanical Garden and its partners, and with INBio, the Costa Rican Institute for Biodiversity, were expanded to support screening at greater depth and breadth in line with the "now or never" theme of the time. The relatively high throughput of mechanism of action screens enabled these expansions.

1.5 Natural Products Chemistry in the Boom Years

By the mid 1990s, natural products chemistry was a mature discipline. HPLC had extended the chemists' ability to isolate active compounds in very small quantities, while countercurrent chromatography had evolved from the Craig apparatus through droplet countercurrent chromatography (DCC) and rotational locular countercurrent chromatography (RLCC) to centrifugal countercurrent chromatography (CCC) and centrifugal partition chromatography (CPC), facilitating isolation of labile molecules not amenable to the more traditional chromatographic techniques. High field NMR and mass spectrometry allowed identification of compounds in sub-milligram quantities. NMR spectrometers of 500 MHz had become routinely available in industrial labs while early 600 MHz instruments had more limited availability. Many 2D-NMR experiments were already in routine use enabling rapid structure determination. While mass spectrometry already had more than enough sensitivity for analyzing sub-milligram quantities of material, its utility was greatly expanded with the introduction of LC-MS in the early 1990s. With LC-MS came the ability to obtain reliable mass spectra from samples that were not homogeneous, opening the door to performing structure determinations on samples that were not completely purified, potentially saving days or weeks of isolation time.

With higher screening throughput came a greater need for dereplication of active extracts. Dereplication is the process by which a researcher determines whether he/she has encountered the same active constituent previously. For some groups of organisms, taxonomic relationships and chemotaxonomy could predict the probability of a known compound being present in a new organism. This approach was far from perfect as the knowledge of the distribution of secondary metabolite families was far from complete. While LC-MS is well suited to this kind of study, it too is an imperfect tool, especially for use in dereplicating crude extracts of complex organisms like higher plants. While the mass spectrometric detection of metabolites was certainly adequate for the purposes of dereplication, the ability to separate the full range of compounds represented in a typical plant extract in a single chromatographic experiment, even with HPLC, was still lacking. Some organizations chose to delay dereplication until after an initial isolation step in the hope of sufficiently simplifying the active fraction to allow LC-MS analysis. Other organizations chose to pre-fractionate every extract prior to screening. Analysis of the fractions by LC-MS or LC with photodiode array detection could then be used for dereplication. Of course, this latter approach greatly magnified the cost of the initial screening. Again, there was no single "correct" way of performing dereplication.

1.6 Production Issues

One area that received much attention during the Boom period was the topic of production of natural products on an industrial scale. While this was never an issue for microbial products, it was a major consideration for macro-organisms such as plants or marine invertebrates. If a real lead or a real potential product were found, how would a corporation make tons of it each year for the life of the product? Pharma has often been portrayed as being ready to over-exploit the environment by collecting vast amounts of plant material needed to support production from wild populations. While this may have been the opinion of potential collectors, it was and is the least likely scenario for industrial scale production of a new natural product. Such an extractive process would continuously diminish the available supply of the source of the product. Relying on natural populations, especially those under the control of someone other than the corporation, simply does not make business sense.

Agriculture was another possibility. While potentially applicable for fast growing herbaceous plants, it is, at best, a long term effort for slower growing organisms such as trees. Given the 17 year patent life for a new chemical entity, major investments in agricultural production of a slow growing organism seems unlikely unless there was no other way of making a real blockbuster product.

Perhaps a step closer to reality for complex products is production by plant cell or tissue culture of either the desired product or a related metabolite that could be readily transformed into the desired product. This was an approach I advocated in 1994. Once established and optimized, a cell culture process removes the need for collections or farms and eliminates the potential adverse environmental impact. This approach was successfully employed for the production of paclitaxel, for example (Onrubia et al. 2013).

Ultimately, for a variety of reasons, the most likely method that would be used for the production of all but the most complex of natural products is total synthesis. The scientific community has repeatedly shown that even complex natural products can be made by total synthesis, with paclitaxel (Wang et al. 2011) and tacrolimus (Ireland et al. 1996) as examples of marketed products successfully synthesized. Arguably, synthesis offers the best prospects for establishing and maintaining a patent position for the life of the product.

1.7 The Demise of Natural Products in Big Pharma

In the decade following the Earth Summit, it was easy to be excited and bullish about the future of natural products research in the pharmaceutical industry. Researchers had a seemingly endless array of potential targets, numerous collection projects were popping up and an incredible array of chemical and biochemical tools were available for pursuing new discoveries. By the end of the next decade, however, most, if not all, of the in-house natural products programs in Big Pharma in the United States had ceased to exist. Merck, for example, shut down its entire natural products research effort in the USA and in Spain in the Spring of 2008 after a very productive history spanning over 50 years. What happened?

A confluence of several factors resulted in the demise of natural products research in Big Pharma in the United States. During the Biodiversity Boom, the value of natural products was grossly overhyped. Without naming any names, on the bioprospecting conference circuit in the 1990s, is was not unusual to hear someone banter about the opinion that there are X number of drugs on the market that were found after studying only a few thousand species of plants, and there are over 300,000 plant species, so there must be 1000 times X drugs in the remaining unstudied species. Any responsible scientist sees immediately the fallacy in this argument. Unfortunately, this argument had the dual effect of driving up the expectations of source countries regarding the potential financial returns for use of their natural resources (and price of using those resources) while simultaneously inflating the expectations of the business community, including Pharma's senior management. At the same time, despite an investment of many millions of dollars by Pharma and by the US National Cancer Institute, no new chemical entities in clinical development had come from recent bioprospecting activities. Other competing technologies, e.g. combinatorial chemistry and library screening, offered more compounds for less money. From the perspective of a businessperson, it is not hard to come to the conclusion that this relatively unproductive endeavour was too costly to continue. Indeed, costs were cited as the reason why Merck's program was closed.

On a more macroscopic scale, the decision to end in house natural products research could be viewed as one manifestation of a larger shift in the constantly evolving business model for the industry. Functions that were important enough to continue, but not perceived as valuable enough to require headcount in house, could be outsourced to other organizations with lower expenses. Frequently this meant relocating these functions to other parts of the world where labor was cheaper. It is a short jump from this point to eliminating the function altogether, in favor of monitoring the activities of academic groups and smaller companies still active in the field, for the discovery of any promising development candidates which could then be purchased or licensed for internal development. For a risk-averse industry, this is an attractive, lower-risk approach to research. The ultimate expression of this approach *will be* a company that is no longer involved in any basic research, relying only on licensing and acquisition for its new product pipeline.

1.8 The Evolution of Natural Products Chemistry in the Twenty-First Century

While corporate business plans have continued to evolve, science has not stood still. Many of the tools used in natural products research have experienced major enhancements in capabilities. HPLC has been a routine tool for many years in most chemistry labs. Using 5 and 10 micron particles, one could separate most kinds of mixtures in a reasonable length of time within the 400 bar pressure limits of most commercially available pumps. Driven by the desire for faster separations, greater resolution and better sensitivity, column manufacturers introduced newer particles with mean diameters extending below 2 microns, dramatically increasing speed, resolution and sensitivity but requiring UHPLC pump technology capable of delivering mobile phase at pressures of 600–900 bar or higher. The higher speed of separation required detectors and data systems with higher sampling rates, able to properly digitize the narrower peaks. Of course, to embrace all of these new capabilities requires a significant investment in new equipment. More recently, column manufacturers have re-introduced pellicular packings on small particles, offering separations comparable to UHPLC but at pressures accessible to the large base of users with traditional HPLCs.

Nuclear Magnetic Resonance spectrometry has also changed dramatically in the last 20 years. The gigahertz spectrometers and cryoprobes capable of achieving signal to noise ratios on 10,000:1, now commercially available, would have been considered science fiction at the Rio Conference. While these gigahertz instruments are not widely deployed (for obvious reasons of cost and upkeep), there is now a substantial user base of instruments in the 700–800 MHz range. These instruments have made it possible to obtain usable data on samples substantially less than 100 micrograms. The increases in hardware capabilities have been matched by a dizzying array of new experiments easily implemented on the modern spectrometers, further simplifying the task of structure determination.

Similarly, mass spectrometry has made major strides in this time period. While mass spectrometry has always been an extremely sensitive method, current generation ion trap, Orbitrap and quadrupole time of flight (qTOF) instruments now boast sensitivities down to femtogram level samples with high mass accuracy. Tandem MS methods add fragmentation information for each ion in the spectrum. When coupled to an HPLC or UHPLC, such an instrument makes it possible to collect mass spectra for every compound eluting from an HPLC column, truly enabling the fingerprinting of an extract. The application of these techniques to dereplication is obvious and becomes even more powerful with the addition of statistical (chemometric) methods such as principle component analysis (PCA). These statistical tools are already available in bundled software packages from the instrument manufacturers. One can easily envision developing a system that uses a standardized, gentle, initial fractionation step (preferably orthogonal to HPLC) such as countercurrent chromatography, followed by assay and LC-MS/MS analysis using PCA to elucidate the peaks of interest. While perhaps not universally applicable, I would be hard pressed to produce an example where it was not.

1.9 The Evolution of Screening in the Twenty-First Century

In biology, what's old is new again. Over the last several years there has been a resurgence in interest in phenotypic screening. This is not to say that interest has resurfaced for using animals in the screening of samples in drug discovery, rather the advent of technologies for imaging cellular processes, coupled with cells engineered to overexpress metabolic pathways, has enabled the development of high content screening. Molecular biologists have studied cellular processes for years using imaging tools such as laser confocal microscopy, but these experiments were laborious and time consuming. High content imaging systems have taken these experiments, adapted them to microplate formats (96, 384 and perhaps 1536 well formats) and automated acquisition and processing of the data. As the development of microplate technology allowed the miniaturization and automation of biochemical assays, enabling high throughput target-based screening with high content imaging, now allows miniaturization and automation of screening at the biological level in intact cells. Phenotypic screening addresses some of the perceived problems involved with target based screens, namely target relevance and bioavailability among others. Observing change in the morphology or behavior of cells is a direct indication of the biological activity of an applied substance. Elicitation of that activity is predicated on the ability of the applied substance to penetrate the cell and access the underlying metabolic process. The ability to work at this level in a high throughput manner is a real game changer.

For decades, the mantra of drug discovery in Big Pharma has been, "one drug, one target". Compounds displaying more than one bioactivity (or interacting with more than one target) were considered inadequately selective for development as products. Recently, interest in polypharmacology has grown in the research community. Polypharmacology accepts that a compound may interact with more than one step in a metabolic pathway and that the net overall effect on the entire pathway is the relevant indicator of biological activity, not just the effect on a single step. Phenotypic assays allow the researcher to observe the net overall effect, positive or negative, of a compound's interaction with an intact metabolic pathway. While polypharmacology may not be the optimal way for finding the most potent HMG-CoA reductase inhibitor, it may well allow for the discovery of safe, effective and novel modulators of cholesterol metabolism.

While most of the world's population still relies on traditional medicines to meet their healthcare needs (World Health Organization 2013), relatively few "modern" medicines have been developed from traditional medicines. Clearly some traditional medicines have some level of efficacy. Why else would their use have persisted for hundreds or even thousands of years? Oftentimes, traditional medicines are comprised of multiple ingredients prepared in a formulaic manner that may be difficult to reproduce accurately in the laboratory. In labs constrained by the "one drug, one target" philosophy of drug discovery, a reductionist approach has been taken, evaluating each component ingredient individually against an individual target. While this approach has yielded many biologically active compounds, few, if any, blockbusters have emerged. Why not? The complex nature of traditional medicines suggests that the observed effects may be the result of multiple biological activities. This is not to suggest that true synergism occurs at a single target, rather that inhibition of multiple steps in a pathway may lead to a net effect that is greater than the sum of the inhibition of each step. Modulation of multiple steps in a metabolic pathway is an example of polypharmacology in action.

1.10 New Frontiers

Over the years, I have heard many administrators suggest that natural products may be played out as a source of new leads or drugs and that some novel approach would surely increase the pace of new drug discovery. Thus were born *de novo* rational drug design, combinatorial chemistry and the all inclusive approach to compound library screening. Like natural products, each of these approaches has had its ups and downs. Natural products can be played out only in the mind of a person lacking imagination and curiosity. During the boom years, studies focused primarily on terrestrial plants, microorganisms and marine invertebrates, each taken out of the context of the environment in which they lived. Each of these groups was known to produce a wide range of secondary metabolites with little apparent overlap with the other groups. More recently, endophytic fungi such as *Taxomyces* (Stierle et al. 1993) have been found that are capable of producing some of the same metabolites produced by their host organisms. Similarly, some of the metabolites ascribed to marine invertebrates have been found to be produced also (or solely) by their microbial symbionts (Gerwick and Fenner 2013), and poison arrow frog toxins actually come from the ants the frogs eat (Daly 1998). Perhaps the ants are eating the true producing organisms. It is not surprising therefore, that marine microbes and terrestrial endophytes have become hotspots of interest.

One further area of interest pushes the boundaries of the definition of natural products. That is metagenomics. We have long been told that the vast majority of microorganisms cannot be cultured, and culturable microbes may represent less than 1 % of the species in the environment (Rappe and Giovannoni 2003). Metagenomics attempts to isolate DNA directly from the environment, without first isolating a producing organism. It then seeks to identify genes and gene clusters analogous to the genes in known biosynthetic pathways, isolates them, and then inserts them into the genome of a suitable heterologous host, hopefully thereby expressing the biosynthetic potential of the isolated genes ((Charlop-Powers et al. 2014); (Brady et al. 2009)). These methods hold great potential for tapping the biological and biochemical diversity of this largely unexplored resource. Perhaps a limiting factor is that the sequence based metagenomics approach relies on our existing knowledge of the genetics of biosynthesis, and is thus more likely to provide variations on known groups of compounds rather than truly novel chemotypes. Nonetheless, as our understanding of biosynthesis is continuously expanding, so too are the opportunities that will be afforded by metagenomic small molecule discovery.

Looking back with the clarity of 20/20 hindsight, many of the advances of the last 20 years could be considered predictable. Certainly, improvements to the quality and capabilities of laboratory equipment happen continuously. Why else would researchers flock to the Pittsburgh Conference (Pittcon) every year? While some of the details may have been surprising, the general trend is not. The resurgent interest in phenotypic screening and polypharmacy are likely a response to years of dealing with the limitations and vagaries of target-based discovery that have led to the questioning of the validity of the "one drug, one target" paradigm. The real breakthroughs in sourcing are really only in their infancy. Aside from the dearth of new blockbusters based on natural products, the science is pretty much right where it would be expected.

1.11 Into the Future

The only thing certain about the road forward is that it will be a bumpy ride. Pharma's interest in natural products research has always been a cyclical phenomenon, but the recent wholesale abandonment of the field suggests that the current cycle may have a longer period than the last few. Pharma's business plan of outsourcing research and relying on acquisitions to fill the pipeline may be myopic, but it does open the door to an expanded emphasis on innovation and the entrepreneurial drive of academia and smaller business entities. It must be noted that American Pharma's business plan is not the *de facto* standard in other parts of the world. There is still substantial interest in natural products in places such as China and Japan where natural remedies had been a part of the culture and tradition long before Europeans laid claim to the New World, and still are in widespread use today. One major discovery, one blockbuster, may be all it takes to start the next cycle of interest in the United States.

There are a few confounding factors. Pharma's abandonment of natural products research has dramatically reduced the number of attractive (i.e. lucrative) jobs in the field. Predictably, this will act as a disincentive for attracting students to become the next generation of natural product researchers. Smaller companies may be able to generate a significant number of positions for these students, but it unlikely that they will be able to offer the compensation packages we have come to expect from multibillion dollar multinationals. Second is the lack of stable, significant, government funding for natural product research in academia. As anyone in Pharma R&D management will tell you, natural products research is expensive and is a long term endeavor. Without stable funding at a level sufficient to maintain this kind of research, the pipeline of well trained and qualified young scientists will surely dwindle.

Will natural products research disappear completely from pharmaceutical R&D? I think not. There is still too much value, too much potential, for this resource to no longer be of relevance. It appears more likely that speciality discovery organizations will evolve to service the discovery needs of Big Pharma via its existing business model. Whether such organizations evolve as broad-based discovery groups or groups focusing on specific niches (sources, disease targets, etc.) or some combination of the two remains to be seen.

Will natural products ever regain the prominence it once enjoyed in Big Pharma? Probably not. It must be understood that Pharma's departure from natural products research was a business decision, not a scientific one, and is part of a much larger business plan. Pharma, as an industry, continues to evolve. Evolution is a prospective, progressive phenomenon. While it is important to learn from the past, it is equally important not to dwell on it. Pharma will eventually see again the wisdom of using natural products to modulate the inherently natural functions of metabolism to meet medical needs. When will this happen? To quote Niels Bohr, "Prediction is difficult, especially about the future."

References

Borris RP (1996) Natural products research: Perspectives from a major pharmaceutical company. J Ethnopharmacol 51:29–38

Brady SF, Simmons L, Kim JH, Schmidt EW (2009) Metagenomic approaches to natural products from free-living and symbiotic organisms. Nat Prod Rep 26:1488–1503

Charlop-Powers Z, Milshteyn A, Brady SF (2014) Metagenomic small molecule discovery methods. Curr Opin Microbiol 19:70–75

Cragg GM, Newman DJ, Snader KM (1997) Natural products in drug discovery and development. J Nat Prod 60:52–60

Daly JW (1998) Thirty years of discovering arthropod alkaloids in amphibian skin. J Nat Prod 61(1):162–172

Gerwick WH, Fenner AM (2013) Drug discovery from marine microbes. Microb Ecol 65(4):800–806

Ireland RE, Gleason JL, Gegnas LD, Highsmith TK (1996) A total synthesis of FK-506. J Org Chem 61(20):6856–6872

Newman DJ, Cragg GM (2007) Natural products as sources of new drugs over the last 25 years. J Nat Prod 70:461–477

Newman DJ, Cragg GM (2012) Natural products as sources of new drugs over the 30 years from 1981 to 2010. J Nat Prod 75:311–335

Newman DJ, Cragg GM, Snader KM (2003) Natural products as sources of new drugs over the period 1981–2002. J Nat Prod 66:1022–1037

Onrubia M, Cusido RM, Ramirez K, Hernandez-Vazquez L, Moyano E, Bonfill M, Palazon J (2013) Bioprocessing of plant in vitro systems for the mass production of pharmaceutically important metabolites: paclitaxel and its derivatives. Curr Med Chem 20(7):880–891

Rappe MS, Giovannoni SJ (2003) The uncultured microbial majority. Annu Rev Microbiol 57:369–394

Stierle A, Strobel G, Stierle D (1993) Taxol and taxane production by *Taxomyces andreanae*, an endophytic fungus of the Pacific yew. Science 260(5105):214–216

Wang Y-F, Shi Q-W, Dong M, Kiyota H, Gu Y-C, Cong B (2011) Natural taxanes: developments since 1828. Chem Rev 111(12):7652–7709

Wilhelm SM, Adnane L, Newell P, Villanueva A, Llovet JM, Lynch M (2008) Preclinical overview of sorafenib, a multikinase inhibitor that targets both Raf and VEGF and PDGF receptor tyrosine kinase signaling. Mol Cancer Ther 7(10):3129–3140

World Health Organization (2013) WHO Traditional Medicine Strategy: 2014–2023. WHO Press, Geneva, p. 16

Chapter 2
Legal and Ethnoecological Components of Bioprospecting

Morten Walløe Tvedt

Abstract Innovation in the use of biological material has surged ahead over the past decades, closely followed by additions and amendments to the regulatory environment. The aim of this chapter is to explore the most important legal tools relating to the use of genetic resources and to identify the significant changes and additions. This article looks at the critical aspects for a successful implementation of the Nagoya Protocol. Three topics are especially covered: the need for (a) developing sound access and benefit sharing (ABS) contracts; (b) establishing provider and user legislation or regulation in all countries; and (c) avoiding fragmentation at the international legal level concerning ABS. The regulatory environment is increasingly focused on strengthening intellectual property rights (patents and plant breeders' rights) and access to and sharing of benefits arising from the use of genetic resources in terms of ABS. Whereas the regulation of use depends on the laws enacted in each country where bioprospecting occurs, the international arena has become increasingly important for law-making in the last 30 years.

Morten Walløe Tvedt is senior research fellow at the Fridtjof Nansen Institute and assistant professor 2 in law, P.O.Box 326, 1326 Lysaker, Norway. Tvedt has published extensively in the area of biological resources law and intellectual property in recent years (see www.fni.no for a complete list of publications). The research on which this article is based was partly funded by the Norwegian Research Council under the ELSA Program and forms part of the three-year project "Exploring Legal Conditions and Framework for Marine-based Bioprospecting and Innovation – International Law, Open Innovative Systems and Proposals for SMTA for Marbank", project number 208543/O10.

M.W. Tvedt (✉)
Fridtjof Nansen Institute, 326, 1326 Lysaker, Norway
e-mail: mwt@fni.no

R. Paterson, N. Lima (eds.), *Bioprospecting*, Topics in Biodiversity and Conservation 16, DOI 10.1007/978-3-319-47935-4_2

2.1 Setting the Scene – Sovereign Rights and Privatisation Through Patents

Innovation in the use of biological material has surged ahead over the past decades, closely followed by additions and amendments to the regulatory environment. The aim of this chapter is to explore the most important legal tools relating to the use of genetic resources and identify the most importance changes and additions. The regulatory environment is increasingly focused on strengthening intellectual property rights (patents and plant breeders' rights) and access to, and sharing of, benefits arising from the use of genetic resources (ABS). Whereas the regulation of use depends on the laws enacted in each country where bioprospecting occurs, the international arena has become increasingly important for law-making in the last 30 years.

The 1992 Convention on Biological Diversity (CBD) established standards of conservation and sustainable use of biological diversity while enabling the fair and equitable sharing of benefits arising from the use of genetic resources. Under international law states enjoy a sovereign right to their natural resources (Schrijver 2008). This was confirmed by the 1992 Convention on Biological Diversity (CBD), which recognizes in Art. 3 a principle of international law:

> ***States have, in accordance with the Charter of the United Nations and the principles of international law, the* sovereign right to exploit their own resources *pursuant to their own environmental policies, and the responsibility to ensure that activities* within their jurisdiction or control *do not cause damage to the environment of other States or of areas beyond the limits of national jurisdiction*.** (Emphasis added)

The CBD confirms a principle of international law that endows states with a sovereign right to regulate the genetic resources under their control. The CBD did not introduce a new world order in 1992, of course; *permanent sovereignty* over natural resources was already a well-established principle within the UN (ibid.). The references to the UN Charter and principles of international law inform the general law governing questions concerning *genetic resources*.

During the 1980s, developed countries sought to establish within the UN Food and Agriculture Organization (FAO) a conception of plant genetic resources as the 'common heritage of mankind'.[1] The open access system suggested for plant genetic resources by the 1983 International Undertaking, was, however, abandoned in

[1] *International Undertaking on Plant Genetic Resources* [International Undertaking 8/83], Rome, FAO, Conference Resolution 8/83, entered into force 23 November 1983, adapted at the Twenty-second Session of the FAO Conference. The term 'common heritage of mankind' had already been introduced in the UN Law of the Sea regulating the legal status of exploitation of minerals in the deep sea bed, in the sense that "No State shall claim or exercise sovereignty or sovereign rights over any part of the Area or its resources, nor shall any State or natural or juridical person appropriate any part thereof." Art. 136–137 of *United Nations Convention on the Law of the Sea* [UNCLOS], United Nations, 10 December 1982, 1833 UNTS 3, entered into force 16 November 1994. FAO's understanding of the term was less developed than UNCLOS's, which provided for the creation of a governing authority to oversee the Common Heritage regime.

favour of a new resolution in 1991.[2] The CBD did not alter any general legal principles; the FAO's rather narrow (non-binding) International Undertaking had already been abandoned by the FAO members themselves. The CBD is the first legally binding treaty to specify rules concerning the enforcement of states' sovereign rights to genetic resources. However, a treaty needs national implementation to become a functional tool changing the behaviour of relevant actors.

Parallel to the discussions on sovereign rights to genetic resources, the US, Japan and Europe were changing their practices with regard to granting patents for inventions based on biological material. Increasingly, patents were awarded for biotechnological inventions. While the practice began in a few states, it became a global obligation in 1994 with the enactment of the Agreement on Trade-Related Intellectual Property Rights (TRIPS) which required all members of the World Trade Organization (WTO) to allow patent protection of all types of invention, save certain very specific categories, one being 'plant varieties'. Intellectual property right protection of plant varieties was harmonised by member states and became the International Convention for the Protection of New Varieties of Plants (UPOV).[3] Regulation of access and benefit sharing should be seen as an attempt to counter this trend of privatization of innovation in the biodiversity area with the aim of creating a system for sharing parts of the benefits patents create to their owners back to conservation purposes.

2.2 Principles on Access to Genetic Resources in the Convention on Biological Diversity

Sovereign rights, according to the CBD, include mechanisms set forth in Art. 15 para. 1–6, which can be applied by states for (a) regulating access (to) and (b) Art. 15.7, according to which all states must implement rules to ensure fair and equitable benefit sharing (arising from) the utilization of 'genetic resources'. In treaty law, the CBD contains the first definition of 'genetic resource' as an subject matter of regulation. '*Genetic resources* means genetic material of actual or potential value.' And '*Genetic material* means any material of plant, animal, microbial or other origin containing functional units of heredity.' Thus, a definition of a new subject matter of legal regulation became legally binding.[4]

---

[2] International Undertaking 8/83, *Agreed Interpretation of the International Undertaking*, Rome, FAO, Conference Resolution 4/89, entered into force 29 November 1989, adapted at the Twenty-fifth Session of the FAO Conference, and finally *Agreed Interpretation of the International Undertaking*, Rome, FAO, Conference Resolution 3/91, entered into force 25 November 1991, adopted at the Twenty-sixth Session of the FAO Conference.

[3] *International Convention for the Protection of New Varieties of Plants 1991* [UPOV-1991], UPOV, 2 December 1961, UPOV/INF/6/1, as amended on 10 November 1972, 23 October 1978 and 19 March 1991. For a discussion of the UPOV and farmers' rights see Christinck and Tvedt (2015).

[4] For a discussion of the term 'genetic resources', see Tvedt and Schei. "The Term 'Genetic Resources': Flexible and Dynamic while Providing Legal Certainty?" in Global Governance of Genetic Resources Access and Benefit Sharing after the Nagoya Protocol. Edited by Oberthür and

With regard to the plant sector, the principle of regulated access has been judged to be unsatisfactory by some. Open access, as a means of exercising their sovereign rights, was adopted by 145 members of the International Treaty on Plant Genetic Resources for Food and Agriculture (ITPGRFA) for certain specific accessions of plant genetic resources for defined food and agriculture purposes. Elements within the plant sector called for specific solutions when the parties were at the point of agreeing to the CBD. When agreeing to the CBD, the final Resolution from recognised that the global collections of plant genetic resources that existed prior to the CBD instigated a special regulation for them in international law with the agreement of CBD:[5]

> ***Further recognizes the need to seek solutions to outstanding matters concerning plant genetic resources within the Global System for the Conservation and Sustainable Use of Plant Genetic Resources for Food and Sustainable Agriculture, in particular: (a) Access to ex-situ collections not acquired in accordance with this Convention; and (b) The question of farmers' rights.***

This call for a special regulatory system for collections was considered by the Commission on Genetic Resources on Food and Agriculture of the FAO. For a specific list of plant genetic resources, the 2001 International Treaty on Plant Genetic Resources for Food and Agriculture (ITPGRFA) Art. 11–14 established a Multilateral System for Access and Benefit Sharing (MLS).[6] Accession of plant genetic resources included in the MLS is collectively managed by ITPGRFA member states, and can be called a global common for certain plant genetic resources and certain uses. The argument is often made that it is not possible to identify which country each of the special properties of a plant is developed in, and consequently not possible to allocate sovereign rights to any particular state.[7] Whether this system can be used for other sectors of genetic resources than plants is an open question (Drankier et al. 2012; Greiber 2011; Elferink 2007; Arico 2010).

Slow implementation along with lack of a functional benefit sharing system prompted the parties to the CBD to negotiate a protocol to the convention containing specific rules on access and benefit sharing. In 2010, the parties to the CBD adopted the Nagoya Protocol (NP). The NP specified rules of access and benefit sharing more closely. From the perspective of international law, the obligations according to the CBD are still valid side by side the new regulations in the Nagoya

---

Rosendal. New York, Routledge, 2014 and Schei and Tvedt. "Genetic Resources" in the CBD: The Wording, the Past, the Present and the Future. Lysaker, Fridtjof Nansens Institutt, 2010. (FNI Report, no. 4/2010)

[5] *Resolution 3 - The Interrelationship between the Convention on Biological Diversity and the Promotion of Sustainable Agriculture* [Nairobi Resolution 3], entered into force 22 May 1992, the Nairobi Conference for the Adoption of the Agreed Text of the Convention on Biological Diversity.

[6] *International Treaty on Plant Genetic Resources for Food and Agriculture* [ITPGRFA], Food and Agriculture Organization, 3 November 2001, 2400 UNTS 303, entered into force 29 June 2004.

[7] Nevertheless, plant variety protection and plant patents are assuming that it is possible to identify the individual and unique contribution of individuals to new traits in the plant varieties. See UPOV-1991 and *Agreement on Trade-Related Aspects of Intellectual Property (Annex 1C of the Agreement establishing the World Trade Organization)* [TRIPS Agreement], WTO, 15 April 1994, especially Art. 27.

Protocol, as there is no indication of intent to change them in the new text. While negotiating and adopting the Nagoya Protocol a politically difficult question concerning the rights to genetic resources that are held in collections outside the countries of origin and their new uses remained unresolved. The compromise solution was NP Art. 10:

> ***Parties shall consider the need for and modalities of a global multilateral benefit-sharing mechanism to address the fair and equitable sharing of benefits derived from the utilization of genetic resources and traditional knowledge associated with genetic resources that occur in transboundary situations or for which it is not possible to grant or obtain prior informed consent. The benefits shared by users of genetic resources and traditional knowledge associated with genetic resources through this mechanism shall be used to support the conservation of biological diversity and the sustainable use of its components globally.***

This Art. 10 provision mandates NP members to consider whether there is a need for a new global multilateral benefit-sharing mechanism. This mechanism shall be considered in respect of two situations in particular: (1) genetic resources in transboundary situations; and (2) when it is not possible to grant or obtain prior informed consent for their use.

In 2011, shortly after agreement was reached on the draft text of the Nagoya Protocol, other forums discussing ABS in relation to specific groups of genetic resources also made rapid progress. The World Health Organisation (WHO) agreed in 2011 to adopt two standard material transfer agreements concerning exchange and use of viral genetic resources with pandemic potential for humans. In these two standard contracts, globally negotiated terms and conditions for rapid access and benefit sharing are pre-set. For the exchange of viral, human pandemic material, speed and unhindered access are crucial to successfully combating outbreaks. One interesting observation is that they also include benefit-sharing clauses that previously were difficult to agree on in the WHO.

For almost a decade, the question of access and benefit sharing arising from the use of genetic resources in areas beyond national jurisdiction (ABNJ) has been on the agenda of the UN Convention on the Law of the Sea (UNLCOS). Negotiations are currently underway on a mandate for future talks relating to a special regime for this category of marine genetic resources. It could include, for example, genetic resources taken from the seabed and/or the high seas. Discussions under the auspices of the Antarctic Treaty are also probing how to regulate genetic resource material from one of the world's most remote, yet biologically unique areas.

In addition, there are large collections of foreign genetic material in genebanks. These collections are seen by some scholars and lawyers as outside the scope of the CBD since they were collected prior to its entry into force. Whether these collections will be subject to rules on benefit sharing and with whom the benefits shall be shared is currently an unresolved question.

Also in 2011, the Commission on Genetic Resources for Food and Agriculture (CGRFA) under the Food and Agriculture Organization of the UN (FAO) agreed to intensify work on access and benefit sharing for genetic resources for food and agriculture (GRFA). These negotiations are particularly interesting in a publication like this with contributions dealing with the different sectors of users of genetic resources. Section 2.5 below explains the current work of the CGRFA in more detail.

2.3 Patent Law and Plant Breeders' Rights

To round out the discussion of ABS and regulation of genetic resources we need to include intellectual property right systems, such as patents and plant breeders' rights. The increase in the patenting of inventions based on genetic material created a global imbalance where powerful companies could establish time-limited exclusive rights to technologies based on genetic material, while the countries conserving biological diversity were left without the legal tools to participate in this value creation. A patent creates a time-limited exclusive right to an "invention". It is the individual claiming to have made the invention who describes what he claims to have invented in the patent claims. In many countries, practice regarding the granting of patents was increasingly aimed at permitting patent protection of microorganisms, genes and methods in combination with biological material. If an invention meets the criteria of novelty, sufficient inventive step and industrial application, then the applicant is awarded a patent. This is why the objectives of the CBD about sustainable use of biological diversity and those of the FAO about food security are interrelated in the granting of patents to bio-inventions.

The CBD almost never discusses patents and plant breeders' rights, but there are some discussions at the COP about the relationship between ABS and patents. Under the CGRFA, intellectual property rights are never discussed, while IPRs are only discussed by ITPGRFA's Governing Body as a trigger for the benefit-sharing obligation. The positive or negative effect from patent on food production is never on the agenda. Thus, there is a lack of seeing these legal systems in conjunction.

The use of patents and plant breeders' rights features to varying degrees in the different areas of innovation based on biological material. For example, the use of intellectual property rights is particularly intensive in the pharmaceutical and enzyme industry, and in the plant sector. In the animal sector, aquaculture sector and forest tree breeding sector, intellectual property rights play a smaller role.

2.4 The Challenges and Potential of the Nagoya Protocol

Access and benefit sharing (ABS) as instruments of law and policy entered into a new phase with the October 2014 entry into force of the NP, and the first Meeting of the Parties (MOP) at the COP to the CBD. The October meeting was a watershed moment in the decades of work to make ABS a functional mechanism for raising funds for the conservation and sustainable use of biological diversity. A lot of effort was put into the negotiations that led to the NP and now the challenge is to make the mechanisms work by functional implementation (Tvedt 2014; Oberthür and

Rosendal 2014). The idea behind ABS is that the world's most biodiverse regions, usually located in developing countries and dubbed "providers", shall partake in the benefits created by "users", located more traditionally in richer economies. The obligations under CBD Art. 15 have not been sufficiently clear, however; as of writing, provider countries of biodiversity under ABS arrangements have only received a limited amount of the monetary benefits. Following the *modus operandi* of the Nagoya Protocol, users of genetic resources should be ploughing some of their profits and other non-monetary benefits drawn from the commercialisation of genetic resources into conservation and sustainable use of biodiversity.

Three issues remain to be resolved to make the ABS regime more functional: (1) contractual mechanisms for access and for benefit-sharing need to be created; (2) domestic legislative, policy and administrative measures in both user and provider countries need to be put in place; and 3) the possibility of unregulated genetic resources in certain arenas needs clarifying at the international level.

2.4.1 Making Genetic Resources Contracts Work

Article 15 of the Convention on Biological Diversity (CBD) prescribes two contractual mechanisms. A contract can be made either at the time of access to the genetic material or at the point of time of utilisation to ensure sharing of benefits arising from its use. According to both CBD Article 15 and the Nagoya Protocol, the principal method of enforcing a country's sovereign rights is by invoking private law contracts – mutually agreed terms – between the provider country and/or country of origin and user. The user is often thought of as a private company from another country. ABS therefore relies on private contracts as the relevant means of regulating exchange and sharing returns (Young 2013). Nevertheless, little research has been done systematically to explore the contractual mechanisms with a view to making the contractual system functional. Companies need clearer incentives to enter into ABS contracts and fulfil their obligation to share benefits accruing from their research and development fairly and equitably. Neither does the Nagoya Protocol resolve the challenges thrown up by these contractual mechanisms.

ABS contracts will be negotiated and enforced as commercial contracts. This raises a number of challenges. ABS contracts must be drafted in such a manner as to make them legally viable in the jurisdiction and under the legal system of the user. Since the Nagoya Protocol does not prescribe a uniform or standardised system of user country legislation, the contracts need to resolve complex legal questions, which will typically vary among countries. Since ABS contract law is a relatively new and unexplored area, background jurisprudence is limited. This raises a number of technical and difficult challenges in contract law. Existing global legal contract tools do not overcome these challenges and international private law has only a limited potential in this area as well.

Another contract-related challenge is how to regulate the subject matter that is being transferred. An ABS contact regulates what is sometimes a highly dynamic situation with a high degree of scientific and commercial potential. "Genetic resources", as they are defined in the CBD and Nagoya Protocol, rarely constitute a commercial product as such and their connection with a product on the market will therefore be more or less remote (Schei and Tvedt2010). For a contract to embody the creation of value arising from the use of genetic resources, it needs to some extent to foresee future developments of the material. The degree of change and uncertainty, however, will vary among the genetic material's users and uses. Often this is presented as a matter of tracing, though it is perhaps more complex, as it is also about understanding the relative contribution of the genetic resource, research, development and other investments. The careful drafting of the subject matter of the contract and the actions allowed by the contract will become crucial to the functionality of this type of right. It would be advisable to avoid as far as possible the term genetic resources to define the subject of the contract (Tvedt 2013). Instead, parties should spell out in detail what actions the contractual partner has an explicit right to perform with the biological material. In addition a number of legal issues of commercial contracts must be observed and resolved in these contracts. For a further analysis of these questions, Tomme, Young and Tvedt are publishing a monography on this topic in 2017.

2.4.2 *Domestic ABS Legislation as a Core Tool*

Well-drafted ABS contracts will work better if supported by ABS laws in the different countries. Both the CBD and the Nagoya Protocol are based on a perception that it is the providing countries that have the primary responsibility for regulating ABS at the point of access. During the negotiations that eventually led to the new instrument, an understanding of "utilisation of genetic resources" gradually gained momentum (Tvedt and Young 2007; see also Hendrickx et al. 1993). The Nagoya Protocol builds on CBD Art. 15.7 in defining what exactly utilisation constitutes under Arts 2(c) and (d). During the process of pinning down the Nagoya Protocol a number of countries likely halted their processes of regulating access to their genetic resources in expectation of the new international framework. Several issues remain to be considered now as states continue to implement ABS access-side legislation. For example, should they set up a national ABS system to avoid the use of biological diversity without a full ABS contract, or should governments encourage users to enter into a contract in a more deliberative way? Related to this, some governments are attempting to strengthen the incentives for users to enter into such contracts.

Australian ABS legislation is viewed as pioneering in this respect, with its simplified mandatory permits for all types of bioprospecting, including for non-commercial use (Prip et al. 2014). The law also includes a built-in clause to cover "change of intent": if activities change from purely scientific or non-commercial to commercial, the user must return to the authorities and have the necessary changes made to the contract. Even more important, this requirement is based on an existing legal instrument known as the statutory declaration, which binds the user to

Australian criminal law, although admittedly it has limited force if genetic material is transferred to third parties. Furthermore, even though the Australian system has been in place for a while, almost none of the initial bioprospecting agreements has resulted in the user coming back to enter into a benefit sharing contract. This demonstrates the scale of the challenge facing countries when it comes to surveying and tracing the use and commercialisation of products based on their genetic resources.

To facilitate the tracing of uses and genetic resources through to final products on the market, steps must be taken by all CBD countries, not only those currently party to the Nagoya Protocol. All CBD countries are already obliged to make ABS functional, as demonstrated by Art. 15.7. The latter places a clear obligation on all parties to take measures to implement ABS both on the user and provider side. Provider countries would be advised in this sense to require all user countries to submit reports to the next CBD COP on all relevant ABS measures put in place.

2.4.3 Avoiding Fragmentation in the International Arena

Among the more polarising questions in the NP negotiations was the relationship between the ABS in the CBD and other international legal regimes touching on genetic resources. The debate circled around rules already in place and possible new regimes. These concerns led to the inclusion of Article 4 recognizing that the Nagoya Protocol "does not apply for the Party or Parties to the specialized instrument in respect of the specific genetic resource covered by and for the purpose of the specialized instrument." The scope of the other regimes will therefore be crucial to define which genetic resources are covered by the Nagoya Protocol.

The International Treaty on Plant Genetic Resources for Food and Agriculture (ITPGRFA) has been in force since 2004. It is still however not clear which plant resources that are within its mandatory scope (Tvedt 2015a). There are substantial differences in the views expressed by Correa (2013), Halewood et al. (2013: 70–96), and Medaglia et al. (2013) on the interpretation of mandatory scope. These differences have also surfaced in the so-called "tandem meetings" between the CBD and ITPGRFA, focal points organised in the past year by the ABS Capacity Building Initiative (2014), and also at a September 2014 Nordic meeting on access and benefit sharing (FNI 2014). In effect, concern for food security under the ITPGRFA trumps the greater emphasis on equity found in the ethos of the CBD ABS. Beyond these unresolved grey zone questions, work is ongoing under the Plant Treaty to explore the conditions for expanding the scope of the list of crops that are covered by the MLS (Tvedt 2015a, b). Expanding the scope of the multilateral system under the Plant Treaty will necessarily lead to a narrowing of the scope of the Nagoya Protocol. A working group under the Governing Body of the ITPGRFA is looking at both the scope for access and how to make the benefit-sharing mechanism more functional. If the functionality of the benefit-sharing mechanism is not improved, the lack of payment by agro-businesses for using plant genetic resources might undermine the open access regime.

2.5 The Challenges for ABS from the Commission on Genetic Resources for Food and Agriculture

The ABS is in danger of fragmenting further. The work by the Commission on Genetic Resources for Food and Agriculture (CGRFA) under the UN Food and Agriculture Organisation (FAO) may fragment ABS in the future. The CGRFA is discussing questions related to access and benefit sharing for six groups of genetic resources, namely, animals; aquatic; invertebrates; plants; forest; and microbial. During the tenth regular session of the CGRFA in 2005, a recommendation was made that the FAO and CGRFA to do more work on ABS issues to improve its support of agriculture considering that all components of biological diversity are essential for improving food and agricultural systems.

In July 2011, the CGRFA ABS discussions began to explore ABS issues related to specific sectors of agriculture. According to Chiarolla et al., there was a debate within the CGRFA on whether to develop 'specialized instruments' for GRFA (Chiarolla et al. 2013). In a background paper prepared for FAO, Schloen et al. (2011) enumerate direct and indirect impacts ABS measures can have on GRFA. While legal certainty and clarity on measures that govern exchange and use of GRFA and transaction costs, time and capacities needed to implement the ABS provisions form the direct impacts; the incentives for exchange and use form the indirect impacts. They also recommend reducing administrative bottlenecks, aggregating and standardizing ABS processes and decoupling benefit sharing from individual providers and individual genetic resources. While these options are worth discussing in a critical light, this cannot be done in a vacuum given the critical role of the nature of the resource, the holding and ownership of the resource, type of use, potential for commercial utilization, risk of resource privatization through IPR and accrual of benefits to the country and communities. As Schloen et al. (2011) argue, if GRFA is left within the scope of the general ABS regulations, several problems will need to be faced. Since the ABS systems under the NP and CBD are not fully developed, nothing more needs to be said on these prospective challenges.

In January 2015, the Commission urged FAO to adapt the "Elements to facilitate domestic implementation of access and benefit-sharing for different subsectors of genetic resources for food and agriculture", included in Annex B to the Report.

The Commission recalls the 2000 CGRFA in calling attention to the GRFA's "special features":

> ***5. The special nature of GRFA, which are included in agricultural biodiversity, its distinctive features and problems needing distinctive solutions, is widely acknowledged. The Conference of the Parties to the CBD, at its fifth meeting in 2000, considered the distinctive features of agricultural biodiversity to include the following:***
>
> ***(a) Agricultural biodiversity is essential to satisfy basic human needs for food and livelihood security;***
>
> ***(b) Agricultural biodiversity is managed by farmers; many components of agricultural biodiversity depend on this human influence; indigenous knowledge and culture are integral parts of the management of agricultural biodiversity;***
>
> ***(c) There is a great interdependence between countries for the genetic resources for food and agriculture;***

(d) For crops and domestic animals, diversity within species is at least as important as diversity between species and has been greatly expanded through agriculture;

(e) Because of the degree of human management of agricultural biodiversity, its conservation in production systems is inherently linked to sustainable use;

(f) Nonetheless, much biological diversity is now conserved ex situ in gene banks or breeders' materials;

(g) The interaction between the environment, genetic resources and management practices that occurs in situ within agro-ecosystems often contributes to maintaining a dynamic portfolio of agricultural biodiversity. (FAO** 2015**)

In 2000, CGRFA was mainly concerned with plant genetic resources, so these features mainly concern the situation for plants. It is an open but complex question whether these features apply to an equal degree to the other groupings of genetic resources. These distinctive features are used as arguments for CGRFA to regulate genetic resources for food and agriculture in a different manner than prescribed in the CBD and NP. There seems to be a tacit assumption here that implementing the NP will reduce access and in turn hinder efforts to ensure food security. The is little systematic or empirical evidence that a careful implementation of the Nagoya Protocol will have this effect, but to prevent it hindering the achievement of these objectives, there is a need to discuss and better understand the gene flow and potential in access to genetic resources for food and agriculture outside the plant sector. *When perusing discussions of the sectors in the other chapters in this volume, this author would encourage the reader to reflect on the whether these characteristics are accurate for other sectors than plants.*

The extent to which these assumptions apply to all six sectors (animals, aquaculture, forest, plants beyond the ITPGRFA, micro-organisms and invertebrates) needs to be further explored before using them as reasons for differentiating legal regulations. Thus a country needs to assess whether any of these GRFA groupings have these features or not. An assessment is therefore necessary to establish whether these sectors have specific needs with respect to ABS regulation, and what they are. Certain general features of these branches using GRFA can be identified even without having to explore the special situation of the sectors in each country.

The Nagoya Protocol itself foresees two important mechanisms for handling ABS in relation to special branches of genetic resources. Article 19 urges the development and updating of information on model contracts. The essence here is that sectoral and cross-sectoral model contracts can be negotiated under the auspices of the Nagoya Protocol to serve special purposes. This is one potential tool in preventing ABS from becoming fragmented by international organisations negotiating separate systems for access and benefit sharing.

2.6 Conclusions and Challenges for Sector Approaches

Access and benefit sharing of the dividends from genetic resources have now entered a critical phase following the entry into force of the Nagoya Protocol. One would expect to see examples of functioning benefit sharing contracts within relatively short time if the Nagoya system is not to lose momentum and the CBD is not to lose credibility. New genetic variations could possibly help plants adapt to a warmer climate. Collecting activities and collections could therefore prove very useful by making material available that can ease efforts to adapt to a changing climate.

Business representatives often claim that genetic resources are of limited value. At the same time, however, the business community is vocal in calling for secure access to genetic resources. This creates something of a paradox for access and benefit sharing. If genetic resources have limited value – actual or potential – why should access to genetic resources be important to business? Furthermore, the fact that patents are taken out on bio-innovation outcomes, the value for business created from the utilisation of genetic resources would appear to be not wholly insignificant.

If the current system of access and benefit sharing in relation to genetic resources does not provide funds for conservation and sustainable use of biological diversity, one of the raisons d'être of the CBD is in jeopardy. It is therefore increasingly urgent for the CBD to get the ABS to work as intended. The entry into force of the Nagoya Protocol represents a step in this direction. The new instrument, however, cannot solve these issues alone and much will therefore rely on functional implementation moving forward.

In light of the following chapters in this volume, it is interesting to reflect on whether the general regulation of the CBD and NP will apply in a fruitful manner to each sector or whether there will be a need for an open access regime where access is combined with complete freedom to privatise the inventions based on the genetic material while benefit sharing objective is not prioritised. In this scenario, those with the expertise of and in the different sectors need to inform law makers and legislative processes at the international level.

References

ABS Capacity Development Initiative (2014). http://www.abs-initiative.info/countries-and-regions/global/italy/the-international-treaty-and-the-nagoya-protocol-a-tandem-workshop/

Arico S (2010) Marine genetic resources in areas beyond national jurisdiction and intellectual property rights. In: Vidas D (ed) Law, technology and science for oceans in globalisation: IUU fishing, oil pollution, bioprospecting, outer continental shelf. Martinus Nijhoff, Leiden, pp. 383–396

Chiarolla C, Louafi S, Schloen M (2013) An analysis of the relationship between the nagoya protocol and instruments realted to genetic resources for food and agriculture and farmers' rights. In: Morgera E, Buck M, Tsioumani E (eds) The 2010 nagoya protocol on access and benefit-sharing in perspective – implications for international law and implementation challenges. Martinus Nijhoff, Leiden, pp. 83–122

Christinck A, Tvedt MW (2015) The UPOV convention, farmers' rights and human rights – an integrated assessment of potentially conflicting legal frameworks. Deutsche Gesellschaft für Internationale Zusammenarbeit (GIZ) GmbH, Bonn

Correa CM (2013) Plant genetic resources under the management and control of the contracting parties and in the public domain: how rich is the ITPGRFA's multilateral system? In: Halewood M, López Noriega I, Louafi S (eds) Crop genetic resources as a global commons – challenges in international law and governance. Routledge, New York, pp. 177–186

Drankier P, Elferink AGO, Bert Visser B, Takács T (2012) Marine genetic resources in areas beyond national jurisdiction: access and benefit-sharing. Int J Mar Coas Law 27(2):375–433

Elferink AGO (2007) The regime of the area: delineating the scope of application of the common heritage principle and freedom of the high seas. Int J Mar Coas Law 22(1):143–175

FAO (2015). Report of the Commission on Genetic Resources for Food and Agriculture. FAO, 15th Session, Rome, 19–23 January 2015 (CGRFA-15/15/Report)

FNI (2014). http://www.fni.no/

Greiber T (2011). Access and benefit sharing in relation to marine genetic resources from areas beyond national jurisdiction: a possible way forward. Bonn: IUCN. (BfN-Skripten, no. 301)

Halewood M, Andrieux E, Crisson L, Rwihaniza Gapusi J, Wasswa Mulumba J, Kouablan Koffi E, Yangzome Dorji T, Raj Bhatta M, Balma D (2013) Implementing 'mutually supportive' access and benefit sharing mechanisms under the plant treaty, convention on biological diversity, and nagoya protocol. Law, Environment and Development Journal 9(2):70–96

Hendrickx F, Koester V, Prip C (1993) The convention on biological diversity – access to genetic resources: a legal analysis. Environ Policy Law 23(6):250–258

Medaglia JC, Tvedt MW, Perron-Welch F, Jørem A, & Phillips F-K (2013). The interface between the Nagoya Protocol on ABS and the ITPGRFA at the international level – potential issues for consideration in supporting mutually supportive implementation at the national level. FNI Report, no. 1/2013. Fridtjof Nansens Institutt, Lysaker

Oberthür S and Rosendal GK, (2014). Global Governance of Genetic Resources: Background and Analytical Framework. In: Oberthür S, Rosendal GK (eds) Global governance of genetic resources access and benefit sharing after the nagoya protocol. Routledge, New York, pp. 1–18

Prip C, Rosendal GK, Andresen S, & Tvedt MW (2014). The Australian ABS framework – a model case for bioprospecting? FNI Report, no. 1/2014. Lysaker; Fridtjof Nansens Institutt

Schei PJ, & Tvedt MW (2010). 'Genetic resources' in the CBD: the wording, the past, the present and the future. FNI Report, no. 4/2010. Fridtjof Nansens Institutt, Lysaker

Schloen M, Louafi S & Dedeurwaerdere T (2011). Access and benefit-sharing for food and agriculture – current use and exchange practices, commonalities, differences and user community needs – a report from a multi-stakeholder expert dialogue. FAO Commission on Genetic Resources on Food and Agriculture, CGRFA, Rome (no. 59)

Schrijver N (2008) Sovereignty over natural resources: balancing rights and duties. Cambridge University Press, Cambridge

Tvedt MW (2013). Disentangling rights to genetic resources illustrated by aquaculture and forest sectors. Law Environ Dev J 9(2): 127–141. http://www.lead-journal.org/content/13127.pdf

Tvedt MW (2014) Beyond nagoya: towards a legally functional system of access and benefit-sharing. In: Oberthür S, Rosendal GK (eds) Global governance of genetic resources access and benefit sharing after the nagoya protocol. Routledge, New York, pp. 158–178

Tvedt MW (2015a) Access to plant genetic resources – legal questions for material on its way into the multilateral system of the plant treaty. Law Environ Dev J 11(1):3–16

Tvedt MW (2015b) Changes in the plant treaty – how can benefit sharing happen and the link to intellectual property rights – assessing the mutually supportiveness. Law Environ and Dev J 11(1):3–14

Tvedt MW, Schei PJ (2014) The term 'Genetic resources': flexible and dynamic while providing legal certainty? In: Oberthür S, Rosendal GK (eds) Global governance of genetic resources access and benefit sharing after the nagoya protocol. Routledge, New York, pp. 18–33

Tvedt MW & Young TR (2007). Beyond access: exploring implementation of the fair and equitable sharing commitment in the CBD. IUCN Environmental Policy and Law Paper, no. 67/2. Gland: IUCN

Young TR (2013) An international cooperation perspective on the implementation of the nagoya protocol. In: Morgera E, Buck M, Tsioumani E (eds) The 2010 nagoya protocol on access and benefit-sharing in perspective – implications for international law and implementation challenges. Martinus Nijhoff, Leiden, pp. 451–506

Agreements and Conventions

Agreed Interpretation of the International Undertaking, Rome, FAO, Conference Resolution 4/89, entered into force 29 November 1989, adapted at the Twenty-fifth Session of the FAO Conference.

Agreed Interpretation of the International Undertaking, Rome, FAO, Conference Resolution 3/91, entered into force 25 November 1991, adapted at the Twenty-sixth Session of the FAO Conference.

Agreement on Trade-Related Aspects of Intellectual Property (Annex 1C of the Agreement establishing the World Trade Organization) [TRIPS Agreement], WTO, 15 April 1994.

International Convention for the Protection of New Varieties of Plants 1991 [UPOV-1991], UPOV, 2 December 1961, UPOV/INF/6/1, as amended on 10 November 1972, 23 October 1978 and 19 March 1991.

International Undertaking on Plant Genetic Resources [International Undertaking 8/83], Rome, FAO, Conference Resolution 8/83, entered into force 23 November 1983, adapted at the Twenty-second Session of the FAO Conference.

International Treaty on Plant Genetic Resources for Food and Agriculture [ITPGRFA], Food and Agriculture Organization, 3 November 2001, 2400 UNTS 303, entered into force 29 June 2004.

Resolution 3 - The Interrelationship between the Convention on Biological Diversity and the Promotion of Sustainable Agriculture [Nairobi Resolution 3], entered into force 22 May 1992, the Nairobi Conference for the Adoption of the Agreed Text of the Convention on Biological Diversity.

United Nations Convention on the Law of the Sea [UNCLOS], United Nations, 10 December 1982, 1833 UNTS 3, entered into force 16 November 1994.

Chapter 3
Current Status and Perspectives in Marine Biodiscovery

Daria Firsova, Nipun Mahajan, Hiren Solanki, Christine Morrow, and Olivier P. Thomas

Abstract In recent years the field of marine biodiscovery has experienced profound changes. The initial approach based on the identification of small molecules from marine organisms through processes guided by the search for the targeted bioactivity has led to a number of successes, especially in the field of pharmaceutical research. By contrast, we would like to highlight the benefits of integrating a slightly different approach, mostly based on the construction of chemical libraries and strong collaboration with other fields of marine science including ecology, biology, taxonomy, microbiology, biochemistry, and chemistry in order to better meet the expectations of this approach in a context of sustainability.

3.1 Introduction

During the twentieth century the global population has increased from 1.65 billion to 7.4 billion (http://www.worldometers.info/). The demands of a rising population has resulted in devastating changes to many of the world's ecosystems, the depletion of natural resources and changes to the global climate (http://www.worldwatch.org/). Half of the world's forests have been cleared for human land use and the area of land used for cultivation has increased by approximately 13 % since measurements began in 1961 (www.unep.org).

With the increasing demand for terrestrial ecosystems to be used for cultivation and development, scientists have turned their attention to the exploration of the marine environment for sources of new biomolecules. The advent of SCUBA diving, submersibles, remotely operated vehicles (ROV's) and other technologies, have enabled the sampling of previously unexplored marine habitats. Only a tiny percentage of marine species have been investigated by scientists for diverse applications.

D. Firsova • N. Mahajan • H. Solanki • C. Morrow • O.P. Thomas (✉)
Marine Biodiscovery, School of Chemistry, National University of Ireland Galway, University Road, Galway, Ireland
e-mail: olivier.thomas@nuigalway.ie

R. Paterson, N. Lima (eds.), *Bioprospecting*, Topics in Biodiversity and Conservation 16, DOI 10.1007/978-3-319-47935-4_3

Many marine organisms have adaptations that enable them to live in habitats which we would consider hostile, such as the high pressure, low temperatures, absence of sunlight and in some cases, low oxygen saturation associated with the deep sea. The opportunity to exploit molecular strategies that have evolved in marine organisms, has had and will have, significant societal benefits. This is exemplified by the green fluorescent protein (GFP) which was first isolated from the bioluminescent hydromedusa *Aequorea victoria* (Shimomura et al. 1962). In 2008 the Nobel Prize in Chemistry was awarded to Osamu Schimomura, Martin Chalfie and Roger Tsien for their discovery and development of GFP, that is now one of the most important tools in cell and molecular research (Phillips 2001).

In addition to these primary biomolecules, a rapid interest has emerged in studying the diversity of small molecules produced by so-called specialized or secondary metabolism. Similar to terrestrial species, marine organisms biosynthesize unique small molecules such as terpenes, steroids, polyketides, peptides, alkaloids and porphyrins. The potential of novelty is huge in this context. In their review of new marine natural products from invertebrates, Leal et al. (2012) reported that of the 11 phyla that have so far been investigated, and which contain approximately 170,000 valid species, less than 1 % of them have been the subject of biodiscovery studies. Many of these compounds are derived from biosynthetic pathways that are uniquely marine (Garson 1993). The marine environment is unlike terrestrial habitats in that it contains an extremely high diversity of sessile invertebrates that rely on bioactive compounds to deter predators, prevent fouling, compete for space or paralyse prey (Paul et al. 2011).

Since the early 1960s, a race commenced involving the random harvesting of marine organism and analysing them for useful biomolecules. This area of research is referred to as 'marine bioprospecting'. No clear definition was given, but rather a broad concept was embraced involving the collection of organisms from a given region with the aim of identifying those that could be exploited for human use. The Food and Agriculture Organisation workshop on marine bioprospecting suggested that the term bioprospecting should be divided into two terms: 'biodiscovery' as the first phase of scientific research into a region's biodiversity and 'bioprospecting' as the second and subsequent phases of re-collection of biological resources for further investigation, with a view to commercial exploitation (http://www.fao.org/docrep/009/a0337e/A0337E15.htm).

Intensive marine bioprospecting began in the early twenty-first century in seas around Australia and New Zealand (Capon 2008) and later spread to all regions of the world, including Ireland (Rae et al. 2013), Scotland and Norway (Svenson 2013). The Biodiscovery Act published in 2004 in Queensland, Australia is without doubt a landmark in the development of this approach. In the same time, Ireland launched a large programme in marine biodiscovery called the Beaufort project (http://www.qub.ac.uk/research-centres/MarineBiodiscovery/). A Marine Biodiscovery Centre was established at Aberdeen, Scotland (http://www.abdn.ac.uk/ncs/departments/chemistry/marine-biodiscovery-centre-112.php) and a research centre in Marine Biotechnology and Biodiscovery by GEOMAR-Biotech at Kiel, Germany (http://www.geomar.de/en/research/fb3/fb3-mn/geomar-biotech/)

was also established. One of the most renowned centres in this area is located at the Scripps Institute of Oceanography, San Diego, USA (https://scripps.ucsd.edu/cmbb).

Due to the novelty of the chemical structures of marine metabolites, the main and most lucrative interest was focused on the search for new medicines (Baker 2015). After some initial screening of marine macro-invertebrates, most of the current research carried out by large consortia around the world is dedicated to the isolation and culture of marine micro-organisms (Jaspars et al. 2016; Reen et al. 2015; Rocha-Martin et al. 2014; Fuerst 2014; Muehling et al. 2013; Kurtboeke 2012; Capon 2012; Joint et al. 2010; Heidelberg et al. 2010). The main justification for using micro-organisms over macro-organisms, is the ability to perform large-scale culture of microbes in the laboratory which enables control over the production of particular compounds. In this context, several pharmaceutical companies, often in collaboration with academic laboratories, have organized expeditions throughout the world's oceans in a search for new drugs. The most striking outcomes of this research culminated in the marketing of several drugs that are used today to treat human diseases. In the first part of this chapter we will detail (a) the successes of this approach, but also (b) some of the limitations that have been identified.

In recent decades there has been a growing awareness of the negative impacts we are having on the environment. Governmental, non-governmental and international organizations have emphasised the need to address the problem of environmental damage caused by uncontrolled economic development. Initially, international legislations permitted the exploitation of resources without the requirement to share benefits with the people living in close vicinity to the natural resources being exploited. The European Environment Agency have identified climate change, pollution, ocean acidification, over-fishing and invasive species as serious threats to marine bioresources and diversity (http://www.eea.europa.eu/highlights/marine-biodiversity-life-in-seas).

It is timely therefore that we consider a new approach to marine biodiscovery that is centred upon a more sensitive and sustainable use of our marine resources. The Nagoya Protocol and the United Nations Convention on the Law of the Sea today provide a legal framework for helping to ensure benefit sharing when a bioresource leaves the country of origin (https://www.cbd.int/abs/text/; http://www.un.org/Depts/los/index.htm). In the second part of this chapter we will propose some perspectives for a more enlightened use of marine bioresources.

3.2 The Bioguided Approach

The main strategy used to identify new compounds of marine origin for cosmetic or pharmaceutical applications is the bioguided screening process (Fig. 3.1). Companies that invest in the marine environment to provide their products and incomes have developed their own screening facilities on specific targets. Expensive expeditions have been organized mostly by companies to investigate the world's oceans and collect a large diversity of marine species (Bhatia and Chugh 2015). However, in the 1990s benefit sharing agreements were not fully in place and the

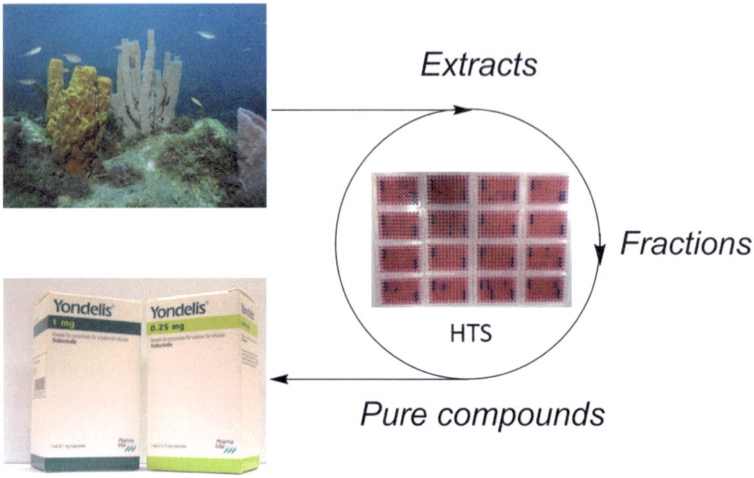

Fig. 3.1 Marine biodiscovery in the context of bioguided and high throughput screening (HTS)

expeditions usually did not involve the participation of the local population or scientists. Collecting permits were granted because the authorities were attracted by putative future benefits. Most of the collections contained a random assortment of marine organisms. Some culture collections of marine samples were then built especially in the laboratories of the companies which financially supported the expeditions (https://www.pharmamar.com/). Whilst some of the academic researchers who were involved in obtaining the original collecting permits were associated with parts of the studies, most if not all of the intellectual property rights remain with the companies that funded the initial collecting expeditions. This wealth is today mostly concentrated in the private sector as the awareness of the economic value was long to be recognized at national levels. In most cases some academic chemists participated in the expeditions and collections, but often marine biologists and specialist taxonomists were not associated with this work, because documenting the biodiversity and species identification were not considered as priorities. Different types of extracts were produced from all kinds of marine organisms, including microbes, algae and a large number of animals, and screened for a specific target after building the culture collection. The screens have focused on anticancer activities, mainly because marine compounds are often more cytotoxic than terrestrial ones, but also because cancers have a major impact on human health. Below we review the major successes in marine biodiscovery, most of which are in the pharmaceutical sector.

3.2.1 Fishes

Omega-3-Acids (Anti-Hypertriglycerimedia): Ethyl esters of omega-3-acids, isolated mainly from fish oils are used as anti-hypertriglyceridemic drugs (Fig. 3.2). Lovaza® is the formulation which chiefly contains the ethyl esters of

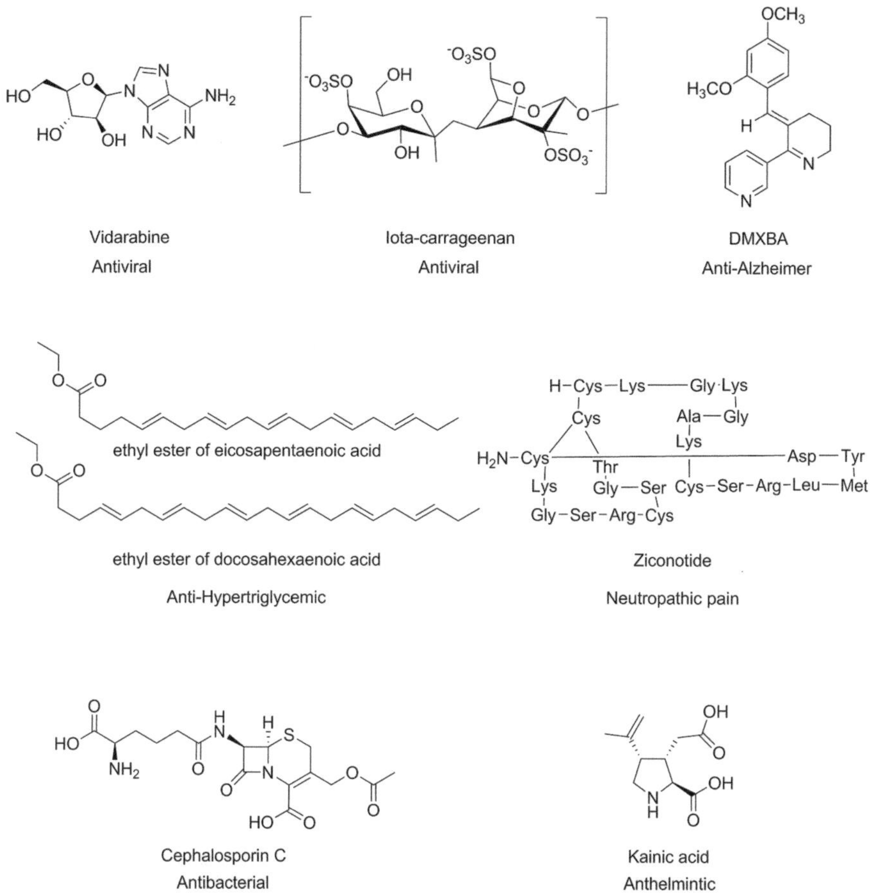

Fig. 3.2 Marine natural products with applications (except anticancer)

eicosapentaenoic acid and docosahexaenoic acid, along with other omega fatty acids esters. They are proven to reduce triglycerides and low density cholesterols and to increase high density cholesterols in the blood. They are used therapeutically, along with other statins and/dietary supplements, to lower triglycerides levels (Davidson et al. 2012; Koski 2008; Lovaza: GlaxoSmithKline 2014)

3.2.2 *Porifera*

Tethya Crypta

Cytarabine (Cytosar-U®): Cytarabine (Ara-C) is the first marine anticancer molecule and was discovered in 1945 by Werner Bergmann (Fig. 3.3). This active small molecule is a nucleoside isolated from the Caribbean sponge *Tethya crypta* (Newman et al. 2009) and is used for the treatment of leukemia. Intracellularly it is converted

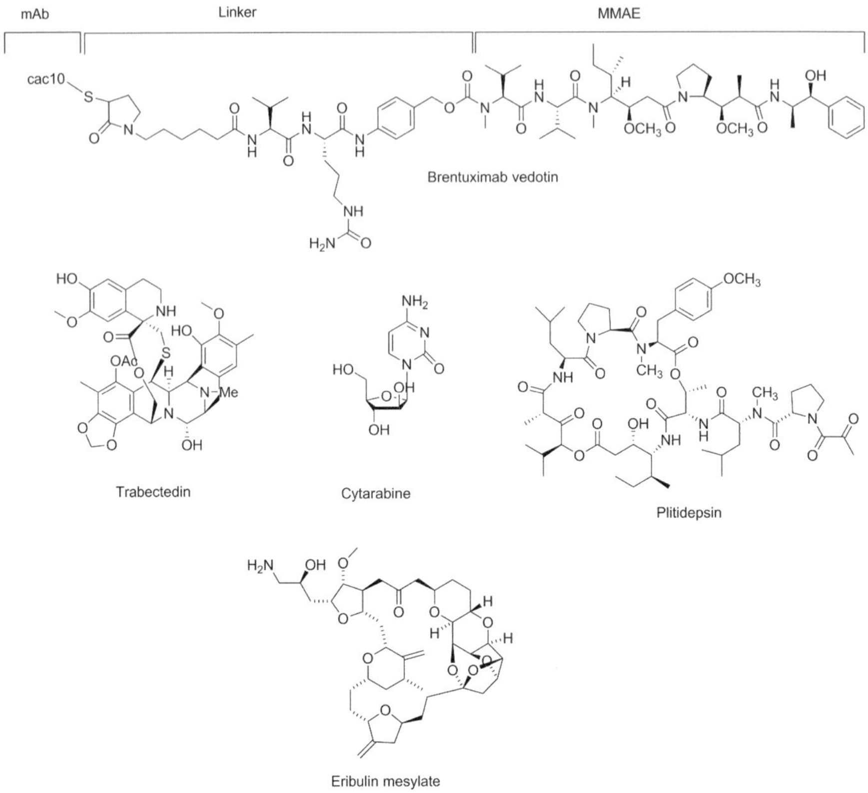

Fig. 3.3 Marine natural products as anticancer/antitumoral agents

into cytosine arabinoside triphosphate that acts as a cytotoxic agent inhibiting DNA polymerase and DNA synthesis (Mayer et al. 2010). Due to its discovery being 70 years ago, extensive studies and trials have been conducted and the drug has been in use for over 40 years. Cytarabine is a simple molecule with similarity to RNA bases.

Tethya Crypta **and** *Streptomyces Antibioticus* **(Actinobacteria)**
Vidarabine (Fig. 3.2) (Antiviral): Vidarabine (Ara-A) is a synthetic antiviral purine nucleoside derived from the natural product spongouridine. Currently obtained from *Streptomyces antibioticus*, it was isolated initially from the Caribbean sponge *Tethya crypta.* Being structurally similar to adenine, it functions as an antimetabolite. After being metabolically converted into its triphosphate form, it inhibits viral DNA polymerase and thus DNA synthesis. Vidarabine was approved by FDA in 1976. However due to its lower therapeutic window in comparison to other antiviral drugs in the market, it was discontinued in the USA in June 2001. Ophthalmic ointment is still in use in the EU against acute keratoconjunctivitis and recurrent epithelial keratitis. It is useful against rhabdoviruses, hepadnarviruses, herpes virus, poxvirus and shingles (Hayden and Douglas 1995; De Clercq 1993; Chabner and Glass 1996; Schabel Jr 1968).

Halichondria Okadai
Eribulin mesylate (Fig. 3.3) (Halaven®): Eribulin mesylate is a marine derived microtubule-targeted agent that was approved in 2010 for use against metastatic breast cancer in patients who have already undergone chemotherapy (Huyck et al. 2011). It is a structurally simplified synthetic analogue of halichondrin B which was originally isolated from the marine sponge *Halichondria okadai* (Dumontet and Jordan 2010). As there was a limited supply of the natural source, there was a clear need for a synthetic analogue. Simplified analogues were discovered, which included eribulin, while searching for the best synthetic route toward halichondrin B. Eribulin inhibits the growth of microtubules hence leading to G2/M cell cycle arrest and ultimately apoptotic cell death.

3.2.3 Chordata, Tunicata

Ecteinascidia Turbinata
Trabectedin (Fig. 3.3) (Yondelis®): Trabectedin was approved by the European Commission in 2007 for the treatment of patients with soft tissue sarcoma (SAS) and in 2009 for treatment of ovarian cancer. In the USA the drug was approved only in 2015 for treatment of SAS. Trabectedin is only advised to be used if previous chemotherapy has failed since severe, negative effects on human health are possible (D'Incalci and Galmarini 2010). The compound is an alkaloid isolated from the tunicate *Ecteinascidia turbinata*. Its chemical structure allows the molecule to bind to DNA *via* two tetrahydroisoquinoline rings, while the third ring is proposed to protrude and interact with DNA binding proteins. It is known that trabectedin also affects transcription factors and DNA repair pathways.

PM01183 (Fig. 3.4): This is another synthetic analogue of ecteinascidins, a potent antitumor compounds isolated from the tunicate *Ecteinascidia turbinata* (Rinehart et al. 1990). Two fused tetrahydroisoquinoline rings of PM01183 play a role of recognition and binding to DNA of tumor cells, while a tetrahydro-β-carboline protrudes from the DNA groove ultimately causing DNA damage *via* double strand breaks. This alkaloid is used for treatment of solid tumours and is currently being used in Phase II clinical trials.

Aplidium Albicans
Plitidepsin (Fig. 3.3) (Aplidin®): this highly potent depsipeptide plitidespin was isolated from the Mediterranean tunicate *Aplidium albicans* but is produced currently through synthesis by the Spanish company Pharmamar (Mayer et al. 2010). Plitidepsin is active against multiple myeloma. The drug causes apoptosis and inhibition of protein synthesis and acts as an angiogenesis inhibitor. Current clinical trials are promising and indicate that whether alone, or in combination with other active agents, plitidepsin will be useful in combating multiple myeloma (Mitsiades et al. 2008). At the chemical level, plitidepsin is a macrocycle that is constituted of 6 subunits and one side chain.

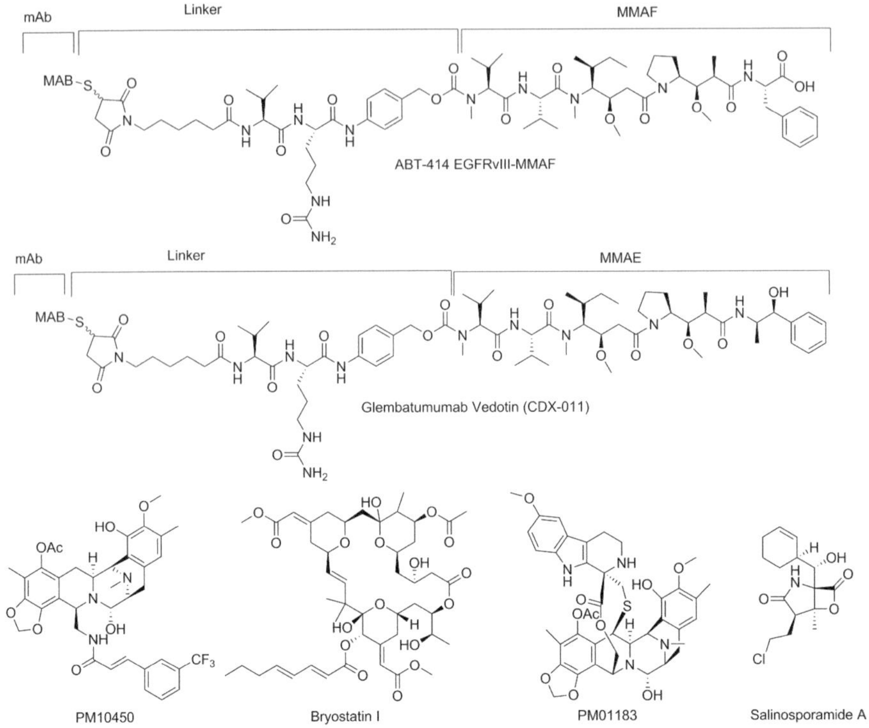

Fig. 3.4 Additional marine natural products as anticancer/antitumoral agents

3.2.4 *Bryozoa*

Bugula Neritina

Bryostatin: Bryostatin-1 (Fig. 3.4) is a macrocyclic lactone with antineoplastic activity isolated from a marine invertebrate bryozoan *Bugula neritina*. The main mechanism of action is modulation of the regulatory domain of protein kinase C. Transient contact to bryostatin-1 promotes activation of PKC, whereas long-lasting exposure advances huge down regulation of PKC. In various haematological and solid tumor cell lines, bryostatin-1 restrains multiplication, impels separation, and advances apoptosis. Stage I studies recommended noteworthy antineoplastic action against a few types of tumor and defined the main dose-limiting toxicity as myalgia. Bryostatin-1 has later been examined broadly in stage II clinical trials as a single agent in patients with malignancies including lymphoma, leukaemia, and melanoma. Although there is nominal single-agent activity, combination chemotherapy with standard medications is giving exceptionally reassuring results and indicates a new direction in cancer therapy. The major toxicities are myalgias, nausea, and vomiting.

However, because of practical difficulty in harvesting it from the naturally occurring bryozoan, and long chemical syntheses that were too cumbersome for drug makers, the compound has fallen out of favor as drug candidates. Recently, the cloud that was hanging over the bryostatins has begun to lift. Animal tests show that bryostatin 1 augments memory and could be used to treat Alzheimer's disease and strokes, and some preliminary studies show it could help eliminate HIV. Based upon a number of Blanchette Rockefeller Neurosciences Institute (BRNI) pre-clinical and autopsy-validated human tissue studies, PKCε deficits have been implicated as a potential cause of Alzheimer's disease.

Neurotrope Bioscience has an exclusive license to develop and commercialize bryostatin for cognitive disorders. The Company is approaching the treatment of Alzheimer's disease through the activation of PKCε. Bryostatin is a potent modulator of an enzyme protein kinase C, epsilon (PKCε). Activation of PKCε isozymes improves learning and memory by (a) inducing synthesis of proteins required for long-term memory, (b) increasing brain neurotropic factors, (c) reducing neurotoxic amyloid accumulation and tau protein hyperphosphorylation and (d) inducing synaptogenesis. Currently Neurotrope Bioscience is conducting Phase II clinical trials on bryostatin for Alzheimer's disease (Pettit et al. 1982; Kraft et al. 1986; Berkow and Kraft 1985; Kortmansky and Schwartz 2003).

3.2.5 Mollusca

Conus Magnus

Ziconotide (Fig. 3.2) (Prialt®, Neutropathic pain): This is a synthetic form of ω-conotoxin, a toxin isolated from the cone snail *Conus magus*. It was made by solid phase synthesis in 1987 and was approved by FDA in 2004 and EMEA in 2005. It acts by blocking *N*-type voltage-sensitive calcium-channels at neuromuscular junctions and thus inhibits the release of pro-nociceptive chemicals. After morphine, ziconotide was the first intrathecal analgesic drug to be approved. However, because of the invasive procedure involved, it is only recommended for the management of severe chronic pain, which otherwise cannot be managed by orally available analgesics. It is preferred over morphine, because of non-opioid related side effects. However its low dose titration, in combination with opioid based therapy, is recommended for neuropathic pain in cancer patients (de la Calle Gil et al. 2015; Staats et al. 2004; Atanassoff et al. 2000).

Jorunna Funebris **and** *Neopetrosia* **sp. (Porifera)**

PM10450 (Zalypsis®): PM10450 (Fig. 3.4) is a synthetic derivative produced by Pharmamar and derived from the original alkaloids Jorumycin and Renieramycin J, isolated from the nudibranch *Jorunna funebris* and the sponge *Neopetrosia* sp. respectively (Petek and Jones 2014). Like plitidepsin, zalypsis acts against multiple myeloma by causing cell cycle arrest, apoptosis and DNA double stranded breaks. The IC_{50} values of the antimyeloma agent zalypsis are outstanding – within the picomolar or low nanomolar range. Moreover, zaplypsis showed activity against resistant cells, which means that the drug can potentially be used for patients already treated for multiple myeloma.

Dolabella Auricularia **and** *Symploca* **sp. (Cyanobacteria)**

Brentuximab vedotin (Adcetris®): Brentuximab vedotin is an anticancer agent that originated from the highly potent dolastatin 10 molecule. Dolastatin 10 (Fig. 3.3) was first isolated from the sea hare *Dolabella auricularia* (Pettit et al. 1987). Subsequently, the cyanobacteria *Symploca* sp. was identified as the true source of the molecule (Luesch et al. 2001). This compound found in this mollusk/cyanobacterium complex has led to a large diversity of drugs in preclinical or clinical trials. Due to a high interest in dolastatin 10, several analogues were synthesized including molecules of the class auristatin (Pettit et al. 1998). Dolastatin 10 entered into Phase II clinical trials, but due to a very high toxicity it was substituted by the synthetic analogue – monomethyl auristatin E (MMAE) coupled to a monoclonal antibody (mAb) (Perez et al. 2005). The concept of Antibody Drug Conjugate (ADC) (Senter and Sievers 2012) is one of a drug composed of coupled cytotoxic molecules to tumor-selective antibodies *via* a linker. ADCs are highly selective due to tumor-seeking antibodies that preferentially bind to tumor cells over normal cell. The antibody used in this particular ADC is the one targeting CD30, which is an antigen highly expressed in Hodgkin lymphoma cells. The linker used with this ADC is valine-citrulline (Val-Cit) dipeptide linker that releases MMAE after proteolysis. Brentuximab vedotin was approved quickly and is currently in the market for treatment of Hodgkin lymphoma.

ABT-414 EGFRvIII-MMAF: ABT 414 is an ADC that links the anti-Epidermal Growth Factor Receptor (EGFR) antibody ABT-806 to monomethylaurisatin F (MMAF) *via* a linker (Fig. 3.4) (Reilly et al. 2016). MMAF conjugate is as much a potent antimitotic agent as MMAE, described previously, but the main difference between them is the presence of phenylalanine at the C-terminus of MMAF. ABT 414 is designed in such way that it binds to cancer cells that are overexpressing EGFRde2-7 (EGFRvIII) and EGFR, causing major recovery even in the most difficult cases. Moreover it is possible to use this ADC in conjugation with radiation and temozolomide treatment causing significant positive effects in glioblastoma models.

Glembatumumab Vedotin (CDX-011): Glembatumumab Vedotin is an ADC that includes in its structure the glycoprotein nonmetastatic B (GPNMB) mAb linked to MMAE (Fig. 3.4) (Vahdat and Chan 2015). In fact GPNMB is an attractive target because it is overexpressed in 85 % of patients with metastatic melanoma including triple-negative breast cancer that is particularly challenging to treat. Glembatumumab vedotin is currently under development by Celldex therapeutics for treatment of metastatic breast cancer and stage III and IV melanoma.

PSMA-ADC: Prostate-specific membrane antigen (PSMA)-ADC is a drug that consists of fully human PSMA mAb linked to MMAE *via* a valine-citrulline dipeptide linker (Ma et al. 2006). ADC is highly specific since PSMA is a typical cell surface marker of prostate cancer (Davis et al. 2005). The company responsible for the production and development is Progenics Pharmaceuticals (Tarrytown, NY, USA), and the drug is currently in Phase II of clinical trials.

GSK2857916: GSK2857916 is constituted by an antagonistic anti-B-cell maturation antigen (BCMA) mAb linked to MMAF *via* a non-cleavable linker (Tai et al.

2014). In this particular ADC, the mAb was afucosylated, which improved the binding affinity of ADC towards BCMA overexpressed in multiple myeloma cells. GSK2857916 causes G2/M cell cycle arrest and apoptosis, the effect of which is enhanced by the ability of the drug to (a) cause multiple myeloma cell lyses, (b) eliminate these vastly distributed cells and (c) initiate their macrophage-mediated phagocytosis. The drug is currently under development by GlaxoSmithKline (GSK), Brentford, UK.

DNIB0600A: This ADC consists of a IgG1 anti-multi-transmembrane, sodium-dependent phosphate transporter (NaPi2b) mAb linked to MMAE through an undisclosed linker (Burris et al. 2014). Currently Phase I and Phase II trials are on-going on patients with non-small cell lung cancer, platinum resistant ovarian cancer and others. Genentech/Roche (San Francisco, CA, USA) produces DNIB0600A.

Pinatuzumab vedotin (DCDT-2980S): Pinatuzumab vedotin is another ADC that is composed of humanized anti-CD22 IgG1 mAb connected to MMAE through maleimidocaproyl-valine-citrulline-p-aminobenzoyloxycarbonyl linker (Li et al. 2013). CD22 is an attractive target since this antigen is overexpressed solely in B cells and non-Hodgkin lymphomas. *In vitro* and *in vivo* studies showed tumor regressions in mouse and cynomolgus monkey models. Current development and production is by Genentech/Roche (San Francisco, CA, USA).

Polatuzumab vedotin (DCDS-4501A): this ADC is produced by Genentech. Its structure is composed of the antimitotic MMAE coupled to anti-CD79b mAb (Health 2016b). Polatuzumab vedotin exhibits a selectivity towards the transmembrane protein CD79b expressed on B cell lymphoma. Upon binding to CD79B, the linker connecting mAb and MMAE ceases through a proteolytic cleavage and MMAE causes G2/M cell cycle arrest and apoptosis through tubulin polymerization. This particular drug is in Phase II clinical trials as an alternative to pinatuzumab vedotin for treatment of follicular B cell lymphoma and Phase I trials against different types of lymphomas.

AGS-16C3F: AGS-16C3F is an ADC that includes in its structure AGS-16C, a human IgG2k mAb specific to ectonucleotide pyrophosphatase/phosphodiesterase family member 3 (ENPP3), attached to MMAF *via* a maleimido-caproyl linker (Newman and Cragg 2014). This drug specifically binds with high affinity to AGS-16 that is overexpressed in renal and liver carcinoma, after which the MMAF is released via proteolytic cleavage of the linker. MMAF then acts by inhibition of tubulin polymerization that causes G2/M cell cycle arrest and apoptosis (Health 2016a). The drug is currently produced by Agensys & Astellas Pharma (Northbrook, IL, USA) and the phase I clinical trials are ongoing.

AGS-67E: AGS-67E is an ADC that consists of the human IgG2 anti-CD37 antibody attached to MMAE *via* a maleimidocaproyl – valine – citrulline – p – aminobenzyloxycarbonyl linker (Pereira et al. 2015). CD-37 is a tetraspanin transmembrane protein overexpressed on B cells. It is involved (a) in the role of a death receptor (Lapalombella et al. 2012), (b) possibly in the regulation of B/T-cell interaction and proliferation and (c) in immune responses (van Spriel et al. 2004; Knobeloch et al. 2000). AGS-67E exhibited highly potent activity against several non-Hodgkin lymphoma and chronic lymphocytic leukemia cell lines *via* apoptosis events and cell-cycle alternations. Current development is conducted by Agensys & Astellas Pharma (Northbrook, IL, USA).

ASG-15ME: this particular ADC was developed for treatment of advanced urothelial cancer (Morrison et al. 2016). The drug is constituted by the human gamma 2 antibody (Igγ2) coupled to MMAE *via* a protease-cleavable linker. The mAb is directed to SLITRK6, the overexpressed antigen in bladder tumor cells. The *in vitro* and *in vivo* studies by Morrison et al. (2016) revealed excellent anti-tumor activity of ASG-15ME against bladder and lung models (Morrison et al. 2016); it is produced by Seattle Genetics (Bothell, WA, USA).

ENFORTUMAB VEDOTIN: ENFORTUMAB VEDOTIN (ASG-22ME, formerly AGS-22M6E) is an ADC sourced under a Phase 1 clinical trial supported by Astellas Pharma Inc and Seattle Genetics Inc. ASG-22ME contains a human monoclonal antibody, AGS-22 focusing on the cell adhesion molecule nectin-4 which is a tumor related antigen, and is over expressed in variety of cancers including breast, bladder, lung and pancreatic malignancy. The antibody is conjugated to the engineered cytotoxic agent MMAE, by means of an exclusive catalyst cleavable linker (AGS-22CE, Seattle Genetics' proprietary innovation), with potential anti-neoplastic activity. The monoclonal counter-acting agent moiety of AGS-22CE specifically ties to nectin-4. After internalization and proteolytic cleavage, MMAE ties to tubulin and hinders its polymerization, which brings about G2/M stage capture and instigates apoptosis in nectin-4 over communicating tumor cells. The linker framework is intended to be steady in the circulatory system and discharges the cell-killing specialists once inside target malignancy cells. This methodology is expected to diminish considerably much of the poisonous impacts patients might encounter amid treatment with customary chemotherapy while improving the antitumor activity. Currently, ASG-22ME is in a phase 1 clinical trial to evaluate the safety and antitumor activity of increasing doses of ASG-22ME in patients with solid tumors (Perez et al. 2014; Roberts et al. 2013; Trail 2013).

DEDN6526A: DEDN6526A is an antibody–drug conjugate (ADC) sourced from the same mollusk/cyanobacterium complex and under Phase – 1 clinical trial supported by Genentech Inc/Roche Inc. In preclinical studies, DEDN6526A confirmed dose-dependent antitumor activity in ETBR-expressing tumor xenografts. The preparation comprises the antimitotic agent MMAE combined to humanized immunoglobulin G1 anti-endothelin B receptor (ETBR) monoclonal antibody via a protease labile linker. ETBR is a G-protein-coupled receptor that can activate RAF/MEK signaling, and is over expressed in more than 50 % of metastatic melanomas. It is allied with malignant conversion of melanocytes and with potentiation of metastatic spread, which could explain its role in the development of melanoma. Clinical benefits were observed starting at a dose of 1.8 mg/kg and the most common adverse events were fatigue, chills, diarrhea, alopecia, nausea, headache, decreased appetite, infusion-related reaction, peripheral sensory neuropathy, asthenia, and vomiting. The drug is predicted to be a potent option for monotherapy and also viable for combinational chemotherapy (Infante et al. 2014).

DMUC5754A: DEDN6526A or Sofituzumab Vedotin is an Antibody–drug conjugate (ADC) under Phase – 1 clinical trial supported by Genentech Inc/Roche Inc.

In preclinical cancer studies, DMUC5754A established selective targeting of MUC16 and antitumor activity. DMUC5754A consists of a humanized monoclonal antibody against MUC16, a transmembrane glycoprotein that is over expressed in ovarian cancers, linked via labile linker to monomethyl auristatin (MMAE), a potent anti-mitotic drug that inhibits cancer cells' ability to form microtubules. When the drug binds to MUC16, a linker molecule releases MMAE into the cancer cell and destroys it. Clinical pharmacokinetic results from phase -1 study in patients with advanced, recurrent, platinum-resistant ovarian cancer confirmed that 2.4-mg/kg dose is the potentially clinical relevant dose with fatigue nausea, vomiting, decreased appetite, diarrhoea, and peripheral neuropathy as side effects. These results are promising and viable representing a novel type of therapy for ovarian cancer, with effectiveness in platinum-resistant ovarian cancer, which is the hardest type of ovarian cancer to treat (Liu et al. 2013; Wang et al. 2015; Weekes et al. 2016).

DSTP3086S: DSTP3086S, or Vandortuzumab vedotin, is an Antibody–drug conjugate (ADC) under Phase – 1 clinical trial supported by Genentech Inc/Roche Inc. DSTP3086S is composed of a humanized IgG1 mAb against Six Transmembrane Epithelial Antigen of the Prostate (STEAP1), linked *via* a protease-cleavable peptide linker to MMAE. It is a potent anti-mitotic drug that inhibits cancer cells' ability to form microtubules. A phase -1 trial demonstrated an acceptable safety and tolerability profile with minimal adverse side effects. Anti-tumor activity was detected in prostate cancer with acceptable pharmacokinetics that was linear and predictable. With these promising results DSTP3086S is expected to be a viable option to treat metastatic castration-resistant prostate cancer (Danila et al. 2013).

HuMax-CD74: HuMax-CD74 is an ADC under Phase 1 clinical trials supported by GenMab in collaboration with Seattle Genetics Inc. This compound is combined with MMAE, a potent anti-mitotic drug that inhibits the ability of cancer cells to form microtubules. This ADC uses HuMax-CD74, an antibody that targets the HLA class II histocompatibility antigen gamma chain (CD74), which is expressed on a wide range of haematological malignancies and solid tumors. The selective targeting of cancer cell approach by HuMax-CD74 is expected to establish new trends in treatment options for ovary, cervix, endometrium, bladder, prostate head and neck cancers (Van Berkel et al. 2012).

Indusatumab vedotin: MLN-0264 (Indusatumab vedotin) is an Antibody–drug conjugate (ADC) under Phase – 1 clinical trial supported by Takeda/Millennium Pharmaceuticals in collaboration with Seattle Genetics Inc. MLN-0264 contains a monoclonal antibody directed against guanylyl cyclase C (GCC or GUCY2C), a transmembrane receptor normally found on intestinal cells and dopamine neurons in the brain. It is also over expressed on the surface of gastrointestinal cancers conjugated *via* a labile linker to MMAE, an auristatin derivative and a potent microtubule inhibitor, with potential antineoplastic activity. The monoclonal antibody moiety of MLN0264 selectively binds to GCC and upon internalization and proteolytic cleavage, MMAE binds to tubulin and inhibits its polymerization, resulting in G2/M phase arrest and tumor cell apoptosis in GCC-expressing tumor cells. Currently a Phase -1 clinical trial is ongoing to assess the efficacy, safety and tolerability of MLN0264 in patients with recurrent or metastatic guanylyl cyclase C

(GCC)-positive adenocarcinoma of the stomach or gastro-oesophageal junction. It is also predicted to be a suitable treatment option for pancreatic cancer as a monotherapy agent or in combinational chemotherapy (Cruz Zambrano et al. 2014; Zhang et al. 2013; Teng et al. 2014).

3.2.6 Nemertea

DMXBA/GTS-21 (Alzheimer): DMXBA (GTS-21) (Fig. 3.2) is a synthetic analogue of anabaseine, which is currently in phase II studies. GTS-21 promotes the cognitive functions in experimental animals and humans. Anabaseine is present in various species of carnivorous marine worms, belonging to the Phylum Nemertea. GTS-21 selectively stimulates various types of animal nicotinic acetylcholine receptors such as $\alpha 12\beta 1\gamma\delta$ (embryogenic) or $\alpha 12\beta 1\gamma\varepsilon$ (adult) and $\alpha 7$ AChRs (Kem et al. 2006).

3.2.7 Rhodophyta

Digenea Simplex
Kainic acid (Anthelmintic): Kainic acid (Fig. 3.2) was originally isolated in 1953 from the red seaweed *Digenea simplex*, commonly known as "Kainin-Sou" (Davies 2007). It is used for nematodal infections in Japan. Additionally, it acts as a potent neuroexcitatory amino acid agonist to kainite receptors. Based on a dose response, it acts as a CNS stimulant in low doses, whereas as a neurotoxin at 10–30 mg/kg in mice. Kainic acid is used in laboratory animal models to study epilepsy and Alzheimer's disease (Baker 2015).

Rhodophyta
Iota-carrageenan (Antiviral): Iota-carrageenan (Fig. 3.2) is a linear sulphated polymer isolated mainly from red edible seaweeds belonging to the Rhodophyta. Also available as an Over-The-Counter (OTC) drug, it is a broad spectrum anti-viral molecule, which prevents the virus attaching to the cell by acting as physical barrier. It is proven to be effective for the prevention and treatment of (a) the common cold and influenza and (b) adenoviruses causing conjunctivitis (Eccles et al. 2010; Grassauer et al. 2008; Girond et al. 1991).

3.2.8 Fungi

Acremonium Chrysogenum **(Ascomycota)**
Cephalosporin C (Anti-infective): Cephalosporin C, one of the well-known β-lactam antibiotics, was isolated initially from the marine derived fungus, *Acremonium chrysogenum*, previously known as *Cephalosporium* sp. Cefalotin is a synthetic analogue of cephalosporin C, which is the first marketed cephalosporin antibiotic,

used because of its high potency compared to cephalosporin C (Fig. 3.2) (Abraham and Newton 1961).

3.2.9 *Bacteria*

Salinispora Tropica **(Actinobacteria)**
Salinosporamide A: Salinosporamide A (Marizomib®; NPI-0052) (Fig. 3.4) is a structurally and pharmacologically unique β-lactone-γ-lactam discovered from the marine actinomycete, *Salinispora tropica* and is a potent proteasome inhibitor. It is currently under phase-1 trials sponsored by Triphase Corporation. Marizomib, is established as an efficient single agent and in combination with biologics, chemotherapeutics and targeted therapeutic agents in models for (a) multiple myeloma, (b) Waldenstrom's macroglobulinemia, (c) chronic and acute lymphocytic leukemia, (d) mantle cell lymphoma and (e) glioma, colorectal and pancreatic cancer models through extensive preclinical evaluation in a variety of hematologic and solid tumor models. Marizomib has also exhibited synergistic activities in tumor models in combination with bortezomib, various histone deacetylase inhibitors and the immunomodulatory agent lenalidomide (Revlimid®) (Nett et al. 2009; Fenical et al. 2009).

Treatment of multiple tumor cell lines with marizomib and the histone deacetylase inhibitor, vorinostat, resulted in a highly synergistic antitumor activity. The combination of full dose marizomib with vorinostat is reported to be tolerable in patients with safety findings and consistent with either drug alone. These studies were granted the framework for continued clinical trials in patients with multiple myeloma, leukemias, lymphomas, and solid tumors. The compound is also expected to be an alternative for those patients who have failed bortezomib treatment for multiple myeloma, and in those where other proteasome inhibitors have not demonstrated significant efficacy (Potts et al. 2011; Millward et al. 2012; Ling et al. 2010). See also:

- *Streptomyces antibioticus* (Actinobacteria) (and *Tethya crypta*) above
- For the work on Vidarabine (Antiviral): Vidarabine (Ara-A) see Sect. 3.1.1 Porifera

3.3 Screening Marine Chemical Libraries

The results obtained over the decades in the search from new drugs from the sea, show several successes that have had major benefits to society. However some of the methods employed in the discovery of these compounds could be considered questionable. At the beginning of the twenty-first century several countries realized the huge potential of their marine bioresources and the benefits that could accrue. They demanded that legislation was put in place to protect their genetic resources, e.g. the

Convention on Biological Diversity and in particular the Nagoya Protocol. Given the potential marine resources that remain untapped in many countries, applications of these recent legislations are required to ensure the rights of each country are properly safeguarded (Lallier et al. 2014). Very importantly, the ABS Capacity Development Initiative was established in 2006 to support the development and implementation of national regulations on "Access and Benefit Sharing" (ABS) and also involves the marine environment especially in the Pacific and Caribbean Islands (http://www.abs-initiative.info/about-us/). Due to the broad remit of marine bioprospecting, legislation on bioprospecting in waters that are outside of any particular countries territorial waters might also be necessary.

In order to ensure maximum societal benefits from bioprospecting, each maritime country should be able to organize expeditions with the combined goals of documenting and describing their biodiversity and collecting samples for biodiscovery. The convention on biological diversity and the Nagoya Protocol on Access to Genetic Resources and the Fair and Equitable Sharing of Benefits Arising from their Utilization (Nagoya Protocol) have established a strong basis for a more sustainable and equitable use of marine bioresources (Lallier et al. 2014). Marine biodiscovery and prospecting should be performed with the assistance of the international scientific community if the expertise does not exist in that country. Well organized biorepositories should be set up in all countries or in international facilities and legislations on the use of this resource established (Fig. 3.5). In this sense the examples of Australia (WA Marine Bioresources Library), Norway (Marbank – National Marine BioBank) and Ireland (Repository of the Marine Institute) should inspire other countries and especially those with access to a rich marine biodiversity. The richness of our oceans is a resource that should be protected and used for the benefit of all humanity.

One of the principal shortfalls of the previous approaches was the lack of correct taxonomic identification of the taxa discovered. In most cases only the species leading to bioactive compounds were identified. It is time to support the proper taxonomic identification of marine species including the use of DNA sequencing. In addition to taxonomic identification, much more work is required in describing common macro-invertebrates as conservative estimates are that only one-third to one-quarter of marine species have been formally described (Appeltans et al. 2012). Both approaches are needed to better understand and sustainably exploit, our marine biodiversity.

The second step of the process should include the study of the metabolome of all collected species. Whilst this is a long term aim, it is necessary to help describe the chemical richness of our oceans as much as its biological diversity. The role of natural product chemists is therefore essential towards this end and the benefits are numerous: (i) the construction of a chemical library will accelerate the process of finding hits through a more efficient dereplication processes; (ii) the ecological role of these compounds could be assessed; (iii) biochemicals as used in chemotaxonomy, will contribute essentially to an integrative approach to systematics hence assisting the search for novel sources of active compounds; (iv) more solid hypotheses would be proposed for metabolic pathways hence facilitating biological synthesis; and (v) a

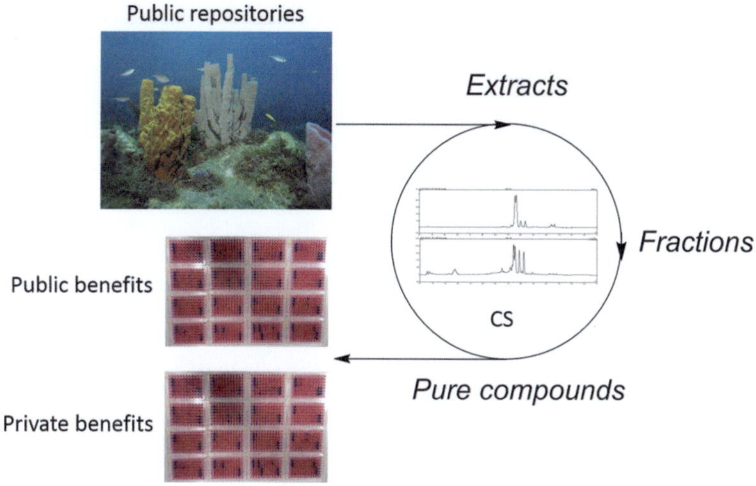

Fig. 3.5 Marine biodiscovery in the context of the construction of chemical libraries

large array of biological activities may be screened. Finally, we have to shift from a bioguided approach whereby a large number of extracts derived from marine organisms are investigated randomly in order to identify a small number of bioactive compound to a more constructive and environmentally sustainable approach whereby one compound is screened against a large array of bioassays in order to find its best target.

These goals will be obtained through strong collaboration and interdisciplinary research, where natural product chemists are key partners in this process. Since large pharmaceutical companies usually hold the intellectual rights to the research, the chemists will have to choose between interacting solely with companies and pharmacologists in the search for potentially large financial benefits, or to work in strong collaborations with marine biologists, microbiologists and ecologists for fewer monetary rewards. Although the latter approach may be more time consuming, it will enable a better understanding of our oceans and the resources they contain, which in the long-term will be of much greater benefit for the sustainable use of the ocean's resources and ultimately the future of humanity.

References

Abraham EP, Newton GGF (1961) The structure of cephalosporin C. Biochem J 79(2):377–393

Appeltans W, Ahyong Shane T, Anderson G, Angel Martin V, Artois T, Bailly N, Bamber R, Barber A, Bartsch I, Berta A, Błażewicz-Paszkowycz M, Bock P, Boxshall G, Boyko Christopher B, Brandão Simone N, Bray Rod A, Bruce Niel L, Cairns Stephen D, Chan T-Y, Cheng L, Collins Allen G, Cribb T, Curini-Galletti M, Dahdouh-Guebas F, Davie Peter JF, Dawson Michael N, De Clerck O, Decock W, De Grave S, de Voogd Nicole J, Domning Daryl P, Emig Christian C, Erséus C, Eschmeyer W, Fauchald K, Fautin Daphne G, Feist Stephen W,

Fransen Charles HJM, Furuya H, Garcia-Alvarez O, Gerken S, Gibson D, Gittenberger A, Gofas S, Gómez-Daglio L, Gordon Dennis P, Guiry Michael D, Hernandez F, Hoeksema Bert W, Hopcroft Russell R, Jaume D, Kirk P, Koedam N, Koenemann S, Kolb Jürgen B, Kristensen Reinhardt M, Kroh A, Lambert G, Lazarus David B, Lemaitre R, Longshaw M, Lowry J, Macpherson E, Madin Laurence P, Mah C, Mapstone G, McLaughlin Patsy A, Mees J, Meland K, Messing Charles G, Mills Claudia E, Molodtsova Tina N, Mooi R, Neuhaus B, Ng Peter KL, Nielsen C, Norenburg J, Opresko Dennis M, Osawa M, Paulay G, Perrin W, Pilger John F, Poore Gary CB, Pugh P, Read Geoffrey B, Reimer James D, Rius M, Rocha Rosana M, Saiz-Salinas José I, Scarabino V, Schierwater B, Schmidt-Rhaesa A, Schnabel Kareen E, Schotte M, Schuchert P, Schwabe E, Segers H, Self-Sullivan C, Shenkar N, Siegel V, Sterrer W, Stöhr S, Swalla B, Tasker Mark L, Thuesen Erik V, Timm T, Todaro MA, Turon X, Tyler S, Uetz P, van der Land J, Vanhoorne B, van Ofwegen Leen P, van Soest Rob WM, Vanaverbeke J, Walker-Smith G, Walter TC, Warren A, Williams Gary C, Wilson Simon P, Costello Mark J (2012) The magnitude of global marine species diversity. Curr Biol 22(23):2189–2202

Atanassoff PG, Hartmannsgruber MWB, Thrasher J, Wermeling D, Longton W, Gaeta R, Singh T, Mayo M, McGuire D, Luther RR (2000) Ziconotide, a new n-type calcium channel blocker, administered intrathecally for acute postoperative pain. Reg Anesth Pain Med 25(3):274–278

Baker BJ (2015) Marine biomedicine: from beach to bedside. CRC Press/Taylor & Francis Group, Boca Raton

Berkow R, Kraft A (1985) Bryostatin, a non-phorbol macrocyclic lactone, activates intact human polymorphonuclear leukocytes and binds to the phorbol ester receptor. Biochem Biophys Res Commun 131(3):1109–1116

Bhatia P, Chugh A (2015) Role of marine bioprospecting contracts in developing access and benefit sharing mechanism for marine traditional knowledge holders in the pharmaceutical industry. Glob Ecol Conserv 3:176–187

Burris HA, Gordon MS, Gerber DE, Spigel DR, Mendelson DS, Schiller JH, Wang Y, Choi Y, Kahn RS, Wood K, Maslyar DJ, Infante JR A phase I study of DNIB0600A, an antibody-drug conjugate (ADC) targeting NaPi2b, in patients (pts) with non-small cell lung cancer (NSCLC) or platinum-resistant ovarian cancer (OC). In: ASCO annual meeting, 2014.

Capon RJ (2008) Australian marine biodiscovery. Planta Med 74(9):930–931

Capon RJ (2012) Microbial biodiscovery: back to the future. Curr Top Med Chem 12(14):1471–1478

Chabner BA, Glass J (1996) Cytidine analogues. In: Chabner BA, Londo DL (eds) Cancer chemotherapy and biotherapy: principles and practice. Wolters Kluwer, Philadelphia, pp. 213–233

Cruz Zambrano C, Almhanna K, Messersmith WA, Rodon Ahnert J, Ryan DP, Faris JE, Jung J, Fasanmade A, Wyant T, Kalebic T (2014) MLN0264, an investigational antiguanylyl cyclase C (GCC) antibody-drug conjugate (ADC), in patients (pts) with advanced gastrointestinal (GI) malignancies: phase I study. J Clin Oncol 32(15_suppl):3546

D'Incalci M, Galmarini CM (2010) A review of trabectedin (ET-743): a unique mechanism of action. Mol Cancer Ther 9(8):2157–2163

Danila DC, Szmulewitz RZ, Higano CS, Gilbert H, Kahn RS, Wood K, Agarwal P, Lin K, Kabbarah O, Fine BM (2013) A phase I study of the safety and pharmacokinetics of DSTP3086S, an anti-STEAP1 antibody-drug conjugate (ADC), in patients (pts) with metastatic castration-resistant prostate cancer (CRPC). In: ASCO annual meeting 2013, vol 15_suppl. pp 5020

Davidson MH, Johnson J, Rooney MW, Kyle ML, Kling DF (2012) A novel omega-3 free fatty acid formulation has dramatically improved bioavailability during a low-fat diet compared with omega-3-acid ethyl esters: The ECLIPSE (Epanova® compared to Lovaza® in a pharmacokinetic single-dose evaluation) study. J Clin Lipidol 6(6):573–584

Davies JA (2007) Kainic acid A2 In: SJ Enna, Bylund DB (ed) xPharm: the comprehensive pharmacology reference Elsevier, New York, pp 1–3. doi:http://dx.doi.org/10.1016/B978-008055232-3.61978-6

Davis MI, Bennett MJ, Thomas LM, Bjorkman PJ (2005) Crystal structure of prostate-specific membrane antigen, a tumor marker and peptidase. Proc Natl Acad Sci U S A 102(17):5981–5986

De Clercq E (1993) Antivirals for the treatment of herpesvirus infections. J Antimicrob Chemother 32(suppl A):121–132

de la Calle Gil AB, Pena Vergara I, Cormane Bornacelly MA, Pajuelo Gallego A (2015) Intrathecal ziconotide and morphine for pain relief: a case series of eight patients with refractory cancer pain, including five cases of neuropathic pain. Neurol Ther 4(2):159–168

Dumontet C, Jordan MA (2010) Microtubule-binding agents: a dynamic field of cancer therapeutics. Nat Rev Drug Discov 9(10):790–803

Eccles R, Meier C, Jawad M, Weinmüllner R, Grassauer A, Prieschl-Grassauer E (2010) Efficacy and safety of an antiviral Iota-Carrageenan nasal spray: a randomized, double-blind, placebo-controlled exploratory study in volunteers with early symptoms of the common cold. Respir Res 11(1):1–10

Fenical W, Jensen PR, Palladino MA, Lam KS, Lloyd GK, Potts BC (2009) Discovery and development of the anticancer agent salinosporamide A (NPI-0052). Bioorg Med Chem 17(6):2175–2180

Food and Drug Administration (2014) Lovaza: GlaxoSmithKline

Fuerst JA (2014) Diversity and biotechnological potential of microorganisms associated with marine sponges. Appl Microbiol Biotechnol 98(17):7331–7347

Garson MJ (1993) The biosynthesis of marine natural products. Chem Rev 93(5):1699–1733

Girond S, Crance J, Van Cuyck-Gandre H, Renaudet J, Deloince R (1991) Antiviral activity of carrageenan on hepatitis a virus replication in cell culture. Res Virol 142(4):261–270

Grassauer A, Weinmuellner R, Meier C, Pretsch A, Prieschl-Grassauer E, Unger H (2008) Iota-Carrageenan is a potent inhibitor of rhinovirus infection. Virology J 5(107)

Hayden FG, Douglas RGJ (1995) Antiviral agents. In: Benneth JE, Dolin R, Blaser MJ (eds) Principles and practice of infectious diseases. Churchill Livingstone Inc, New-York, p. 411

Health NCIatNIo (2016a) anti-ENPP3 antibody-drug conjugate AGS-16C3F. http://www.cancer.gov/publications/dictionaries/cancer-drug?cdrid=739766.

Health NCIatNIo (2016b) polatuzumab vedotin. http://www.cancer.gov/publications/dictionaries/cancer-drug?cdrid=695317.

Heidelberg KB, Gilbert JA, Joint I (2010) Marine genomics: at the interface of marine microbial ecology and biodiscovery. Microb Biotechnol 3(5):531–543

http://www.abdn.ac.uk/ncs/departments/chemistry/marine-biodiscovery-centre-112.php. Accessed 30th March 2016a

http://www.abs-initiative.info/about-us/. Accessed 30th March 2016b

http://www.eea.europa.eu/highlights/marine-biodiversity-life-in-seas. Accessed 30th March 2016c

http://www.fao.org/docrep/009/a0337e/A0337E15.htm. Accessed 30th March 2016d

http://www.geomar.de/en/research/fb3/fb3-mn/geomar-biotech/. Accessed 30th March 2016e

http://www.qub.ac.uk/research-centres/MarineBiodiscovery/. Accessed 30th March 2016f

http://www.un.org/Depts/los/index.htm. Accessed 30th March 2016g

http://www.worldometers.info/. Accessed 30th March 2016h

http://www.worldwatch.org/. Accessed 30th March 2016i

https://scripps.ucsd.edu/cmbb. Accessed 30th March 2016j

https://www.cbd.int/abs/text/. Accessed 30th March 2016k

https://www.pharmamar.com/. Accessed 30th March 2016l

Huyck TK, Gradishar W, Manuguid F, Kirkpatrick P (2011) Eribulin mesylate. Nat Rev Drug Discov 10(3):173–174

Infante JR, Sandhu SK, McNeil CM, Kabbarah O, Li C, Zhong W, Asundi J, Wood K, Chu Y-W, Hamid O (2014) Abstract CT233: a first-in-human phase I study of the safety and pharmacokinetic (PK) activity of DEDN6526A, an anti-endothelin B receptor (ETBR) antibody-drug con-

jugate (ADC), in patients with metastatic or unresectable melanoma. Cancer Res 74(19 Supplement):CT233–CT233
Jaspars M, De Pascale D, Andersen JH, Reyes F, Crawford AD, Ianora A (2016) The marine biodiscovery pipeline and ocean medicines of tomorrow. J Mar Biol Assoc UK 96(1):151–158
Joint I, Muehling M, Querellou J (2010) Culturing marine bacteria – an essential prerequisite for biodiscovery. Microb Biotechnol 3(5):564–575
Kem W, Soti F, Wildeboer K, LeFrancois S, MacDougall K, Wei D-Q, Chou K-C, Arias H (2006) The nemertine toxin anabaseine and its derivative DMXBA (GTS-21): chemical and pharmacological properties. Mar Drugs 4(3):255–273
Knobeloch KP, Wright MD, Ochsenbein AF, Liesenfeld O, Lohler J, Zinkernagel RM, Horak I, Orinska Z (2000) Targeted inactivation of the tetraspanin CD37 impairs T-cell-dependent B-cell response under suboptimal costimulatory conditions. Mol Cell Biol 20(15):5363–5369
Kortmansky J, Schwartz GK (2003) Bryostatin-1: a novel PKC inhibitor in clinical development. Cancer Investig 21(6):924–936
Koski RR (2008) Omega-3-acid ethyl esters (Lovaza) for severe hypertriglyceridemia. Pharm Ther 33(5):271–303
Kraft AS, Smith JB, Berkow RL (1986) Bryostatin, an activator of the calcium phospholipid-dependent protein kinase, blocks phorbol ester-induced differentiation of human promyelocytic leukemia cells HL-60. Proc Natl Acad Sci U S A 83(5):1334–1338
Kurtboeke DI (2012) Biodiscovery from rare actinomycetes: an eco-taxonomical perspective. Appl Microbiol Biotechnol 93(5):1843–1852
Lallier LE, McMeel O, Greiber T, Vanagt T, Dobson ADW, Jaspars M (2014) Access to and use of marine genetic resources: understanding the legal framework. Nat Prod Rep 31(5):612–616
Lapalombella R, Yeh YY, Wang L, Ramanunni A, Rafiq S, Jha S, Staubli J, Lucas DM, Mani R, Herman SE, Johnson AJ, Lozanski A, Andritsos L, Jones J, Flynn JM, Lannutti B, Thompson P, Algate P, Stromatt S, Jarjoura D, Mo X, Wang D, Chen CS, Lozanski G, Heerema NA, Tridandapani S, Freitas MA, Muthusamy N, Byrd JC (2012) Tetraspanin CD37 directly mediates transduction of survival and apoptotic signals. Cancer Cell 21(5):694–708
Leal MC, Puga J, Serôdio J, Gomes NCM, Calado R (2012) Trends in the discovery of new marine natural products from invertebrates over the last two decades – where and what are we bioprospecting? PLoS One 7:e30580
Li D, Poon KA, SF Y, Dere R, Go M, Lau J, Zheng B, Elkins K, Danilenko D, Kozak KR, Chan P, Chuh J, Shi X, Nazzal D, Fuh F, McBride J, Ramakrishnan V, de Tute R, Rawstron A, Jack AS, Deng R, Chu YW, Dornan D, Williams M, Ho W, Ebens A, Prabhu S, Polson AG (2013) DCDT2980S, an anti-CD22-monomethyl auristatin E antibody-drug conjugate, is a potential treatment for non-Hodgkin lymphoma. Mol Cancer Ther 12(7):1255–1265
Ling T, Potts BC, Macherla VR (2010) Concise formal synthesis of (−)-salinosporamide A (marizomib) using a regio-and stereoselective epoxidation and reductive oxirane ring-opening strategy. J Org Chem 75(11):3882–3885
Liu J, Moore K, Birrer M, Berlin S, Matulonis U, Infante J, Xi J, Kahn R, Wang Y, Wood K (2013) Targeting MUC16 with the antibody-drug conjugate (ADC) DMUC5754A in patients with platinum-resistant ovarian cancer: A phase I study of safety and pharmacokinetics. Cancer Res 73:LB-290
Luesch H, Moore RE, Paul VJ, Mooberry SL, Corbett TH (2001) Isolation of dolastatin 10 from the marine cyanobacterium Symploca species VP642 and total stereochemistry and biological evaluation of its analogue symplostatin 1. J Nat Prod 64(7):907–910
Ma D, Hopf CE, Malewicz AD, Donovan GP, Senter PD, Goeckeler WF, Maddon PJ, Olson WC (2006) Potent antitumor activity of an auristatin-conjugated, fully human monoclonal antibody to prostate-specific membrane antigen. Clin Cancer Res 12(8):2591–2596
Mayer AMS, Glaser KB, Cuevas C, Jacobs RS, Kem W, Little RD, McIntosh JM, Newman DJ, Potts BC, Shuster DE (2010) The odyssey of marine pharmaceuticals: a current pipeline perspective. Trends Pharmacol Sci 31(6):255–265
Millward M, Price T, Townsend A, Sweeney C, Spencer A, Sukumaran S, Longenecker A, Lee L, Lay A, Sharma G (2012) Phase 1 clinical trial of the novel proteasome inhibitor marizomib with the histone deacetylase inhibitor vorinostat in patients with melanoma, pancreatic and

lung cancer based on in vitro assessments of the combination. Investig New Drugs 30(6):2303–2317
Mitsiades CS, Ocio EM, Pandiella A, Maiso P, Gajate C, Garayoa M, Vilanova D, Montero JC, Mitsiades N, McMullan CJ, Munshi NC, Hideshima T, Chauhan D, Aviles P, Otero G, Faircloth G, Mateos MV, Richardson PG, Mollinedo F, San-Miguel JF, Anderson KC (2008) Aplidin, a marine organism-derived compound with potent antimyeloma activity in vitro and in vivo. Cancer Res 68(13):5216–5225
Morrison K, Challita-Eid PM, Raitano A, An Z, Yang P, Abad JD, Liu W, Ratay Lortie D, Snyder JT, Capo L, Verlinsky A, Avina H, Donate F, Joseph IB, Pereira DS, Morrison K, Stover DR (2016) Development of ASG-15ME, a novel antibody drug conjugate targeting SLITRK6, a new urothelial cancer biomarker. Mol Cancer Ther 15:1301–1310
Muehling M, Joint I, Willetts AJ (2013) The biodiscovery potential of marine bacteria: an investigation of phylogeny and function. Microb Biotechnol 6(4):361–370
Nett M, Gulder TA, Kale AJ, Hughes CC, Moore BS (2009) Function-oriented biosynthesis of beta-lactone proteasome inhibitors in Salinispora tropica. J Med Chem 52(19):6163–6167
Newman D, Cragg G (2014) Marine-sourced anti-cancer and cancer pain control agents in clinical and late preclinical development. Mar Drugs 12(1):255–278
Newman DJ, Cragg GM, Battershill CN (2009) Therapeutic agents from the sea: biodiversity, chemo-evolutionary insight and advances to the end of Darwin's 200th year. Diving Hyperbar Med 39(4):216–225
Paul VJ, Ritson-Williams R, Sharp K (2011) Marine chemical ecology in benthic environments. Nat Prod Rep 28(2):345–387
Pereira DS, Guevara CI, Jin L, Mbong N, Verlinsky A, Hsu SJ, Avina H, Karki S, Abad JD, Yang P, Moon SJ, Malik F, Choi MY, An Z, Morrison K, Challita-Eid PM, Donate F, Joseph IB, Kipps TJ, Dick JE, Stover DR (2015) AGS67E, an anti-CD37 monomethyl auristatin E antibody-drug conjugate as a potential therapeutic for B/T-cell malignancies and AML: a new role for CD37 in AML. Mol Cancer Ther 14(7):1650–1660
Perez EA, Hillman DW, Fishkin PA, Krook JE, Tan WW, Kuriakose PA, Alberts SR, Dakhil SR (2005) Phase II trial of dolastatin-10 in patients with advanced breast cancer. Investig New Drugs 23(3):257–261
Perez HL, Cardarelli PM, Deshpande S, Gangwar S, Schroeder GM, Vite GD, Borzilleri RM (2014) Antibody–drug conjugates: current status and future directions. Drug Discov Today 19(7):869–881
Petek BJ, Jones RL (2014) PM00104 (Zalypsis): a marine derived alkylating agent. Molecules 19(8):12328–12335
Pettit GR, Herald CL, Doubek DL, Herald DL, Arnold E, Clardy J (1982) Isolation and structure of bryostatin 1. J Am Chem Soc 104(24):6846–6848
Pettit GR, Kamano Y, Herald CL, Tuinman AA, Boettner FE, Kizu H, Schmidt JM, Baczynskyj L, Tomer KB, Bontems RJ (1987) The isolation and structure of a remarkable marine animal antineoplastic constituent: dolastatin 10. J Am Chem Soc 109(22):6883–6885
Pettit GR, Srirangam JK, Barkoczy J, Williams MD, Boyd MR, Hamel E, Pettit RK, Hogan F, Bai R, Chapuis JC, McAllister SC, Schmidt JM (1998) Antineoplastic agents 365. Dolastatin 10 SAR probes. Anticancer Drug Des 13(4):243–277
Phillips GJ (2001) Green fluorescent protein – a bright idea for the study of bacterial protein localization. FEMS Microbiol Lett 204(1):9–18
Potts B, Albitar M, Anderson K, Baritaki S, Berkers C, Bonavida B, Chandra J, Chauhan D, Cusack J Jr, Fenical W (2011) Marizomib, a proteasome inhibitor for all seasons: preclinical profile and a framework for clinical trials. Curr Cancer Drug Targets 11(3)
Rae M, Folch H, Moniz MBJ, Wolff CW, McCormack GP, Rindi F, Johnson MP (2013) Marine bioactivity in irish waters. Phytochem Rev 12(3):555–565
Reen FJ, Gutierrez-Barranquero JA, Dobson ADW, Adams C, O'Gara F (2015) Emerging concepts promising new horizons for marine biodiscovery and synthetic biology. Mar Drugs 13(5):2924–2954
Reilly EB, Phillips AC, Boghaert ER, Vaidya KS, Mitten MJ, Norvell S, Falls HD, DeVries PJ, Cheng D, Meulbroek JA, Buchanan FG, McKay LM, Goodwin NC (2016) ABT-414, an antibody drug conjugate targeting a rumor-selective EGFR epitope. Mol Cancer Ther 15:661–669

Rinehart KL, Holt TG, Fregeau NL, Stroh JG, Keifer PA, Sun F, Li LH, Martin DG (1990) Ecteinascidins 729, 743, 745, 759A, 759B, and 770: potent antitumor agents from the Caribbean tunicate Ecteinascidia turbinata. J Org Chem 55(15):4512–4515

Roberts SA, Andrews PA, Blanset D, Flagella KM, Gorovits B, Lynch CM, Martin PL, Kramer-Stickland K, Thibault S, Warner G (2013) Considerations for the nonclinical safety evaluation of antibody drug conjugates for oncology. Regul Toxicol Pharmacol 67(3):382–391

Rocha-Martin J, Harrington C, Dobson ADW, O'Gara F (2014) Emerging strategies and integrated systems microbiology technologies for biodiscovery of marine bioactive compounds. Mar Drugs 12(6):3516–3559

Schabel F Jr (1968) The antiviral activity of 9-β-D-arabinofuranosyladenine (ara-A). Chemotherapy 13(6):321–338

Senter PD, Sievers EL (2012) The discovery and development of brentuximab vedotin for use in relapsed Hodgkin lymphoma and systemic anaplastic large cell lymphoma. Nat Biotechnol 30(7):631–637

Shimomura O, Johnson FH, Saiga Y (1962) Extraction, purification and properties of aequorin, a bioluminescent protein from the luminous hydromedusan, Aequorea. J Cell Comp Physiol 59:223–239

Staats PS, Yearwood T, Charapata SG et al (2004) Intrathecal ziconotide in the treatment of refractory pain in patients with cancer or aids: a randomized controlled trial. JAMA 291(1):63–70

Svenson J (2013) MabCent: arctic marine bioprospecting in Norway. Phytochem Rev 12(3):567–578

Tai YT, Mayes PA, Acharya C, Zhong MY, Cea M, Cagnetta A, Craigen J, Yates J, Gliddon L, Fieles W, Hoang B, Tunstead J, Christie AL, Kung AL, Richardson P, Munshi NC, Anderson KC (2014) Novel anti-B-cell maturation antigen antibody-drug conjugate (GSK2857916) selectively induces killing of multiple myeloma. Blood 123(20):3128–3138

Teng S-W, Zopf C, Yang J, Stringer B, Zhang J, Shyu WC, Chakravarty A, Veiby P, Mettetal J (2014) Using pharmacokinetic/efficacy modeling to identify the optimal schedule for MLN0264, an anti-guanylyl cyclase C (GCC) antibody-drug conjugate, in a range of xenograft models. Cancer Res 74(19 Supplement):4649–4649

Trail PA (2013) Antibody drug conjugates as cancer therapeutics. Antibodies 2(1):113–129

Vahdat LT, Chan N (2015) The antibody-drug conjugate Glembatumumab Vedotin (CDX-011) and its use in treatment of breast cancer. In: Wang J, Shen W-C, Zaro JL (eds) Antibody-drug conjugates, AAPS Advances in the Pharmaceutical Sciences Series, vol 17. Springer International Publishing, Cham, pp. 225–232. doi:10.1007/978-3-319-13081-1_13

Van Berkel P, Bleeker WK, Van Dijkhuizen R, Lisby S, Overdijk M, Parren P, Verploegen S (2012) Human antibodies and antibody-drug conjugates against cd74. WO 2012104344 A1

van Spriel AB, Puls KL, Sofi M, Pouniotis D, Hochrein H, Orinska Z, Knobeloch KP, Plebanski M, Wright MD (2004) A regulatory role for CD37 in T cell proliferation. J Immunol 172(5):2953–2961

Wang Y, Firestein R, Ryner L, Darbonne W, Guan Y, Lu S, Choi Y, Xiao Y, Polakis P, Suttmann B (2015) Abstract POSTER-THER-1441: biomarker evaluation of phase 1 clinical trials of antibody-drug conjugates (ADCs) in platinum resistant ovarian cancer. Clinical Cancer Research 21(16 Supplement):POSTER-THER-1441-POSTER-THER-1441

Weekes CD, Lamberts LE, Borad MJ, Voortman J, McWilliams RR, Diamond JR, de Vries EG, Verheul HM, Lieu CH, Kim GP (2016) Phase I study of DMOT4039A, an antibody–drug conjugate targeting mesothelin, in patients with unresectable pancreatic or platinum-resistant ovarian cancer. Mol Cancer Ther 15(3):439–447

www.unep.org. Accessed 30th March 2016m

Zhang J, Gallery M, Wyant T, Stringer B, Manfredi M, Danaee H, Veiby P (2013) Abstract PR12: MLN0264, an investigational, first-in-class antibody-drug conjugate (ADC) targeting guanylyl cyclase C (GCC), demonstrates antitumor activity alone and in combination with gemcitabine in human pancreatic cancer xenograft models expressing GCC. Mol Cancer Ther 12(11 Supplement):PR12–PR12

Chapter 4
Microbial Resource Centers Contribute to Bioprospecting of Bacteria and Filamentous Microfungi

Jörg Overmann and David Smith

Abstract The immense diversity of microorganisms has remained largely untapped, especially with regard to bioprospecting. Through their isolation, microbes attain a substantial monetary value which microbial domain Biological Resource Centers (mBRCs) preserve in a highly cost-effective manner. Typically, mBRCs are publicly funded in order to provide quality-controlled, and well-characterized microbial resources and data, at low cost to researchers. The present chapter outlines the preconditions and scenarios for mBRCs to expand their traditional tasks and enter the field of bioprospecting. It appears most promising to generate information on the biosynthetic potential of novel types of microorganisms through extended characterization, metabolic profiling, and genome analyses. Particular challenges are an improved access to the vast uncharted microbial biodiversity, the compliance with new legal requirements, and the efficient linking to private industry as a novel stakeholder. A business plan is developed herein that proposes to join the expertise of different mBRCs to create a platform that provides a "one-stop-shop" with restricted access to a large number of well-characterized, pre-screened microbial resources in a legally compliant manner. As a typical and inherent weakness, the limited public funding of mBRCs often will not permit an expansion of tasks through the existing funding alone. Revenues generated from sales of even high value microbial resources to bioindustry rarely will cover costs. Therefore, alternative funding could be sought from the government agencies in charge of the bioeconomy that traditionally are not stakeholders of mBRC, and through the participation of mBRCs in dedicated funding programs for bioprospecting.

J. Overmann (✉)
Leibniz-Institut DSMZ-Deutsche Sammlung von Mikroorganismen und Zellkulturen GmbH, Braunschweig, Germany

Technische Universität Braunschweig, Braunschweig, Germany
e-mail: joerg.overmann@dsmz.de

D. Smith
CABI, Bakeham Lane, Egham, Surrey TW20 9TY, UK

R. Paterson, N. Lima (eds.), *Bioprospecting*, Topics in Biodiversity and Conservation 16, DOI 10.1007/978-3-319-47935-4_4

4.1 Introduction

Culture collections have been undertaking research with microorganisms for over 100 years. During this time, the public access to live cultures of microorganisms has played a pivotal role for progress in basic microbiological sciences in a number of ways: In microbial systematics, novel isolates need to be compared to existing strains to determine biochemical or physiological characteristics that serve to delineate and define previously unreported taxa (Rosselló-Móra and Amann 2015). Microbial strains also permit the testing of biochemical features sometimes predicted by the analysis of genome sequences, and allow the elucidation of novel metabolic pathways. In microbial ecology, isolated strains serve as models for microorganisms of the same phylotype that occur in complex environments. Isolated microbial strains have been deposited in public collections to ensure their unrestricted availability for subsequent scientific studies for these and other purposes. Live microbial strains also provide numerous opportunities for applications in agriculture, food processing, catalyses, environmental protection, and public health. The potential for application of microbial resources is particularly prominent in the pharmaceutical sector, especially given the problems of resistance in pathogenic bacteria to antibiotics and the lack of novel leads to treat the diseases as discussed elsewhere in this book.

Microbial resource centers (more precisely, microbial domain Biological Resource Centers *sensu* OECD, mBRCs) provide products and services that go far beyond the maintenance and distribution of microbial resources offered by typical culture collections (OECD 2004). In particular, mBRCs rely on specific and approved quality assurance procedures, provide profound expertise in microbial systematics, and maintain legal expertise with regard to property rights, biosafety, and biosecurity. They also deliver the associated data and maintain the expert knowledge essential for cultivation and physiological analysis of microbial strains. Whereas mBRCs traditionally support basic research, their holdings and competencies have so far been exploited only rarely for natural compound discovery and the bioeconomy.

The present chapter analyses the current state of mBRCs as they apply to bacteria and filamentous microfungi. The larger macrofungi such as mushrooms are considered elsewhere (www.iucnredlist.org). It features the (a) type and value of their holdings, (b) currently changing legal framework for their acquisition and distribution policies, and (c) future challenges and potential novel functions. A business plan is developed that intends to support mBRCs to cope with these future demands.

4.2 Microbial Resources

A bacterial strain represents the progeny of a single isolated cell and constitutes the basis for subsequent studies. Microbial strains that are authenticated, taxonomically defined, physiologically characterized, quality controlled, and also well-documented, are designated 'microbial resources', in order to distinguish them from mere laboratory isolates that may lack the respective information. The worldwide inventory of microbial strains that are registered in the Culture Collections Information Worldwide (CCINFO) system of the World Data Centre for Microorganisms (WDCM 2015) currently amounts to 2.5 million strains. Of these, 1.05 million, 727,000 and 38,000 are bacteria, fungi and viruses respectively maintained in 692 culture collections from 71 countries. Despite these impressive numbers, however, they represent the existing microbial diversity rather poorly since most, or all, bacterial isolates belong to only 10,693 bacterial and archaeal species that are recognized to date (LPSN 2015), or to the 100,000 fungal species (Kirk et al. 2008). In contrast, current estimates of total prokaryotic diversity range between 10^7 and 10^9 species (Dykhuizen 1998; Curtis et al. 2002). The described number of species thus may represent only 0.1–0.001 % of the total prokaryotic diversity. Consequently, the worldwide holdings mostly consist of multiple strains of each bacterial species and hence represent an extensive phylogenetic redundancy. Furthermore, most of the currently available bacterial isolates belong to just four of the approximately 90 phyla that are presently recognized (i.e., the *Actinobacteria*, *Bacteroidetes*, *Firmicutes* and *Proteobacteria*) (Overmann 2012). By comparison, representatives of 60 bacterial phyla have rarely or never been isolated to date (Baker and Dick 2013).

Recent estimates based on high-throughput sequencing methods suggest that as many as 5.1 million fungal species exist (Blackwell 2011). Considering that there are currently around 100,000 species described, the majority of fungi are also yet to be discovered and cultured. At least in part, the failure to recover phylogenetically novel microbial lineages of bacteria and fungi can be attributed to current isolation methodology that is often found to be inadequate (Overmann 2012).

Since replication-competent cultures of bacterial type strains (those strains that follow the description and characteristics of novel species) constitute the basis for follow-up systematic comparisons, the deposit of a type strain in two public collections (preferably in two different countries) is mandatory for its valid description as a novel species (Lapage et al. 1992). Accordingly, the acquisition, maintenance and distribution of type strains have been major tasks of bacterial mBRCs. However, unlike the bacterial type strains that are almost completely secured in public collections, only 25.5 % of type strains from the 100,000 recognized fungal species are publicly available. In addition, mBRCs also maintain strains that are deposited in association with the publication of patents. The World Intellectual Property Organization (WIPO) statistics report that over 4500 organisms were deposited in 2012 (WIPO 2015), which allows researchers to fulfil this legal prerequisite for new patent applications.

From the high ratio between the total number of strains and the few type strains hosted by mBRCs and culture collections worldwide, and from the small overlap in the holdings of type strains in different culture collections, it can be deduced that the majority of the worldwide inventory of microbial resources represent non-type strains that were deposited for purposes other than systematics (Overmann 2015). Indeed, of the strains in the WDCM Global Catalogue of Microorganisms that had a registered origin, 48 % were genetically engineered fungi and 25 % were bacteria of human origin (GCM 2015). Thus, isolates obtained through biotechnological projects or isolated during medical studies may represent a considerable fraction of the microbial resources that are currently deposited in mBRCs worldwide. Notwithstanding the rather high numbers of non-type strains, the holdings of mBRCs still represent only a small fraction of the isolates that are actually generated in microbiological laboratories worldwide: in contrast to the bacterial type strains that are almost all accessible, less than 1 % of the microbial isolates obtained through publicly funded research are deposited in mBRCs and hence run a higher risk of being lost before their potential has been explored further (see Sect. 4.5).

Because the uncharted proportion of microbial diversity is vast, it has been suggested that novel biochemical features and biosynthetic pathways are likely to occur in the underexplored bacterial or fungal lineages, which therefore may provide novel solutions for agriculture, biotechnology and public health. This hypothesis has been substantiated by the discovery of natural compound synthesis gene clusters and novel types of natural compounds in several phylogenetic lineages such as members of the candidate bacterial phylum 'Tectomicrobia' that occur in marine sponges (Wilson et al. 2014). Also, they occur in *Acidobacteria* (Quaiser et al. 2003), *Chloroflexi* (Nett et al. 2006), and *Planctomycetes* (Jeske et al. 2013). Recent studies report sampling fungal diversity by applying new isolation techniques and suggest that many species new to science are being discovered in unexplored environments (Blackwell 2011). As much as 13% of the tropical fungi yielded active compounds when screened for antifungal, antibacterial, antiviral, insecticidal, antihelminthic, anti-cancer, anti-diabetes melitus, anti-inflammatory and pro-endocrinological leads in one large screening program (Bills et al. 2002).

Unfortunately, mBRCs (a) have not succeeded in securing a large fraction of the microbial isolates that are generated worldwide, (b) could make only limited contributions towards cultivating uncultured microbial diversity and (c) have played only a small role in attempts to access and explore microorganisms for future applications and these situations needs to change for a more effective bioeconomy.

4.3 Monetary Value of Microbial Resources

mBRCs represent an investment largely made by the public, often through governmental funding over long time periods. As such, it is important to understand what these investments produce in concrete terms and to assess future possibilities for exploitation of these investments for social purposes. There are several ways of assigning a financial value to the existing holdings of mBRCs through the: (1) cost for isolating a microbial strain from a complex sample, (2) cost of acquisition and curation, (3) supply fee, and (4) potential of microbial resources to yield high value marketable products. The following calculations do not consider non-monetary value, such as the scientific, cultural, and educational significance, as no methods exist to quantify them.

Obtaining novel types of microbes from environmental samples involves a series of labour-intensive steps for enrichment and isolation. Microorganisms that grow rapidly in aerobic, high nutrient, predefined complex media require a rather limited effort for cultivation that corresponds to a value of 450 € (Smith 2012). This cost represents an average of the 47 WFCC affiliated collections. In general, slow growing organisms such as filamentous fungi, which may take 7 days to grow fully, will cost more than fast growing bacteria, which may take 24 h. This is from increased (electrical) power requirements, the need for different equipment, greater obsolescence of equipment because they take longer to grow, and more person hours per organism. On the other hand, bacteria may require higher powered and more expensive microscopes because they are smaller.

However, strain characterisation was not included in the above estimations. Recent cost for an accession of a fungus to the CABI collection is ca. € 900 (cf. € 918 for DSMZ, see below). This value is modest (compared to the much higher estimate of the full monetary value of a more fastidious bacterial strain described in the following paragraphs) and does not consider the fact that many isolates occur only occasionally in other mBRCs worldwide. For example, of the 25,611 names of fungi listed in the World Data Centre for Microorganisms (WDCM 2015), almost 50 % are present in only one of the fungal collections listed (Sugawara et al. 1993). While these numbers are inflated as they represent names listed in the individual collections that also include synonyms, anamorph names, and spelling variants, they are still indicative and are supported by the data from the CABI collections. Here, 3360 of the 4541 species of filamentous fungi (i.e. 56 %) are represented by only one strain and if such a unique strain was lost, the recovery of an isolate with similar characteristics from its natural environment may be difficult and would incur substantially higher costs. This would inevitably render strains with specific and particularly desirable phenotypic properties much more valuable than the value of 450 € given above.

Significantly, the readily-culturable bacteria and filamentous fungi that are still being obtained through low-cost cultivation approaches, typically show little biological novelty (Singh et al. 2013). On the other hand, phenotypically novel bacteria often have unknown growth requirements, and are highly fastidious: These require substantially more person hours for enrichment and isolation as exemplified by *Myxobacteria*, *Acidobacteria* and *Dehalococcoides ethenogenes* (Foesel et al. 2013; Maymó-Gatell et al. 1997; Sanford et al. 2002). Based on the detailed compilation of all actual costs that are associated with the isolation and characterization of a more fastidious bacterial strain, a monetary value of 9836 € through cultivation can be obtained (Overmann 2015). This estimate encompasses all work steps and materials comprising of costs for personnel, consumables, and the depreciation of the equipment, but not the costs for sampling itself. However, in countries with an emerging economy (e.g., India), the monetary value attained would be somewhat lower due to lower wages (e.g. 5042 €; Overmann 2015).

In contrast to these high values, the costs for deposit and curation of microbial resources are modest. The acquisition of a bacterial isolate by mBRCs requires the cultivation and initial preservation (liquid nitrogen and freeze-drying) and identification with extensive quality control (e.g., by 16S rRNA gene sequencing, MALDI-TOF, biochemical testing, fatty acid analysis and microscopy). Acquisition of filamentous microfungi would be similarly laborious (in this case, sequencing of the internal transcribed spacers is used instead of 16 S rRNA gene sequencing). Based on recent estimates, these activities incur costs of 918 € at the Leibniz-Institut DSMZ (Overmann 2015) and 900 € at CABI (see above), which is in the same order of magnitude as previous estimates for other mBRCs (Smith 2012). Costs for the curation, i.e. the long-term preservation and maintenance of live cultures, are even lower, with numbers ranging from 7.45 € (CABI) to 3.60 € (DSMZ) (both per strain and year). These expenses for the long-term storage of microbial resources amount to only 0.03–0.07 % of their overall monetary value. Evidently, mBRCs provide a very cost-effective way of preserving the monetary value of isolated strains.

Supply fees (retail prices) for microbial resources are even lower than the costs for acquisition and curation, and vary considerably between individual mBRCs. For example, CABI charges a fee of 209 € per freeze-dried ampoule and 278 € for an active culture, whereas corresponding prices at the DSMZ currently range between 75 and 175 €. Those of the American Type Culture Collection (ATCC) are between 350 and 418 € for most bacterial strains. Based on the feedback of DSMZ customers, researchers at public institutions often do not have the funds to afford the more expensive microbial resources. It becomes evident then, that the revenues generated through microbial resources are not sufficient to cover the concomitant expenditures of strain acquisition by the mBRCs. Thus, mBRCs could not support research and development in a sustainable manner without substantial public funding.

Finally, a monetary value can be assigned to microbial resources through their potential to yield high value marketable products. Annual global sales of pharmaceutical drugs amounted to about 956 billion US$ in 2011 (IMS Health Market Prognosis 2012). Of every 5000 to 10,000 natural products that generate a hit in initial screens, only one eventually will become an approved drug (ten Kate and Laird 1999; PhRMA 2012). If only uncharacterized microbial isolates are available for bioprospecting, they need to be screened entirely in an untargeted approach which typically yields only a low percentage of "talented" strains (strains capable of synthesizing natural compounds) of approximately 10 % (Hindra Huang et al. 2014; Weissmann and Müller 2010; Xie et al. 2014). Taken together, about 100,000 uncharacterized strains would be needed to statistically yield just one single pharmaceutical product that would reach the market place. Consequently, even the entire holdings of the largest mBRCs ($\leq$ 165,000; WDCM 2015) are too small to arouse the interest of the pharmaceutical industry. Equally, providing microbial strains for untargeted screening is economically unattractive for mBRCs, since the market prices of microbial resources used by pharmaceutical companies has been calculated to range between US$ 2–60 per strain and hence barely reached the retail prices of mBRCs (Miyazaki 2006).

Based on the above considerations, mBRCs could strengthen their role in the bioeconomy and public health by providing more attractive microbial resources for research and development. This can be achieved through (1) an increase in holdings of strains from underexplored phylogenetic groups (recovered through their own research projects or a suitable accession policy; see Sect. 4.2), (2) an in-depth characterization of microbial strains and growth experiments that systematically vary growth parameters to release a greater potential for natural compound synthesis or other applications, (3) collecting high quality genomic information that permits genome mining, and (4) providing sufficient documentation of genetic and metabolic properties of microbial strains that would permit a concise, targeted, and sector-specific bioprospecting by the users. Based on the numbers given above, improving the chances of novel discoveries by these measures would increase the monetary value of microbial resources by up to tenfold.

4.4 Current Functions of Microbial Domain Biological Resource Centers

Functions of a mBRC include the *ex situ* conservation of microorganisms, strain identification services, training opportunities and consultancy (WDCM 2015). mBRCs thus represent custodians of national resources providing the living materials to underpin the science base. So far, mostly strains isolated in the course of studies of microbial systematic and diversity and other projects of basic research, are deposited in mBRCs during the publication process. In addition to their public collections, many mBRCs carry out safety deposit (confidential holdings) and patent deposit services for researchers.

The role of mBRCs for basic research extends far beyond microbial systematics and the provision of type strains (see Sect. 4.1). This is exemplified by the fact that two thirds of the 11,020 scientific publications referring to strains of the DSMZ over the past 30 years appeared in journals outside of microbial systematics (Overmann 2015). Scientific articles based on deposited strains are cited more than twice as often as publications on strains that are not publicly available (Furman and Stern 2011). These data emphasize that public collections of mBRCs are a key to future scientific discoveries.

The current demand for microbial strains (including archea, bacteria, filamentous fungi, microalgae, viruses and yeasts) provided by 13 European mBRCs organized within the Microbial Resource Research Infrastructure (MIRRI) (MIRRI 2015) has been determined in the course of preparing the MIRRI business plan. The mBRCs distributed about 198,000 individual strains between 2010 and 2012. These were provided equally on national and international bases; 60 % of the strains were shipped to users located in the non-profit sector and 40 % to for-profit organizations (in the latter case often for non-commercial research, however). MIRRI estimates that microorganisms are utilised by around 1000 institutions in Europe, constituting about 400 universities and 600 research institutes which encompass at least one biological department.

The 13 MIRRI-mBRCs represent less than 10 % of the worldwide holdings and less than 2 % of the collections registered with the WDCM. Half of all CCINFO registered microbial strains are kept in just 10 % of the countries registered, in particular the United States, Brazil, Japan, China, the Republic of Korea, and India (WDCM 2015). Obviously, the global demand for microbial strains is much higher than that in Europe alone and are approximately 0.5 million per year for the WFCC affiliate collections (Smith 2012). Since the strains are distributed by less than half (308 of 692) of the registered culture collections (WDCM 2015), the global average amounts to about 1600 cultures which are supplied per culture collection per year. Larger mBRCs distribute much higher numbers (e.g., the DMSZ provides 37,000 different microbial resources per year).

mBRCs have been little involved in bioprospecting because of their public mission. In the past, large pharmaceutical companies typically established their in-house, proprietary collections through bilateral agreements for sampling and exploitation with partner countries (e.g., Salazar et al. 2002). However, as the pharmaceutical and biotech industries have gone through significant streamlining and cost-cutting measure in the recent decade and especially in the recent global economic crisis, many natural product departments have closed or been downsized. Furthermore, small to medium sized companies often are unable to maintain a significant in-house capacity. Thus, in principle, there is a market niche here that is not currently fully serviced. Through their decades of successful operation, mBRCs have established cultivation and preservation skills, as well as profound knowledge on microbial biochemistry and physiology, which will be required to access a larger fraction of the uncharted microbial diversity and which are particularly important for fastidious or slow growing microbial strains, that require specific cultivation techniques for isolation. It is particularly attractive to participate in the search for, and retrieval of, novel types of microorganisms and to unlock their potential for future research and applications to facilitate the future development of mBRCs.

4.5 Future Demands for Microbial Resources and Microbial Domain Biological Resource Center Services in the Bioeconomy and Biotechnology

A recent analysis of 835 articles of eight European microbiology journals revealed that less than 1 % of the bacterial isolates cited had been secured in public mBRCs (Stackebrandt 2010). By accessing, maintaining and professionally distributing a higher fraction than just 1 % of the microorganisms used in microbiological research, mBRCs could become instrumental in unlocking their potential for basic and applied research. Although many of the existing microbial strains may represent novel isolates of already described species, phylogenetically closely related strains can still exhibit a distinct genetic potential and novel, unknown phenotypic properties (Jaspers and Overmann 2004). Therefore, at least some of these not-yet-publicly-available isolates are of potential relevance for future scientific discovery. The so-called ‘key strain’ concept has recently been established (Stackebrandt et al. 2014) and subsequently amended (Overmann 2015) to aid the prioritization of suitable strains that should be deposited and secured in mBRCs. The substantial monetary value of microbial resources (see Sect. 4.2) could be preserved and future scientific work with these strains promoted if this concept was consistently applied. A similar situation would exist if scientific journals or research funders required the deposit of the strains in mBRCs as a pre-condition to publication or continued funding (as is the case for the deposit of nucleotide and protein sequences in public databases). The present financial budgets of most mBRCs severely constrain their capacities for additional curation, storage and distribution work. Tight networking, moderate increases in budgets, and concerted action of mBRCs would provide a highly cost-effective way to accommodate a significantly larger amount of microbial resources (see Sect. 4.8).

However, future demands for support by mBRCs are likely to extend far beyond securing the microbial strains isolated in academic research laboratories. In particular the profound competence of mBRCs to cultivate fastidious and novel types of microorganisms and thereby render the uncharted fraction of microbial diversity accessible to others, is likely to become increasingly important. One potential area where the existing strengths of mBRCs could be directly coupled to societal needs is in the development of new antibiotics. Indeed, the number of new anti-infective compound classes developed over the past 40 years has been decreasing steadily, while resistance problems caused by current antibiotics treatments have been increasing concomitantly. This increasing gap in anti-infective innovation has rendered the discovery of novel lead compounds a highly pressing issue (Cooper and Shlaes 2011). The present lack of novel lead compounds has been attributed to the small share of bacterial (mostly streptomycetes and myxobacteria) and fungal diversity that have been explored for secondary metabolites (Bérdy 2005; Scannell et al. 2012). Past screening programs by the pharmaceutical industry that aimed at the discovery of novel natural compounds yielded microbial isolates that frequently produced already-known antibiotics and antibiotics classes (Baltz 2006). However, biochemically and physiologically novel bacteria are typically found in underexplored phylogenetic lineages (Wu et al. 2009) in which novel natural compounds have already been detected. The wide swath of uncharted microbial diversity provides mBRCs with huge targets for novel acquisitions.

Approximately, 60 % of all marketed pharmaceuticals are either developed directly from natural compounds or from chemically modified derivatives (Newman and Cragg 2012). More than half of prescriptions filled in the USA in 1993 contained at least one major active compound derived from, or modelled on, natural compounds and 42 % of the sales of the 25 top-selling drugs worldwide are either biologicals, natural products or entities derived from natural products (ten Kate and Laird 1999). Also, the screening and chemical analysis of natural products continues to provide novel chemical scaffolds for the development of novel drugs (Butler 2005; Chin et al. 2006). Given the current decline in expenditures for R&D by biotechnology companies (Ernst and Young 2014) that in part may be attributed to the limited access to novel biotechnological products and processes (EuropaBio 2014), a proprietary biodiscovery strategy of mBRCs would provide specific and novel opportunities for mBRCs and fill a need from bioindustry. Simultaneously, the vast microbial diversity makes it essential to coordinate efforts and focus detailed studies on those microbial resources that have the highest potential for application. As an example, although about 20,000 marine bioactive compounds were discovered since the mid-1960s through successful bioprospecting programs, only about ten have reached the market (Mayer et al. 2010; Rocha et al. 2011). Here, major bottlenecks have been difficulties of synthesizing larger amounts of natural compounds either through their producer organisms or by chemical synthesis (Glaser and Mayer 2009). Therefore, it may be rewarding for mBRCs to (1) develop procedures for selecting suitable microbial resources early during the isolation or acquisition stage, (2) generate the necessary information on biosynthetic pathways and the underlying gene clusters on the induction conditions of non-constitutively expressed biosynthesis pathways, and (3) provide upscaling data on large scale cultivation of suitable producers.

Finally, a growing future demand for scientific services that are technologically demanding and/or require specialized scientific knowledge. With the recent advances in and widespread availability of, sequencing technologies and bioinformatics, standardized and high quality genomic DNA for purchase will become an increasingly attractive resource for the customers of mBRCs, since it alleviates the users of the burden to cultivate fastidious or slow-growing microbial strains. Microbial sequence information is expanding exponentially and large sequencing programs (e.g., the GEBA project of the DOE Joint Genome Institute, CA, USA; Wu et al. 2009) are devoted to generating a large, phylogenetically diverse database of genome sequences for future in-depth studies of bacterial strains. This means that results from comparative genomics will replace the experimental services traditionally offered by mBRCS, such as DNA/DNA hybridization, and determination of the GC-content. Other services, such as taxonomic identification or chemotaxonomic analysis based on phospholipid fatty acid profiles, will probably continue to be requested from mBRCs since the necessary equipment and expertise is not maintained by many research laboratories. Furthermore, isolation of difficult-to-grow pure cultures, optimization of growth conditions, phenotypic characterization, and unique environmental sampling most likely represent services that will be requested at an increased frequency in the future. An integrated spectrum of such expensive and labour-intensive services can be offered most cost-effectively through combining the complementary expertise and instrumentation of individual mBRCs by means of networking.

4.6 Structured Information on Microbial Resources and the Key Role of Databases

In order to harness bacteria for new applications, and as discussed in the preceding sections of this chapter, a sufficient amount of strain-associated (meta)data will be needed for future use and application of the available microbial resources.

As a first step toward structuring information on microbial resources, the World Data Centre for Microorganisms was established and went online in 1997. Of the 692 public culture collections registered in the WDCM CCINFO system to date, only 110 have published their strain holdings online, or in a digitalized catalogue (WDCM 2015). Accordingly, the WDCM Global Catalogue of Microorganisms (GCM) has been established to improve the access to strain-related information and to promote scientific and industrial use of the public microbial resources (Wu et al. 2013). The GCM database links strain catalogue information as provided by individual collections to the nucleotide and protein sequences of the strains stored in molecular databases. At the time of writing, this information had become available for about 335,000 strains, including 108,000 bacteria and 171,000 fungi, from 68 culture collections (GCM 2015). The WDCM minimum dataset comprises 15 items, most importantly, the name and number of the individual strains, their source, history of deposit, geographic origin, and growth media or temperature. This information can be searched interactively in the database which also provides a homology and keyword search for the literature linked to the strain.

As a second initiative, the StrainInfo Bioportal (Dawyndt et al. 2005) offers comprehensive aggregated information on the numbers, taxonomic names, and International Nucleotide Sequence Database Collaboration (INSDC) accession numbers of deposited strains and dynamically compiles their exchange history by integrating the catalogues of many culture collections worldwide. In addition, StrainInfo links to literature connected to the strains. StrainInfo thus constitutes an important resource to elucidate the history of the deposit and assist in the exchange of bacterial strains by the different culture collections.

Despite these efforts, the wealth of biochemical, physiological and ecological data available for each microbial strain had remained largely inaccessible until recently and was not systematically searchable, since the strain-associated (meta) data were typically dispersed across a considerable number of publications. In order to overcome the fragmented nature of available information on bacterial strains, Bac*Dive* – the Bacterial Diversity Metadatabase was recently established by the Leibniz-Institute DSMZ (Söhngen et al. 2014). Bac*Dive* covers data on the taxonomy, morphology, physiology, cultivation, geographic origin, application, biological interactions, and the appropriate sources of supply for each strain together with its genome and 16S rRNA sequences deposited at the INSDC. The relational database behind BacDive was constructed by defining more than 400 data fields for each strain. Besides the relevant primary scientific literature, sources for the annotation include detailed internal descriptions of culture collections and expert-compiled compendia on strains which are not publicly available. In Bac*Dive*, the majority of the data is manually annotated and curated. Importantly, the portal offers powerful advanced search functionalities that allow the combination of more than 30 search fields for text and numerical data. Bac*Dive* will provide quick and complete access to information about the cultured microbial biodiversity which is currently not available in any other database; this innovative tool also enables the user to filter the information for all strains according to particular attributes.

An important precondition for improving the usability of microbial resources is the central accessibility and comparability of all existing strain metadata and molecular data. A central, integrated portal providing this combined information based on common standards and quality criteria would constitute a highly desirable resource for academia and industry. This would facilitate locating a particular microbial resource for the customer and is, indeed, one of the core goals of the MIRRI network. By offering additional information on natural products (e.g. antibiotics), genomics, and biosecurity, the integrated portal could also serve as an essential resource for bioindustry and the development of the bioeconomy. An integrated portal would provide a major incentive for participation of individual mBRCs, since it would increase their visibility. The most practical solution to the challenge could be to link the data of the already existing databases (e.g., StrainInfo Bioportal, Bac*Dive*, Silva; Quast et al. 2013). In fact, StrainInfo and Bac*Dive* already have established standards and data fields according to the methods developed in collaboration with the Genomic Standards Consortium (Tuama et al. 2012; Yilmaz et al. 2011). Individual mBRCs will need further improvements to provide their data more dynamically, in addition to providing the data exchange standards. Accordingly, an integrated portal could be realized in a short time period (estimated 2 years) and, importantly, would capitalize on existing expertise and human resources to ensure the future availability of the data.

4.7 Ownership and Legal Constraints for the Exploitation of Microbial Resources

Through the Convention on Biological Diversity (CBD) that entered into force on 29 December 1993, the sovereign rights over genetic resources were transferred to the country of origin (CBD 2015; Article 15.1), instead of being considered a common heritage of mankind. The CBD aims at the (a) conservation of biological diversity, (b) sustainable use of its components, and (c) fair and equitable sharing of benefits (both monetary and non-monetary) arising from the utilisation of genetic resources. Since the latter is expected to provide incentives to provider countries for the conservation of biological diversity within the states of origin, these goals are interdependent.

In the case of microorganisms, which constitute a significant, if not the largest, fraction of biological diversity on the planet ((a) 10^7 to 10^9 and (b) 5.1×10^6 species of (a) bacteria and (b) fungi respectively; Sect. 4.2), the first (conservation) and second (sustainable use) aim of the CBD are less relevant for four biological reasons (Overmann 2015):

1. Microorganisms, especially bacteria, are highly unlikely to become extinct because of the large size of their populations;
2. Microbial field sampling of microscopic organisms is not as invasive or destructive compared to larger organisms since small environmental samples are normally required;

3. Most species of bacteria do not show a definable biogeography due to an efficient dispersal across and between continents;
4. Furthermore, different bacterial and fungal species show a considerable functional redundancy in metabolic functions and may also synthesize identical natural compounds.

Based on the missing threat of extinction of bacteria and also many other microorganisms, it is understandable, that pharmaceutical companies have not been willing to finance the conservation of areas with high natural biodiversity (Simpson 1997). However, it must be pointed out that environmental conditions impact on microbial chemistry and strains from different locations can have different properties, hence preserving these areas could be important.

Thus, the inherent ecological assumptions of the CBD with regard to conservation and sustainable use, do not fit our understanding of microbial ecology and therefore the policy implications of the CBD are misaligned for microbial research. We exclude the larger fungi here, some appearing on the International Union for Conservation of Nature Red List of Threatened Species (www.iucnredlist.org) and the current authors refer generally to the microfungi. By contrast, the third goal of the CBD (Access and Benefit Sharing, ABS) is of much higher relevance for the future work with microbial resources, particularly since the large potential for natural product discovery (Sect. 4.5) renders microorganisms prime targets for exploitation and future utilization.

With the aim to provide the necessary legal framework for the effective implementation of Access and Benefit Sharing (ABS), The *Nagoya Protocol on Access to Genetic Resources and the Fair and Equitable Sharing of Benefits Arising from their Utilization to the Convention on Biological Diversity* (the 'Nagoya Protocol'; CBD 2015) was adopted on 29 October 2010 in Nagoya, Japan and entered into force on 12 October 2014, 90 days after the deposit of the fiftieth instrument of ratification and has been ratified by 59 parties (although the U.S., Canada, Russia, and China are notably absent). In Europe, the implementing EU Regulation No 511/2014 on ABS (EU 2014) was enacted simultaneously. In Europe, individual EU member states will now implement the necessary individual national legislation. Globally, each country signatory to the CBD and the Nagoya Protocol will implement its own controls, ensuring compliance with its requirements.

The Nagoya Protocol addresses all activities related to the utilization of genetic resources in the broadest sense. Besides research and development directly targeting future applications, e.g., through screening of microbial strains for useful natural products, or for the control of pests (including invertebrates, vertebrates, weeds and diseases), the Nagoya Protocol also covers basic research that is not intended to commercially exploit microbial resources, for example biodiversity surveys, or the export for taxonomic identification. However, article 8 does offer countries the opportunity to make special considerations for non-commercial use. The objective is the fair and equitable sharing of benefits arising from the utilization of genetic resources, their biochemical composition, and from traditional knowledge associated with these genetic resources, through the implementation of suitable mechanisms and by balancing regulatory controls and due diligence. Utilization of a genetic resource must be traceable along the value chain of research, development, innovation, pre-commercialization and commercialization. Benefits are not only monetary but also non-monetary, such as the sharing of research results, technology transfer, and training opportunities.

Accordingly, the Nagoya Protocol imposes the following obligations on all users (in the broadest sense) of genetic resources (compare Article 4 of EU Regulation No 511/2014):

- all materials collected or received by the users must have relevant permits or Prior Informed Consent (PIC) and Mutually Agreed Terms (MAT) that include sharing of benefits, if applicable,
- the user is required to provide information on the source of genetic material, and the associated PIC and MAT to the national checkpoint, including available unique identifiers. For materials received from other sources, the user needs to obtain the information relevant to ABS,
- the user is required to transfer to subsequent users all information relevant to ABS,
- records of all the activities related to the use of a genetic resource must be maintained by the user for 20 years after termination of use.

On the level of signatory states, the following specific measures to implement the Nagoya Protocol are currently considered. The ABS Clearing House (ABSCH) represents a key tool for facilitating the implementation of the Nagoya Protocol (ABSCH 2015). The ABSCH is a platform for exchanging information on access and benefit-sharing where each Party to the Nagoya Protocol is required to make available (a) its legislative, administrative and policy measures on ABS, (b) information on the national focal point and competent national authority or authorities, and (c) information on permits or their equivalent, that were issued at the time of access. The National Authorities may choose to issue an Internationally Recognized Certificate of Compliance (IRCC), a record created nationally when a permit or equivalent (PIC and MAT) is made available to the ABS Clearing House in order to facilitate the monitoring of the utilization of genetic resources along the value chain, that extends from (a) isolation, characterization, screening and application trials to (b) commercialization of products. Although 197 party and non-party states have filed information on their national focal points, no single country has yet registered any IRCC (ABSCH 2015). Based on the current legal situation described above, mBRC now face numerous challenges:

1. Information on existing PIC has so far been supplied for only 26 of the ~25,000 bacterial strains deposited in the DSMZ. This very low incidence strongly suggests that many depositors were not aware of the legal restrictions that have been in place since 1993. It is foreseeable that the implementation of the Nagoya Protocol will result in a significantly increased demand of mBRCs customers for legal support and expertise at all stages of microbiological work, starting at the planning phase of research projects.

2. Notably, the Nagoya Protocol clearly states (in Article 8) the need for simplified measures on access for non-commercial research purposes for developing countries. Countries of origin can very efficiently facilitate utilization and capitalize of their microbial resources by rapidly implementing the concept. By making access mechanisms straight-forward, whilst implementing best practice in tracking genetic resource use by commercial companies, it will encourage the use of microbial resources and increase the chances of discovery and long-term benefits. This will have a lasting effect on the future development of science in and the bioeconomy of a country. A delayed implementation of the ABS clearing house will result in users remaining unaware of relevant procedures. Delayed implementation or over complicated and demanding systems for the legal mechanisms, may result in a significant competitive disadvantage for domestic science, research and development. In particular, (a) novel types of microorganisms (due to their presence in other countries) may be isolated, characterized and published by competing researchers, (b) domestic researchers cannot participate in bi- and multilateral research projects and hence will be isolated from technology transfer and capacity building, and (c) identical natural compounds may be isolated more rapidly by competitors in other countries (Overmann 2015).
3. The 115,000 strains in the WDCM Global Catalogue of Microorganisms with known geographic origin records were isolated from 164 different countries. Most of these characterized strains originated in Europe (36 %) and Asia (25 %) (Wu et al. 2013). For the rest of the WDCM registered strains, information is unavailable or has not been entered into the GCM database. Nevertheless, the available data indicate that microbial diversity of the entire African and South American continents is severely underrepresented within culture collections. These regions could unlock the potential of their microbial diversity through implementation of appropriate ABS tools and networking with existing mBRCs and research centres.
4. The workload for scientific users to comply with the Nagoya Protocol can be significantly reduced by sourcing materials from mBRCs that have implemented community best practice and utilizing clusters of expertise and competent institutions. As a suitable tool, EU Regulation No 511/2014 identifies, and stipulates that 'Union trusted collections' or 'Registered collections' be established to apply standardized procedures for (a) exchanging and supplying microbial resources, (b) providing the necessary documentation of compliance with the Nagoya Protocol, (c) monitoring the transfer of microbial resources, and (d) keeping the necessary records. The fact that only 12,147 (i.e., 4 %) of the strains in the GCM have detailed information on their history (Wu et al. 2013), clearly demonstrates the magnitude of the additional effort that mBRCs need to establish the database entries that are now needed. This includes source, geographic coordinates of sampling location and information on the isolator, depositor and ownership.
5. The microbial resources in public collections are used only rarely for commercial applications. Using the Leibniz-Institute DSMZ as an example, 37,000

resources (mostly bacterial, but also fungi, human cell lines, plant viruses, and genomic DNA) are delivered annually, of which only one (i.e., 0.003 %) was used for commercialization. Interestingly, DSMZ each year receives 100 enquiries for commercialization. This means that almost all potential customers eventually abstain from commercial applications because of the public accessibility of the resource and difficulties in clarifying the legal requirements for their use. It is evident that legal certainty combined with an exclusive access to microbial resources, will be critical to their future commercial exploitation. This will require negotiations of appropriate ABS measures with the view to bioprospecting. However, for countries interested in advancing their bioeconomy sector, the fact that most microorganisms do not occur exclusively in their territory, will constitute a particular incentive to participate in multilateral biodiscovery programs. These will enable a rapid and efficiently translation of the results to biotechnological or medical applications.

4.8 Role of Networking Between Microbial Domain Biological Resource Centers

Dedicated, long-term research infrastructures provide the most effective means to (a) preserve the monetary value of living microbial resources, (b) gather and maintain the professional knowledge and data associated with them, and (c) offer the essential scientific services (in particular identification, chemotaxonomy, and physiological characterization). Through these characteristics, mBRCs underpin innovation and discovery and are key to the realisation of the targets for sustainable and inclusive growth in the bioeconomy. However, single mBRCs alone are not capable of covering the vast, so far unexplored molecular and physiological diversity, including the supporting services and data that are of interest to future bioprospecting projects. Furthermore, the current fragmentation of individual holdings, services, strategies and policies result in duplication and gaps in what can be offered. These inherent weaknesses are costly in times of public budget constraints, major demographic changes and increasing global competition. Networking of mBRCs could provide a practical solution to these challenges.

There are various forms in which culture collection networking has taken place in the past (Smith et al. 2013, 2014); efforts to coordinate and develop mBRCs have been undertaken at national, regional and global levels. An example is when collections collaborate in project consortia to answer research programme calls which address specific microbial problems or research areas. These are often short lived, and are usually dictated by the term of the funding: Hence they are insufficient to maintain new structures and competences after termination of the project. Another is the formation of national, regional and international federations to (a) further the activities of collections, (b) facilitate access to their materials, and (c) improve operations. Networking at the national level is quite frequent. Most are loose federations bringing together collection staff and users to discuss common issues and share information. The Belgium Co-ordinated Collections of Microorganisms (BCCM™) is an example of a more defined infrastructure where policy and strategy are set through the Belgian Science Policy Office which coordinates operations, research and development. The U.S. Culture Collection Network (USCCN) is funded by the National Science Foundation and facilitates scientists working with laboratory based collections of microbes. Overall, there are 17 such national federations (Smith 2014).

At the international level, the oldest and largest federation of collections is the World Federation for Culture Collections (WFCC) which was founded in 1968 and currently has over 120 affiliated collections. It promotes and supports the establishment of culture collections and related services. Information networks are established between the collections and their users, while workshops and conferences are organised resulting in publications and newsletters (WFCC 2015). In particular, this effort has resulted in the World Data Centre for Microorganisms with 692 registered collections (WDCM 2015). The European Culture Collections' Organisation (ECCO 2015) followed in 1981 and joins 61 member collections from 22 European countries. ECCO has been an incubator for pan-European activities and has driven the development of collections. It was instrumental in coordinating activities through joint initiatives and projects, e.g., within recent European Community Framework Programmes, including the Common Access to Biological Resources and Information (CABRI – http://www.cabri.org), European Biological Resource Centres Network EBRCN – http://www.ebrcn.net) and the European Consortium of Microbial Resource Centres (EMbaRC – http://www.embarc.eu).

Federations, like the majority of other regional and national networks, are mostly run on a volunteer basis and rely on individuals to carry out the operations using small amounts of funding, often from membership fees. Federations typically have no mandate to change the institutional policy and strategy of their members, although they are more enduring than project consortia. However, to meet modern day challenges, mBRCs must more actively share tasks and strategically coordinate their activities. They need to (a) work under common quality management systems to deliver consistently high quality materials, (b) coordinate their accession policies to arrive at complementary holdings, (c) share facilities, technologies, and expertise and (d) train young researchers for higher cost efficiency.

The Organisation for Economic Cooperation and Development (OECD) Biological Resource Centre initiative (1999–2006) provided the framework for best practice and biological resource networking. It proposed a Global Biological Resource Centre Network (GBRCN) to (a) integrate services and resources, (b) encourage innovative solutions, (c) provide coherence in the application of quality standards, (d) allow homogeneity in data storage and management, and (e) facilitate workload sharing. It recommended that the new generation culture collections (i.e. mBRCs) undertake a proof of concept for the GBRCN that would enhance microbial resource availability and quality. As a consequence, the demonstration project for a GBRCN (http://www.gbrcn.org) commenced in 2008. The German Government, through the German Federal Ministry of Research and Education (Bundesministerium für Bildung und Forschung, BMBF), supported a small

Secretariat to draw national efforts together in developing tools for the establishment of the GBRCN. National and regional efforts have been initiated to establish the GBRCN with the aim to build a structured, long-lasting global network, enabling collections to meet user needs since the project report was published (Fritze et al. 2010). Unnecessary competition between regional networks with similar goals can be reduced if they (a) become partners of GBRCN, (b) become signatories to a membership agreement that will establish a common operational framework, and (c) participate in the decision making processes of the GBRCN. Currently, GBRCN has partners in North- and South America, Africa, Asia and a strong base in Europe.

The success of the GBCRN demonstration project and collaboration with EMbaRC and ECCO, led to the pan-European initiative MIRRI being placed on the European Strategy Forum for Research Infrastructures (ESFRI) road map. This distributed infrastructure currently interlinks 43 public collections and research institutes from 19 countries holding more than 360,000 microbial resources. MIRRI brings together European microbial resource collections and their stakeholders (users, policy makers, potential funders and the plethora of microbial research efforts) and links them to non-European country partners with the aim to add value through:

- a coordinated approach to the implementation of best practice and coherent application of quality standards,
- a coordinated strategy to provide a broader, less redundant coverage of microbiological resources and services,
- a distributed platform for microbial taxonomy to ensure best use of the remaining expertise and to put in place a human resource development program,
- common policies across international boundaries facilitating legitimate access,
- establishing facilities and resources in countries or regions rich in microbial diversity, but without resources and facilities to utilize them in research,
- homogeneity in data storage and management, enabling data mining and targeting of specific microbial resources for specific tasks, through the MIRRI information portal,
- a business model where access to all microbial resources can be found through contact with a single mBRC (Schüngel et al. 2013).

Finally, the Asian Consortium for the Conservation and Sustainable Utilization of Microbial Resources (ACM) has been established (Ando et al. 2014). Based on the principles of the Nagoya Protocol, ACM has established a Network of International Exchange of Microbes under ACM (NIEMA; http://www.acm-mrc.asia/am/acm10.html) which facilitates movement from a single collection (the registering collection) to another network member, but without distribution to outside of the network without permission. Strains exchanged under the NIEMA scheme can only be used for non-commercial purposes whereas a facilitated exchange of microbes for commercial use is not addressed. Thus this particular network scheme requires additional mechanisms to link with other networks for a broader exchange of biomaterials, and cannot legally provide them for a broader use in research and development. mBRC networks have to develop novel mechanisms that conform to existing laws to allow a broader application of bioresources, while limiting additional administrative burdens. Given the persistent growth of the bioeconomical sector and the very limited funding available for taxonomic work, mBRCs are expected to make a much more visible contribution towards the bioeconomy than previously.

4.9 A Business Plan for Microbial Domain Biological Resource Centers

As outlined in Sect. 4.3, mBRCs cannot operate in a cost-covering manner if their mission is to support publicly funded scientific research with high quality, and well characterized, microbial resources. Indeed, the majority of all CCINFO registered culture collections are affiliated to universities or are governmental (WDCM 2015) and the level of public funding of mBRCs is in the range of 65–92 % (Smith et al. 2013). Nevertheless, user needs dictate the structure, activities and future planning of mBRCs which is similar to the private enterprise situation. In this respect, mBRCs are distinct from other scientific institutions in that they are not primarily founded on the success and originality of their scientific research, but need to provide key microbial resources, information, and expertise for research and development. These particular features have to be considered when developing business plans that integrate additional tasks of mBRCs related to bioprospecting.

Traditionally, the main stakeholders of mBRCs are (a) the depositors, academic users, national funding bodies, and umbrella organisations to which mBRCs are affiliated (scientific associations and WFCC), and (b) intergovernmental organisations including OECD. The (a) necessity to safeguard key strains (Sect. 4.5), (b) need for novel types of microbial resources and (meta)data (Sects. 4.5 and 4.6), (c) recent developments regarding the national implementation of the Nagoya Protocol (Sect. 4.7), and (d) necessity to network (Sect. 4.8), will call additional stakeholders into action. In particular, these comprise (a) scientific journals, (b) other mBRCs and collaborating scientific institutes that act as network partners, (c) industrial users such as small and medium enterprises (SMEs) and industrial organizations, and (d) national regulatory authorities (Fig. 4.1). Individual mBRCs will be able to identify the requirements of additional and specific stakeholders through a more tailored stakeholder analysis.

Identifying the most influential and reliable stakeholders, together with raising their interest and engaging their support, will be key issues for mBRCs if they are to enter the field of bioprospecting in a sustainable manner. The establishment of small biotechnology companies can play a decisive role in the initial discovery of, for example, promising marine bioactive compounds, as these enterprises will work closely with academics and governmental agencies when performing the initial steps in the discovery of new natural products (Genilloud 2014). Collaboration between private companies and public institutions will be of paramount importance for financial support in the discovery process. Crude extracts and pure compounds

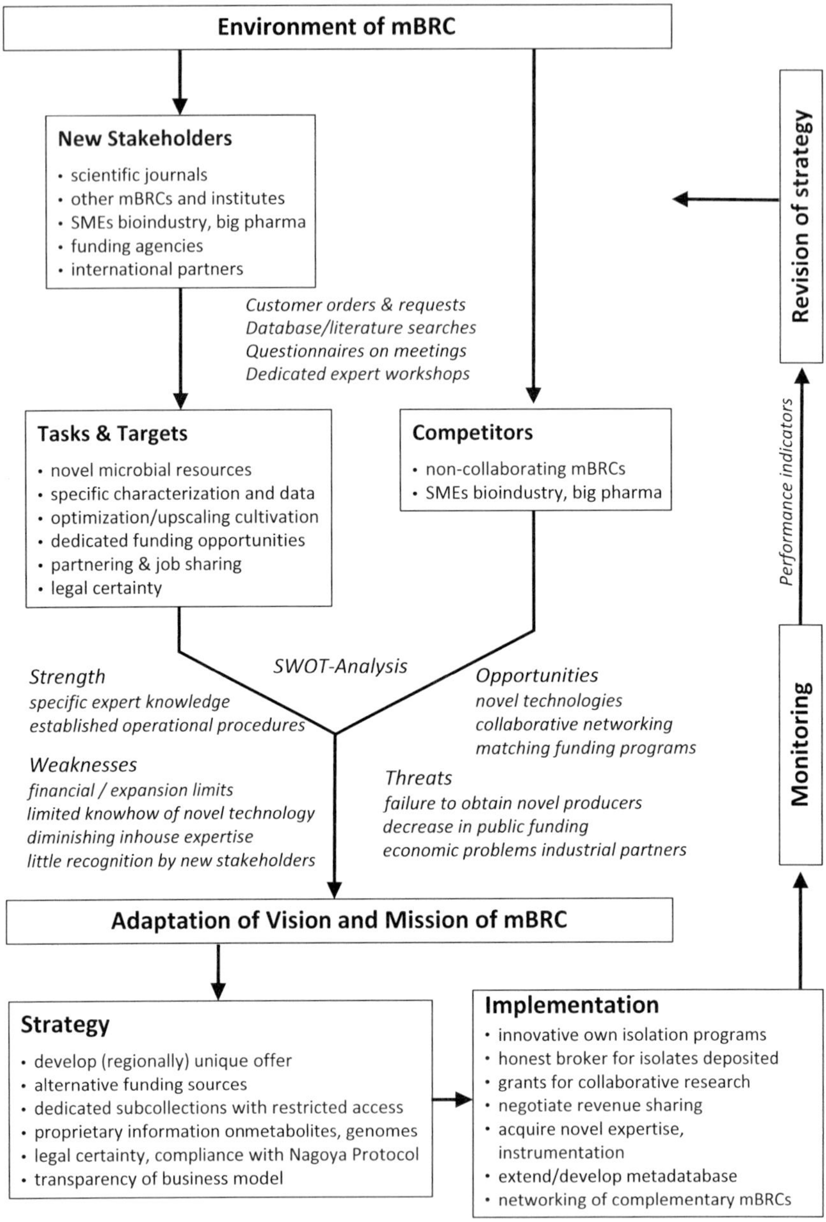

Fig. 4.1 Elements with key aspects (*boxes*) and specific approaches (*in italics*) that constitute a possible business plan for the future development of an mBRCs considering to enter the field of bioprospecting. The envisioned time frame is 5 years

produced by academic laboratories can be used for diverse bioassays as part of broader collaboration programs with private biotech companies worldwide. One challenge for universities is to devise mechanisms that protect intellectual property and simultaneously encourage partnerships with the private sector, by recognizing that the chances of a major commercial pay-off are small, especially if drug discovery is pursued by a single institution (Rocha et al. 2011). In this respect, mBRCs can also support universities by providing the necessary expertise on the true costs and time requirements of isolation, cultivation, upscaling and characterization procedures.

The development of the downstream elements of the business plan is largely determined by the actual needs of depositors and users that are involved in bioprospecting. In contrast, public funders of mBRCs determine the overall mission of most mBRCs, but typically have less specific expectations and directives. Based on the public relations experience of the DSMZ, the analysis of novel customer needs must be performed proactively by the mBRC itself and cannot solely be based on (rather sporadic) requests filed by individual customers. The crucial information on depositor and user needs can be gained through (i) an analysis of customer orders and (ii) the requests for missing materials, (iii) database and literature searches revealing the potential type and number of future deposits and the associated information required, (iv) questionnaires to interrogate delegates of the appropriate scientific meetings or bioindustry events, and (v) expert workshops organized by mBRCs to discuss needs and solutions with international experts (Fig. 4.1).

Traditional products must be complemented by the accession of novel products to meet the needs of new users. Many mBRCs make available DNA, enzymes, secondary metabolites and other derivatives from authenticated strains, or curated databases linked to genome sequences, either as the standard inventory, or on a case-by-case basis. mBRCs can move beyond their traditional services by developing commercial products through the provision of biotechnological solutions to problems, active compounds, and contract research services. These can be funded through public-private investment and spin-off companies.

The non-governmental organization CABI, has been moving in this direction since the 1990's after direct UK Government funding ceased. For example, it identified the need for a rapid test kit to detect fungal contamination in kerosene used as plane fuel because previously available detection methods required as long as 3–10 days for a test result. The company Conidia Bioscience (http://www.conidia.com) was established to develop the FUELSTAT™ detection kit that is changing paradigms for detection of contaminants in fuel and which is now recommended in the Boeing Aircraft Maintenance Manual (http://samtheta.org/id-docs/b/boeing-aircraft-maintenance-manual-pdf-39871.pdf). This demonstrates that culture collection staff can devise solutions to current microbial problems and establish companies. At CABI, the profits are partly used to support biosystematics, biological collections and fundamental research. CABI has also been involved in developing biocontrol agents and one particular success, in collaboration with partners, has been Green Muscle, a fungal product used for control of African Locust (Lomes et al. 2001). The profits fund biodiversity initiatives in Africa. Although neither of these examples are related to bioprospecting per se, the future will reveal whether establishing spin-off companies with proprietary products are as beneficial as mBRCs acting directly as brokers between owners and their commercial users of microbial resources.

In conclusion, mBRCs planning to make a significant contribution towards bio-prospecting, need to gain suitable information on the (a) required microbial resources and associated (meta)data, (b) additional expertise, (c) funding programs that might support additional activities of mBRCs, (d) legal requirements, (e) possible partners, and (f) potential competitors. In this analysis, the internal strengths and weaknesses as well as external opportunities and threats must be considered, e.g. by undertaking a SWOT-analysis (Fig. 4.1). This will support the development of a specific product and an appropriate marketing strategy that promote the unique selling points of the individual mBRCs.

It is likely that individual mBRCs will choose to focus on those groups of organisms for which they already possess the essential skills and operational procedures, given the tremendous taxonomic, biochemical and physiological diversity that remains undiscovered. As an inherent weakness, the limited public funding of mBRCs often will not permit an expansion of tasks into novel diversity. It has to be emphasized that revenues generated through sales of even high value microbial resources and associated metabolomic and genomic data to bioindustry will not cover costs. Therefore, alternative funding requires to be sought from the government agencies in charge of the bioeconomy that traditionally are not stakeholders of mBRC, and through participation in dedicated funding programs for bioprospecting such as the EU Horizon 2020 INFRADEV provide numerous opportunities.

Suitable performance indicators have to be developed and applied when monitoring mBRCs for the success of novel strategies and their implementation (Fig. 4.1). One indicator is the rate of increase in the number of specific target strains deposited (Fig. 4.2). The success of data provision can be monitored through analysis of online visitors to (meta)databases maintained by an mBRC. In addition, the third-party funds acquired need to be considered. Publications relating to biotechnology and bioprospecting are another suitable measure of impact. Standard evaluation procedures ascribe to first or corresponding authorships of scientific publications have a significantly higher value than co-authorships. In the case of mBRCs, however, a large number of co-authorships by curators, or other workers in mBRCs, should be valued highly since it clearly documents that the work and expertise is enabling the scientific success of institutions other than mBRCs.

Even if numerous novel types of microorganisms can be efficiently recovered in the future, an untargeted screening approach to the bioprospecting of thousands of novel isolates would be far too costly (i.e. between 0.5 and 1 billion Euros per approved drug (Overmann 2015). Instead, promising isolates for the future development of pharmaceuticals, or for biotechnological applications could be identified, based on a sufficient knowledge of their biochemistry and physiology, which can be obtained by mBRCs as partners and brokers. Therefore, one key to a successful entry of mBRCs into the field of bioprospecting will be to generate information on the biosynthetic potential of novel types of microorganisms. This can be achieved by extended characterization, metabolic profiling, and genome and transcriptome

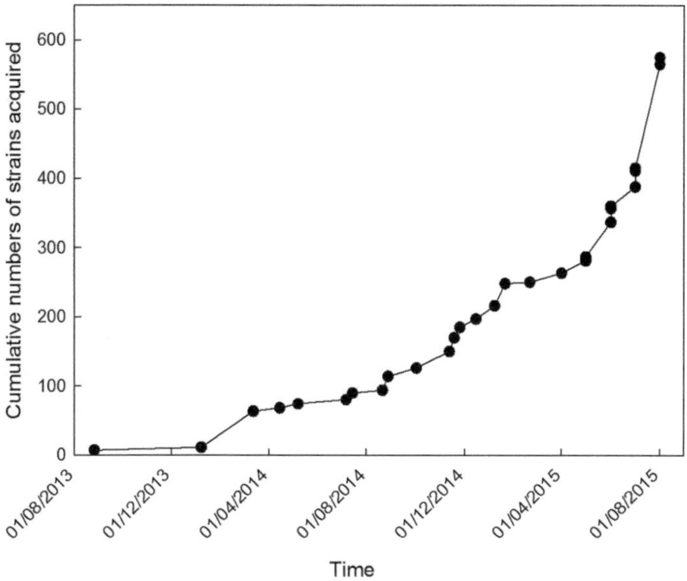

Fig. 4.2 Example for the monitoring of the successful acquisition of novel target microbial resources by the Leibniz-Institute DSMZ. The time course of the increase in strains of pathogenic bacteria deposited in the DSMZ sub-collection "Central Pathogen Repository of the German Centre of Infection Research" is depicted

analyses. Furthermore, information on mass cultivation in bioreactors would be advantagenous as mentioned previously. Pre-screening will probably yield information to substantially increase the rate of discovery of novel compounds and significantly increasing the value of the individual microorganism. Most mBRCs do not possess state-of-the-art genomic and metabolomic technology which enable the generation of the essential high throughput data. However, this could be gained through collaboration with external partners.

Another key issue for a successful entry into the field of bioprospecting, is to establish sustainable strategies for gaining better access to the vast uncharted microbial biodiversity and better quality control, characterization and distribution to users. Particular challenges include (a) compliance with new legal requirements that threaten to impede efficient international collaboration, (b) the difficulties of individual mBRCs to collaborate with industry, (c) the need to increase the attractiveness of mBRCs to bioindustry, and (d) the need for developing countries to gain access to bioindustry.

One possible solution is to join the existing expertise of established mBRCs that attract sufficient funding, to rapidly create a platform providing a single place for a large number of well characterized, pre-screened microbial resources in a legally compliant manner (Fig. 4.3). For example, there are currently 45 major collections supplying cultures in Europe. No single collection would be able to compete with the coverage attained by the MIRRI consortium. Due to its public funding and

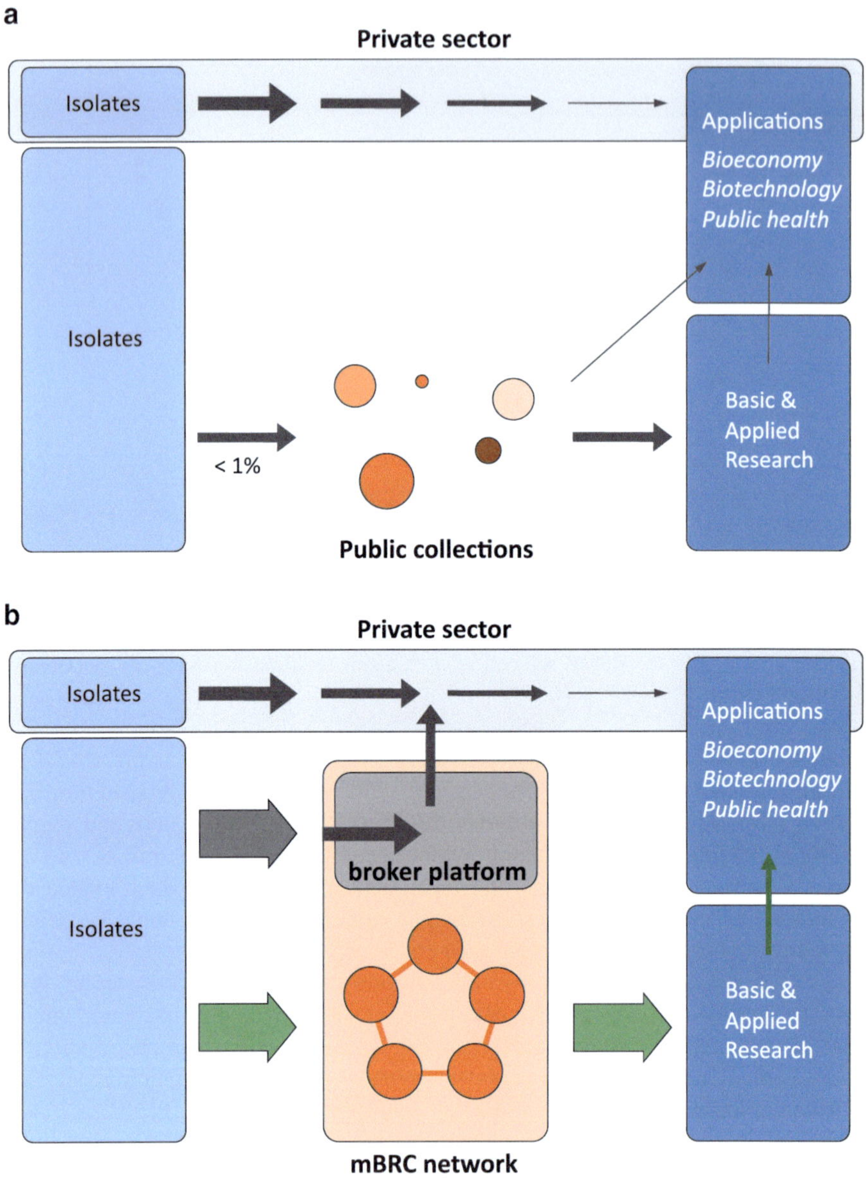

Fig. 4.3 (**a**) Present situation of culture collections and mBRCs that provide microbial isolates for basic research and application. Currently, culture collections and mBRCs acquire less than 1 % of isolates generated through public research. Since they act mostly in an uncoordinated and non-complementary manner, they cannot make major contributions towards bioprospecting. (**b**) Concept to enable mBRCs to make major contributions towards bioprospecting through formation of a lasting, tight network and establishing a platform specifically devoted to broker strains for bioindustry

hence financial independence, this network of mBRCs could act as financially independent, honest broker that would constitute a more visible and attractive partner for private companies than each individual mBRC. Furthermore, such a platform would provide an easy and reliable entry point for developing countries to supply microbial resources at different levels (undefined natural samples/enrichments, microbial isolates, pre-screened target strains with existing metadata), and offer them to bioindustry in geographic regions and markets that would otherwise be inaccessible. On the other hand, if bioindustry needs to be established in a developing country, it could use such a platform to gain access to a much broader choice of microbial resources than the individual country could provide individually. Microbial resources that are particularly enriched in a certain region, e.g. strains related to the bioremediation of specific mine tailings, specific xenobiotica contaminations, or plant-growth-promoting bacteria associated with endemic plants, could be offered most effectively and in a competitive manner by the respective country for future bioprospecting.

Finally, the management of those collections dedicated to bioprospecting has to consider the distinct legal requirements of bioindustry. Accordingly, such subcollections would only become attractive to a user from bioindustry if it was given exclusive access to the microbial resources. The would include the associated data, and if legal certainty is provided, e.g., by pre-negotiated contracts for each bioresource that fix the share of benefits for all partners involved (i.e. country of origin, scientists involved in the detection, isolation and characterization, and mBRCs providing storage, quality control, databases and the distribution system).

4.10 Conclusions

The immense diversity of bacteria and filamentous fungi remains largely untapped. Extending bioprospecting activities to these novel microorganisms with often unknown physiological properties requires specialized skills with respect to the isolation, cultivation and characterization of fastidious strains, dedicated (meta)databases, but also legal competencies with respect to international law. While many of these essential skills and infrastructures are maintained at mBRCs, the latter so far focus on preserving microbial resources for basic scientific studies and providing them to academic and industrial researchers. mBRCs could make a significant contribution towards bioprospecting by (a) improving access to uncharted microbial diversity, (b) extending the characterization, metabolic profiling, and genome analysis of novel microbial resources, (c) establishing proprietary databases with information relevant to bioprospecting, and (d) providing the means for exclusive exploitation of microbial resources. However, since public funding is limited and the revenues generated from sales of microbial resources rarely cost-covering, alternative funding sources are needed by mBRCs to enter the field of bioprospecting.

There is not one single business model that fits all. However, the mBRCs business plan (Fig. 4.1) developed here emphasizes the elements necessary to enter the field of bioprospecting. It goes beyond a simple strategic plan by including specific internal goals and measures for their implementation and providing tools for monitoring and decisions. The business plan would be useful for involving additional new stakeholders and help to efficiently acquire novel bioresources and develop services, improve database structure, novel funding schemes and thereby restructuring a mBRC.

However, for many mBRCs, their business plan will be an ongoing, cyclic process as user needs, technology and funding schemes change (Fig. 4.1). It provides a procedure to improve culture collections by linking them into a network to efficiently underpin the bioeconomy and ultimately make a lasting impact on bioprospecting (Fig. 4.3). Remote mBRCs in, for example, developing countries can also be included in this network. However, it is crucial that mBRCs do not become commercial entities - they must not compromise their public service role.

Acknowledgements This work was funded by grants of the German Federal Ministry of Science and Education (BMBF) through the German Centre of Infection Research (DZIF) (grant 8000-105-3). Additional support was by the European Commission through funding of the preparatory phase of the Microbial Resource Research Infrastructure (MIRRI, No. 312251) and through the FP7-project 'Marine Microorganisms: Cultivation Methods for Improving their Biotechnological Applications' (MaCuMBA, No. 311975) and is gratefully acknowledged. We thank Amber Hartman Scholz for useful comments and suggestions regarding the manuscript.

References

ABSCH, Acess and Benefit Sharing Clearing House (2015). https://absch.cbd.int/

Ando K, Jin TU, Funabiki R, Wu L, Thoetkiattikul H, Lee J-S, Techapattaraporn B, Changthavorn T (2014) Network of international exchange of microbes under the ACM (NIEMA) – a transfer and exchange system of microbes for microbial resource centres for non-commercial puposes according to the CBD and the nagoya protocol. Microbiol Cult Collect 30:85–96

Baker BJ, Dick GJ (2013) Omics approaches in microbial ecology: charting the unknown. Microbe 8:353–360

Baltz RH (2006) Marcel faber roundtable: is our antibiotic pipeline unproductive because of starvation, constipation or lack of inspiration? J Ind Microbiol Biotechnol 33:507–513

Bérdy J (2005) Bioactive microbial metabolites. J Antibiot 58:1–26

Bills G, Dombrowski A, Paláez F, Polishook J, An Z (2002) Recent and future discoveries of pharmacologically active metabolites from tropical fungi. In: Watling R, Frankland JC, Ainsworth AM, Isaac S, Robinson CH (eds) Tropical mycology, Vol. 2, *Micromycetes*. CAB International, Wallingford, pp. 165–194

Blackwell M (2011) The fungi: 1, 2, 3 … 5.1 million species? Am J Bot 98:426–438

Butler MS (2005) Natural products to drugs: natural product derived compounds in clinical trials. Nat Prod Rep 22:162–195

CBD, Convention on Biological Diversity (2015). Nagoya protocol on access to genetic resources and the fair and equitable sharing of benefits arising from their utilization (ABS) to the convention on biological diversity. http://www.cbd.int/abs/doc/protocol/nagoya-protocol-en.pdf

Chin Y-W, Balaunas MJ, Chai HB, Kinghorn AD (2006) Drug discovery from natural sources. AAPS J 8:E239–E253

Cooper MA, Shlaes D (2011) Fix the antibiotics pipeline. Nature 472:32
Curtis TP, Sloan WT, Scannell JW (2002) Estimating prokaryotic diversity and its limits. Proc Natl Acad Sci U S A 99:10494–10499
Dawyndt P, Vancanneyt M, De Meyer H, Swings J (2005) Knowledge accumulation and resolution of data inconsistencies during the integration of microbial information sources. IEEE Trans Knowl Data Eng 17:1111–1126
Dykhuizen DE (1998) Santa Rosalia revisited: why are there so many species of bacteria? Ant Van Leeuwenhoek 73:25–33
Ernst & Young (2014) Beyond borders – biotechnology industry report, http://www.ey.com/GL/en/Industries/Life-Sciences/EY-beyond-borders-unlocking-value
ECCO, European Culture Collection's Organisation (2015). www.eccosite.org
EU (2014). Regulation (EU) No 511/2014 of the European Parliament and of the Council of 16 April 2014 on compliance measures for users from the Nagoya Protocol on Access to Genetic Resources and the Fair and Equitable Sharing of Benefits Arising from their Utilization in the Union. http://eur-lex. europa.eu/legal-content/EN/ALL/?uri=CELEX:32014R0511
EuropaBio™ (2014) Annual Report 2013-2014, http://www.europabio.org/europabio-annual-report-2013-2014
Foesel BU, Rohde M, Overmann J (2013) *Blastocatella fastidiosa* gen. nov., sp. nov., isolated from semiarid savanna soil – the first described species of Acidobacteria subdivision 4. Syst Appl Microbiol 36:82–89
Fritze D, Smith D, Martin D (2010). Third intermediate report on the GBRCN demonstration project. Germany: GBRCN Secretariat. ISBN 978-3-00-031458-2
Furman JL, Stern S (2011) Climbing atop the shoulders of giants: the impact of institutions on cumulative research. American. Econ Rev 101:1933–1963
GCM, Global Catalogue of Microorganisms (2015) http://gcm.wfcc.info
Genilloud O (2014) The re-emerging role of microbial natural products in antibiotic discovery. Ant van Leeuwenhoek 106:173–188
Glaser KB, Mayer AM (2009) A renaissance in marine pharmacology: from preclinical curiosity to clinical reality. Biochem Pharmacol 78:440-448
Hindra TH, Yang D, Rudolf JD, Xie P, Xie G, Teng Q, Lohman JR, Zhu X, Huang Y, Zhao L-X, Jiang Y, Duan Y, Shen B (2014) Strain prioritization for natural product discovery by a high-throughput real-time PCR method. J Nat Prod 77:2296–2303
IMS Health Market Prognosis (2012) http://www.imshealth.com/deployedfiles/ims/Global/Content/Corporate/Press%20Room/Top-Line%20Market%20Data%20&%20Trends/2011%20Top-line%20Market%20Data/Regional_Pharma_Market_by_Spending_2011-2016.pdf
Jaspers E, Overmann J (2004) The ecological significance of "microdiversity": identical 16S rRNA gene sequences represent bacteria with highly divergent genomes and physiology. Appl Environ Microbiol 70:4831–4839
Jeske O, Jogler M, Petersen J, Sikorski J, Jogler C (2013) From genome mining to phenotypic microarrays: planctomycetes as source for novel bioactive molecules. Ant Van Leeuwenhoek 104:551–567
Kate KT, Laird SA (1999) The commercial use of biodiversity: access to genetic resources and benefit sharing . Earthscan Publications Ltd., Londonpp 398
Kirk PM, Cannon PF, Minter DW, Stalpers JA (2008) Dictionary of the fungi, 10th edn. CAB International, Wallingford
SP Lapage, PHA Sneath, EF Lessel, VBD Skerman, HPR Seeliger, WA Clark (1992) International code of nomenclature of bacteria. Bacteriological code, 1990 revision. ASM Press, Washington
Lomes CJ, Bateman RP, Johnson DL, Langewald J, Thomas M (2001) Biological control of locusts and grasshoppers. Annu Rev Entomol 46:667–702

LPSN, List of prokaryotic names with standing in nomenclature (2015) http://www.bacterio.net/-number.html#total
Mayer AM, Glaser KB, Cuevas C, Jacobs RS, Kem W, Little RD, McIntosh JM, Newman DJ, Potts BC, Shuster DE (2010) The odyssey of marine pharmaceuticals: a current pipeline perspective. Trends Pharmacol Sci 31:255–265
Maymó-Gatell X, Chien Y-T, Gossett JM, Zinder SH (1997) Isolation of a bacterium that reductively dechlorinates tetrachloroethene to ethene. Science 276:1568–1571
MIRRI, Microbial Resource Reseach Infrastructure (2015) http://www.mirri.org/home.html
Miyazaki M (2006) Economic value of microbial resources. Microbiol Cult Collect 22:15–19
Nett M, Erol Ö, Kehraus S, Köck M, Krick A, Eguereva E, Neu E, König GM (2006) Siphonazole, an unusual metabolite from *Herpetosiphon* sp. Angew Chem Int Ed 45:3863–3867
Newman DJ, Cragg GM (2012) Natural products as sources of new drugs over the 30 years from 1981 to 2010. J Nat Prod 75:311–335
OECD (2004) Guidance for the operation of Biological Resource Centres (BRCs) ['Biological Research Centres' is erroneously printed on the cover of this document], OECD. http://www.oecd.org/science/biotech/23547743.pdf
Overmann J (2012) Principles of enrichment, isolation, cultivation, and preservation of bacteria. In: Rosenberg E, Stackebrandt E, DeLong EF, Thompson F, Lory S (eds) The prokaryotes, vol 4. Springer, New York. doi:10.1007/978-3-642-30194-0_7
Overmann J (2015) Significance and future role of microbial resource centers. Syst Appl Microbiol 38:258–265
PhRMA (2012) Pharmaceutical research and manufacturers of America, drug discovery and development: understanding the R&D process, www.innovation.org
Rocha J, Peixe L, Gomes NCM, Calado R (2011) Cnidarians as a source of new marine bioactive compounds – an overview of the last decade and future steps for bioprospecting. Mar Drugs 9:1860–1886
Quaiser A, Ochsenreiter T, Lanz C, Schuster SC, Treusch AH, Eck J, Schleper C (2003) Acidobacteria form a coherent but highly diverse group within the bacterial domain: evidence from environmental genomics. Mol Microbiol 50:563–575
Quast C, Pruesse E, Yilmaz P, Gerken J, Schweer T, Yarza P, Peplies J, Glöckner FO (2013) The SILVA ribosomal RNA gene database project: improved data processing and web-based tools. Nucleic Acids Res. 41(Database issue): D590-596
Rosselló-Móra R, Amann RI (2015) Past and future species definitions for *Bacteria* and *Archaea*. Syst Appl Microbiol 38:209–216
Salazar O, González I, Genilloud O (2002) New genus-specific primers for the PCR identification of novel isolates of the genera *Nocardiopsis* and *Saccharothrix*. Int J Syst Evol Microbiol 52:1411–1421
Sanford RA, Cole JR, Tiedje JM (2002) Characterization and description of *Anaeromyxobacter dehalogenans* gen. nov., sp. nov., an aryl-halorespiring facultative anaerobic myxobacterium. Appl Environ Microbiol 68:893–900
Scannell JW, Blnackley A, Boldon H, Warrington B (2012) Diagnosing the decline in pharmaceutical R&D efficiency. Nat Rev Drug Discov 11:191–200
Schüngel M, Stackebrandt E, Bizet C, Smith D (2013) MIRRI – The microbial resource research infrastructure: managing resources for the bio-economy. EMBnet J 19(1):5–8. (http://journal.embnet.org/index.php/embnetjournal/article/view/706)
Simpson DR (1997) Biodiversity prospecting: shopping the wilds is not the key to conservation. Resour Futur 126:12–15
Singh SB, Zhang C, Zink DL, Herath K, Ondeyka J, Masurekar P, Jayasuriya H, Goetz MA, Tormo JR, Vicente F, Martín J, González I, Genilloud O (2013) Occurrence, distribution, dereplication and efficient discovery of thiazolyl peptides by sensitive-resistant pair screening. J Antibiot 66:599–607
Smith D (2012) Culture collections. Adv Appl Microbiol 79:73–118

Smith D (2014) Culture collections. In: Batt CA, Tortorello ML (eds) Encyclopedia of food microbiology, vol 1. Elsevier Ltd/Academic Press, Amsterdam, pp. 546–552

Smith D, Fritze F, Stackebrandt E (2013) Public service collections and biological resource centres of microorganisms. In: Rosenberg E, De Long EF, Lory S, Stackebrandt E, Thompson F (eds) The prokaryotes – prokaryotic biology and symbiotic associations. Springer, Berlin/Heidelberg, pp. 267–304

Smith D, Mc Cluskey K, Stackebrandt E (2014) Culture collection funding models and BRC business plans, SpringerPlus 3, 81. http://www.springerplus.com/content/3/1/81

Söhngen C, Bunk B, Podstawka A, Gleim D, Overmann J (2014) Bac*Dive* – the bacterial diversity metadatabase. Nucleic Acids Res 42:D592–D599

Stackebrandt E (2010) Diversification and focusing: strategies of microbial culture collections. Trends Microbiol 18:283–287

Stackebrandt E, Smith D, Casaregola S, Varese GC, Verkleij G, Lima N, Bridge P (2014) Deposit of microbial strains in public service collections as part of the publication process to underpin good practice in science. Springer Plus 3:208

Sugawara H, J Ma, Miyazaki S, Shimura J, Takishima Y (eds) (1993) World directory of collections of cultures of microorganisms. ISBN 4-930813-56-5, 1152 pp

Tuama EÓ, Deck J, Dröge G, Döring M, Field D, Kottmann R, Ma J, Mori H, Morrison N, Sterk P et al (2012) Meeting report: Hackathon-workshop on Darwin Core and MIxS standards alignment (February 2012). Stand Genomic Sci 7:166–170

WDCM, World Data Center for Microorganisms (2015). Culture collections information worldwide. http://www.wfcc.info/ccinfo/statistics/

Weissmann KJ, Müller R (2010) Myxobacterial secondary metabolites: bioactivities andmodes-of-action. Nat Prod Rep 27:1276–1295

WFCC, World Federation of Culture Collections (2015) http://gcm.wfcc.info/overview/

Wilson MC, Mori T, Rückert C, Uria AR, Helf MJ, Takada K et al (2014) An environmental bacterial taxon with a large and distinct metabolic repertoire. Nature 506:58–62

WIPO, World Intellectual Property Organization (2015) (http://www.wipo.int/ipstats/en/statistics/micros/index.html)

Wu D, Hugenholtz P, Mavromatis K, Pukall R, Dalin E, Ivanova NN et al (2009) A phylogeny-driven genomic encyclopaedia of bacteria and archaea. Nature 462:1056–1060

Wu L, Sun Q, Sugawara H, Yang S, Zhou Y, McCluskey K, Vasilenko A, Suzuki K I, Ohkuma M, Lee Y, Robert V, Ingsriswang S, Guissart F, Philippe D, Ma J (2013) Global catalogue of microorganisms (gcm): a comprehensive database and information retrieval, analysis, and visualization system for microbial resources. BMC Genomics 14:933

Xie P, Ma M, Rateb ME, Shaaban KA, Yu Z, Huang SX, Zhao LX, Zhu X, Yan Y, Peterson RM, Lohman JR, Yang D, Yin M, Rudolf JD, Jiang Y, Duan Y, Shen B (2014) Biosynthetic potential-based strain prioritization for natural product discovery: a showcase for diterpenoid-producing actinomycetes. J Nat Prod 77:377–387

Yilmaz P, Kottmann R, Field D, Knight R, Cole JR, Amaral-Zettler L, Gilbert JA, Karsch-Mizrachi I, Johnston A, Cochrane G et al (2011) Minimum information about a marker gene sequence (MIMARKS) and minimum information about any (x) sequence (MIxS) specifications. Nat Biotechnol 29:415–420

Chapter 5
Bioprospecting Archaea: Focus on Extreme Halophiles

André Antunes, Marta F. Simões, Stefan W. Grötzinger, Jörg Eppinger, Judith Bragança, and Vladimir B. Bajic

Abstract In 1990, Woese et al. divided the Tree of Life into three separate domains: Eukarya, Bacteria, and Archaea. Archaea were originally perceived as little more than "odd bacteria" restricted to extreme environmental niches, but later discoveries challenged this assumption. Members of this domain populate a variety of unexpected environments (e.g. soils, seawater, and human bodies), and we currently witness ongoing massive expansions of the archaeal branch of the Tree of Life. Archaea are now recognized as major players in the biosphere and constitute a significant fraction of the earth's biomass, yet they remain underexplored. An ongoing surge in exploration efforts is leading to an increase in the (a) number of isolated strains, (b) associated knowledge, and (c) utilization of Archaea in biotechnology. They are increasingly employed in fields as diverse as biocatalysis, biocomputing, bioplastic production, bioremediation, bioengineering, food, pharmaceuticals, and nutraceuticals. This chapter provides a general overview on bioprospecting Archaea, with a particular focus on extreme halophiles. We explore aspects such as diversity, ecology, screening techniques and biotechnology. Current and future trends in mining for applications are discussed.

A. Antunes (✉) • M.F. Simões
Department of Biology, Edge Hill University,
St Helens Road, Ormskirk, Lancashire L39 4QP, UK
e-mail: antunesa@edgehill.ac.uk

S.W. Grötzinger • J. Eppinger
Division of Physical Sciences and Engineering (PSE), KAUST Catalysis Center, King Abdullah University of Science and Technology,
Thuwal 23955-6900, Kingdom of Saudi Arabia

J. Bragança
Department of Biological Sciences, Birla Institute of Technology and Science Pilani,
K K Birla, Goa Campus, NH 17B, Zuarinagar, Goa 403 726, India

V.B. Bajic
Division of Computer, Electrical and Mathematical Sciences and Engineering (CEMSE),
King Abdullah University of Science and Technology, Computational Bioscience Research Center (CBRC), Thuwal 23955-6900, Saudi Arabia

R. Paterson, N. Lima (eds.), *Bioprospecting*, Topics in Biodiversity and Conservation 16, DOI 10.1007/978-3-319-47935-4_5

5.1 The Archaea

5.1.1 Archaeal Diversity

Pioneer work in the 1970s first recognized the non-monolithic nature of prokaryotes, and the distinctiveness of "archaeabacteria" (Balch et al. 1977; Fox et al. 1977; Woese and Fox 1977). Ribosomal RNA (rRNA)-based analyses further supported these findings and this uniqueness was eventually formalized with the establishment of the three domains of Life: Eukarya, Bacteria and Archaea (Woese et al. 1990).

The first visual representations of the archaeal branch of the Tree of Life were sparsely populated, and sub-divided into *Euryarchaeota* and *Crenarchaeota* (Woese et al. 1990). They included only cultured representatives, all of which originating from extreme environments. Advances in cultivation-independent techniques, and particularly the use of metagenomics and single amplified genomes (SAGs), revealed several novel phylogenetic groups, which are quickly reshaping our view of the Archaea (e.g. Eme and Doolittle 2015). Indeed, we currently witness a flurry of ongoing additions of new archaeal phyla, which culminated in the creation of two superphyla, commonly referred to as TACK (Guy and Ettema 2011) and DPANN (Rinke et al. 2013).

The TACK superphylum encompasses the *Thaumarchaeota* (Brochier-Armanet et al. 2008), *Aigarchaeota* (Nunoura et al. 2010), *Crenaerchaeota* (Woese et al. 1990), *Korarchaeota* (Barns et al. 1996), and the recently suggested *Bathyarchaeota* (Meng et al. 2014). The DPANN superphylum includes *Diapherotrites*, *Parvarchaeota*, *Aenigmarchaeota*, *Nanohaloarchaeota*, and *Nanoarchaeota* (Rinke et al. 2013). However, there has been debate on the taxonomic ranking of the *Nanoarchaeota* since their discovery (Brochier et al. 2005; Huber et al. 2002; Waters et al. 2003). The rapidly evolving nature of the archaeal branch is further highlighted by the (a) even more recent description of the *Pacearchaeota* and the *Woesearchaeota*, two massive novel groups within the DPANN superphylum (Castelle et al. 2015), (b) description of the putative new order "Altiarchaeales", placed within the *Euryarchaeota* (Probst and Moissl-Eichinger 2015), and (c) discovery of the "Lokiarchaeota" (Spang et al. 2015). The "Lokiarchaeota" represent a candidate novel phylum within the TACK superphylum and seem to bridge the phylogenetic gap between Archaea and Eukarya (Spang et al. 2015). Some authors believe that this discovery might result in the future fusion of both branches of the Tree of Life into a single domain (Fig. 5.1).

5.1.2 Archaeal Ecology

From an ecological perspective, members of the domain Archaea were traditionally split into thermophilic, methanogenic, and halophilic, all of them isolated from extreme environments (e.g. Ferrera et al. 2008). The widespread use of

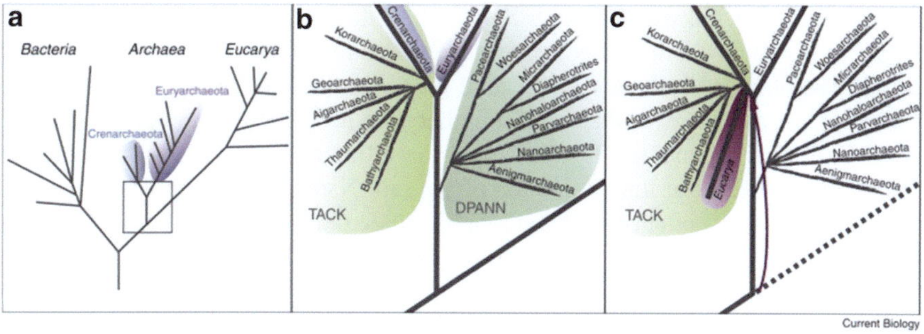

Fig. 5.1 The Tree of Life, as originally envisioned (**a**), and expanded view of latest additions to the Archaea branch (**b**). Recent findings point suggest an alternative placement for the Eukarya branch, and possible fusion with Archaea (**c**). Reprinted from Current Biology, Vol 25 No 6, Eme, L. & Doolittle, F., Microbial Diversity: A Bonanza of Phyla, R228, Copyright (2015), with permission from Elsevier

molecular-based methodologies brought drastic changes, and revealed members of the Archaea to be much more diverse and ubiquitous than previously expected (see Sect. 5.1.1). Archaea populate and thrive in a variety of cold and moderate environments, including landfills, soils, fresh-water sediments, and deep-sea locations, and are involved in symbiotic relationships, e.g. with sponges (e.g. Antunes et al. 2011a; Ferrera et al. 2008; Karner et al. 2001; Leininger et al. 2006; Preston et al. 1996). They contribute to global energy and element cycles, most noticeably via ammonia oxidation in pelagic environments and soils (Ferrera et al. 2008). Archaea might also play important roles in the human body with methanogenic archaea present in the human gut and oral cavities (de Macario and Macario 2009), and abundant *Thaumarchaeota* detected on human skin (Probst et al. 2013).

5.2 Archaea in Saline Environments

5.2.1 *High Salinity Biotopes*

An environment is considered hypersaline when its salt concentration surpasses that of the average seawater (i.e. 3.5 % total dissolved salts). Many of the known hypersaline water bodies derive from simple evaporation of seawater, therefore closely mirroring its ionic composition and proportions. These are known as thalassohaline, as opposed to athalassohaline environments, which are usually derived from inland water bodies, hence the dissolved ions are of non-marine proportions (DasSarma and Arora 2001; Rodríguez-Valera 1988).

Hypersaline biotopes occur in high abundance in arid, coastal and deep-sea locations across the globe. Seawater often penetrates through seepage or narrow inlets near coastal areas creating several small evaporation ponds. Well known examples are (a) the Solar Lake and Gavish Sabkha near the Red Sea coast, (b) Guerrero Negro on the Baja California coast, (c) Lake Sivash near the Black Sea, (d) Sharks Bay in Western Australia, and (e) several locations in Antarctica (e.g. Deep Lake, Organic Lake and Lake Suribati). The number of hypersaline water bodies in coastal areas is further augmented by the numerous artificial solar salterns constructed throughout the ages for the production of sea salt. Natural inland hypersaline lakes have higher salinities than coastal ones and include the Dead Sea (Middle East) and the Great Salt Lake (USA), which are the two largest and best-studied examples. The conjugation of high salinity and alkaline conditions produces unusual alkaline hypersaline soda brines. Some of the better-known examples include the Wadi Natrum lakes, Egypt; Lake Magadi, Kenya; the Great Basin lakes, western United States (Mono Lake, Owens Lake, Searles Lake, and Big Soda Lake), and several others in China, India and throughout the world.

Another type of hypersaline biotopes are the often-overlooked saline soils. These include desolate areas present in, for example, Death Valley (California, USA), Alicante (Spain), dispersed locations across Iraq, and Dry Valleys (Antarctica), among several others (Ventosa et al. 1998). Additional examples of less conspicuous highly saline environments include pickled food, fermented products of oriental cuisine (soy sauce, fish paste), surfaces of salt-excreting desert shrubs, human or animal skin, and other places exposed to periodical drying (Galinsky & Trüper, Galinski and Trüper 1994; Lee 2013).

The least explored hypersaline environments include subterranean brines, evaporite deposits, and brine-filled deep-sea basins. Exploration of such environments has been hampered by the remoteness of such locations, and technical and sampling impediments. The underexplored potential of such locations has attracted considerable attention in the last decades, and resulted in several interesting studies (e.g. Antunes et al. 2008, 2011a, b, 2015; Bougouffa et al. 2013; Daffonchio et al. 2006; Fish et al. 2002; Guan et al. 2015; Joye et al. 2009; Mapelli et al. 2012; McGenity et al. 2000; Siam et al. 2012; van der Wielen et al. 2005; Vreeland et al. 2000; Wang et al. 2011), and some preliminary insights into potential applications of their microbial inhabitants (e.g. Antunes et al. 2011b; Mohamed et al. 2013; Sagar et al. 2013; Sayed et al. 2014).

5.2.2 *The* Halobacteria*: Extremely Halophilic Archaea*

Microbes living in hypersaline environments are called halophiles. Based on their preferred salinity, they can be categorized as slight (0.3–0.8 M or 1.7–4.8 % NaCl), moderate (0.8–3.4M or 4.7–20 % NaCl), or extreme halophiles (above 3.4 M or 20 % NaCl) (Ollivier et al. 1994). Extreme halophiles are traditionally associated with the members of the euryarchaeal class *Halobacteria*. This class contains the single order *Halobacteriales* and its single family *Halobacteriaceae*, although a recent proposal argues for splitting *Halobacteria* into *Haloferacales*, *Natrialbales*, and an emended order *Halobacteriales* (Gupta et al. 2015). The class *Halobacteria* currently includes 177 species with validly published names, placed in 48 genera (LPSN- List of Prokaryotic Names with Standing in Nomenclature 2015; Table 5.1).

Extremely halophilic behaviour is not, however, an exclusive characteristic of the *Halobacteriales*, as it is also observed in *Methanohalobium* and *Methanohalophilus*, of the family *Methanosarcinaceae*, within the *Euryarchaeota* (Oren 2000). No halophiles have, thus far, been described within the kingdom *Crenarchaeota* (Oren 2002).

5.3 Applications of Halophilic Archaea

Mankind has been using halophiles for at least 5000 years. For example, the characteristic red coloration seen in salterns across the globe is imparted mostly by halophilic archaea, and aids in the process of salt crystallization. Other ancient applications include the production of fish sauce, soy sauce and other traditional fermented foods (e.g. Lee 2013).

However, an exponential increase in the applications for halophiles has been observed in the last few decades, particularly after the discovery of extremophiles and extreme-condition adapted enzymes (extremozymes). The industrial application of extremozymes, is clearly the most prominent direct applications for halophiles, and most other extremophiles. However, further exploration is leading to an increase in the number of isolated strains, general knowledge, and number of applications for halophiles and halophilic archaea (Table 5.1; Fig. 5.2).

5.3.1 Archaeal Pigments

Environments with high densities of halophilic archaea frequently have a characteristic red coloration. This is mostly due to the production of C-50 carotenoid pigments (α-bacterioruberin and its derivatives mono-anhydrobacterioruberin (MABR) and bis-anhydrobacterioruberin (BABR), along with small fractions of C-40 carotenoids such as lycopene and β-carotene) (Yatsunami et al. 2014), which are found in the membranes of several halophiles that thrive in such environments. The reddening of the brines contributes to the absorption of light energy, thereby increasing water evaporation and speeding up the process of salt crystallization (Oren 2002). Within these pigments, β-carotene is the most widely used, mainly as a natural food colorant and as an antioxidant, but also as an important additive in cosmetics, multivitamin preparations, and health food products (nutraceuticals) (Margesin and Schinner 2001; Oren 2002, 2010).

Table 5.1 Extremely halophilic archaea: List of genera within the *Halobacteria* and associated patents

Genus	Reference	Patents[a]
Haladaptatus	Savage et al. (2007)	24
Halalkalicoccus	Xue et al. (2005)	41
Halapricum	Song et al. (2014)	0
Haloarchaeobius	Makhdoumi-Kakhki et al. (2012a)	0
Halarchaeum	Minegishi et al. (2010)	4
Haloarcula	Torreblanca et al. (1986)	334
Halobacterium	Elazari-Volcani (1957); Skerman et al. (1980)	560
Halobaculum	Oren et al. (1995)	47
Halobellus	Cui et al. (2011c)	0
Halobiforma	Hezayen et al. (2002)	48
Halococcus	Schoop (1935); Skerman et al. (1980)	266
Haloferax	Torreblanca et al. (1986)	374
Halogeometricum	Minegishi et al. (1998)	80
Halogranum	Cui et al. (2010b)	7
Halohasta	Mou et al. (2012)	0
Halolamina	Cui et al. (2011b)	0
Halomarina	Inoue et al. (2011)	2
Halomicroarcula	Echigo et al. (2013)	0
Halomicrobium	Oren et al. (2002)	51
Halonotius	Burns et al. (2010)	3
Halorientalis	Cui et al. (2011c)	0
Halopelagius	Cui et al. (2010c)	0
Halopenitus	Amoozegar et al. (2012)	0
Halopiger	Gutiérrez et al. (2007)	34
Haloplanus	Bardavid et al. (2007)	8
Haloquadratum	Burns et al. (2007)	76
Halorhabdus	Wainø et al. (2000)	62
Halorubellus	Cui et al. (2012)	0
Halorubrum	McGenity and Grant (1995)	204
Halorussus	Cui et al. (2010a)	0
Halosarcina	Savage et al. (2008)	5
Halosimplex	Vreeland et al. (2002)	25
Halostagnicola	Castillo et al. (2006a)	35
Haloterrigena	Ventosa et al. (1999)	72
Halovenus	Makhdoumi-Kakhki et al. (2012b)	0
Halovivax	Castillo et al. (2006b)	25
Natrialba	Kamekura and Dyall-Smith (1995)	159
Natrinema	McGenity et al. (1998)	47
Natronoarchaeum	Shimane et al. (2010)	1
Natronobacterium	Tindall et al. (1984)	100
Natronococcus	Tindall et al. (1984)	88

(continued)

Table 5.1 (continued)

Genus	Reference	Patents[a]
Natronolimnobius	Itoh et al. (2005)	17
Natronomonas	Kamekura et al. (1997)	102
Natronorubrum	Xu et al. (1999)	36
Salarchaeum	Shimane et al. (2011)	0
Salinarubrum	Cui and Qiu (2014)	0
Salinigranum	Cui and Zhang (2014)	0
Salinarchaeum	Cui et al. (2011a)	0

[a]Number of results obtained from text search of the name of each genus in Google Patents

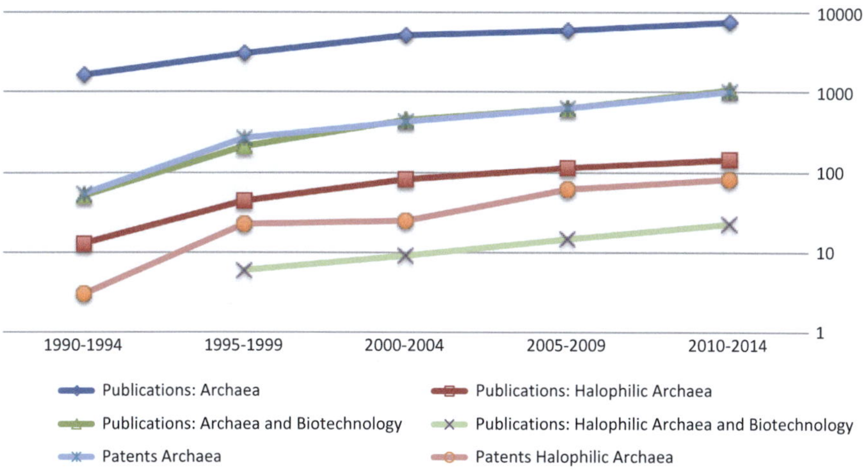

Fig. 5.2 Overview of the evolution in number of publications and patents associated with Archaea and Halophilic Archaea (Publications: data collected using PUBMED; Patents: data collected using Google Scholar)

5.3.2 *Bacteriorhodopsin*

Some of the most interesting uses of halophilic archaea arise from the different proposed applications of bacteriorhodopsin. This molecule, discovered in the early 1970s, is the key protein of the halobacterial photosynthetic system (Hampp 2000; Oren 2002). It is present in *Halobacterium salinarum* and a few other representatives of the *Halobacteriaceae* where it forms a two-dimensional crystal integrated into the cellular membrane in patches (usually referred to as "purple membrane"). Bacteriorhodopsin is involved in the light-driven ejection of protons from the cell, establishing a protonic gradient across the membrane. Cells couple the dissipation of this gradient to the production of energy (i.e. ATP) by a membrane-bound ATPase.

The naturally occurring two-dimensional crystalline structure of bacteriorhodopsin is responsible for its (a) astonishing stability toward chemical and thermal degradation, and (b) photosensitivity and cyclicity to illumination. This favourable combination of properties clearly distinguishes the halophilic protein from synthetic materials and makes it attractive for numerous applications (Hampp 2000). These include holography, spatial light modulators, artificial retina, artificial neural networks, optical computing, and new types of optical memories (Margesin and Schinner 2001). From 2005–2010, over 50 patents were granted associated with different uses of bacteriorhodopsin (Trivedi et al. 2011).

5.3.3 *Bioplastics*

Polyhydroxyalkanoates (PHAs) are a heterogenous family of polyesters, usually used as intracellular carbon storage compounds (most frequently in the form of poly-β-hydroxybutyrate, PHB). The properties of some PHAs are comparable to those of polyethylene and polypropylene with further advantages such as biodegradability, complete water impermeability, and biocompatibility, making them a viable alternative to oil-derived thermoplastics (Divya et al. 2013; Margesin and Schinner 2001; Ventosa and Nieto 1995).

Some halophilic archaea such as *Haloarcula marismortui* and *Haloferax mediterranei* were successfully used to produce high amounts of PHA (Han et al. 2007). *H. mediterranei* can accumulate up to 6 g (60 % of the total biomass dry weight) of PHB per liter of culture using inexpensive starch (DasSarma et al. 2010) or rice bran as carbon source (Huang et al. 2006). The vulnerability of the haloarchaeal cells to pure water (no salt) facilitates isolation of PHA granules by hypoosmotic shock treatment (Quillaguamán et al. 2010). This cheap, straightforward and high yielding harvest procedure reduces downstream processing costs which can account up to 40 % of the total production costs for bacterial PHA production (Choi and Lee 1999).

5.3.4 *Enzymes*

The inability of "normal" enzymes to operate under the harsh conditions imposed by many industrial processes has limited their widespread use. The discovery of extremophiles and their extreme-adapted extremozymes, is revolutionizing this field with an apparently unceasing range of novel industrial applications. Furthermore, as extremozyme discovery is coupled with enzyme tailoring by rational engineering or directed evolution, the development of economical bioprocesses will accelerate and be enabled on larger scales (Demirjian et al. 2001; DasSarma et al. 2010; Liszka et al. 2012).

The special characteristics of halophilic enzymes, which allow them to function properly under high salinities (Reed et al. 2013), are also responsible for their frequently very poor solubility and denaturation at lower salinities, which could limit their applicability (Madern et al. 2000; van den Burg 2003). These same specific properties seem, however, to make them particularly advantageous in aqueous/

organic and non-aqueous media (DasSarma and Arora 2001; Karan et al. 2012; van den Burg 2003). Furthermore, the combination of reverse micelles with halophilic enzymes is further extending the range of applications for these enzymes (van den Burg 2003; Marhuenda-Egea and Bonete 2002).

Relevant enzymes from halophilic archaea include glycosyl hydrolases, proteases, and lipases (Table 5.2). Such enzymes have great potential for biocatalysis in high-salt environments (used in, e.g. the food and detergent industries; Delgado-García et al. 2012; Liszka et al. 2012).

Table 5.2 Selected list of biocatalytically relevant enzymes produced by extremely halophilic archaea (adapted from Demirjian et al. 2001; Ventosa et al. 2005)

Enzyme	Organism	Stability/activity	Reference
β-Galactosidase	*Haloferax lucentense*	Optimal activity at 23 % NaCl	Holmes et al. (1997)
β-Xylanase	*Halorhabdus utahensis*	Optimal activity at 5–15 % NaCl	Wainø and Ingvorsen (2003)
β-Xylosidase	*Halorhabdus utahensis*	Optimal activity at 5 % NaCl	Wainø and Ingvorsen (2003)
Amylase	*Halobacterium salinarum*	Optimal activity at 1 % NaCl	Good and Hartman (1970)
Amylase	*Halorubrum xinjiangense*	Optimal activity at 23 % NaCl	Moshfegh et al. (2013)
Amylase	*Haloferax mediterranei*	Optimal activity at 17 % NaCl	Pérez-Pomares et al. (2003)
Amylase	*Natronococcus amylolyticus*	Optimal activity at 15 % NaCl	Kobayashi et al. (1994, 1992)
Amyloglucosidase	*Halorubrum sodomense*	Optimal activity at 7.5 % NaCl	Oren (1983); Chaga et al. (1993)
Class I fructose aldolase	*Haloarcula vallismortis*	Optimal activity at 2.5 M KCl	Krishnan and Altekar (Krishnan and Altekar 1991)
Lipase	*Natronococcus* sp.	Optimal activity at 23 % NaCl	Boutaiba et al. (2006)
Protease	*Natronobacterium* sp.	Optimal activity at 5.5 % NaCl	Yu (1991)
Protease	*Haloferax mediterranei*	–	Stepanov et al. (1992)
Protease	*Halobacterium salinarum*	Optimal activity at 23 % NaCl	Ryu et al. (1994)
Serine protease	*Halobacterium salinarum*	–	Izotova et al. (1983)
Serine protease	*Natrialba asiatica*	Optimal activity at 10–15 % NaCl	Kamekura and Seno (1990); Kamekura et al., 1992
Serine protease	*Natrialba magadii*	Optimal activity at 6–9 % NaCl	Giménez et al. (2000)
Serine protease	*Natronococcus occultus*	Optimal activity at 6 % NaCl	Studdert et al. (1997)

5.3.5 *Food Industry*

In general, halotolerant and halophilic microorganims (bacteria and archaea) play an essential role in the production of several traditional fermented foods, giving them their characteristic taste, flavor, and aroma. Their salinities range from low to intermediary as present in *Sauerkraut*, pickles or olives, to the concentrated brines used for fermentation of several traditional food products found in the Pacific Rim area. Within the halophilic archaea the importance of *Halobacterium salinarum* and *Halococcus* strains in the production of *nam pla*, a Thai fish sauce, is well recognized (Ventosa and Nieto 1995). Also, *Natrinema gari* and *Halococcus thailandensis*, which were originally isolated from fish sauce, are implicated as important players in the fermentation process (Tapingkae et al. 2008; Namwong et al. 2007), while a protease secreting *Halobacterium* strain was reported to enhance the overall sauce fermentation process (Akolkar et al. 2010). More modern applications include the use of halophilic archaea for the production of food additives (e.g. polyunsaturated fatty acids; Ventosa and Nieto 1995) and pigments (see Sect. 5.3.1).

5.3.6 *Halocins*

Halocins are archaeal bacteriocin-like antimicrobial peptides, produced by many members of the *Halobacteriales*, which inhibit the growth of closely related microbes (Riley and Wertz 2002). According to Kis-Papo and Oren (2000), they could have a role in interspecies competition, particularly on solid substrates.

To name but a few examples, species within *Haloferax*, *Haloarcula* and *Halobacterium* are reported to secrete specific halocins such as S8, H1, H4, C8, H6/H7, and R1 (Salgaonkar et al. 2012; O'Connor and Shand 2002). Despite the almost universal production of these compounds by haloarchaea (Torreblanca et al. 1994), they have been generally overlooked in the ongoing search for new antibiotics (Litchfield 2011). Possible reasons are that many of these purified halocins are not active against the classic group of tested bacteria, and also that many are only active after proteolytic cleavage (Li et al. 2003; Litchfield 2011).

5.3.7 *Metal Bioremediation and Nanoparticles*

Natural and anthropogenic activities such as erosion and mining have resulted in deposition of toxic heavy metals and their derivatives in soils, rivers and oceans (Paula et al. 2013). The use of microbial-based bioremediation attracts considerable interest, and research on the use of halophiles for metal bioremediation is flourishing (Bini 2010). Several taxa of halophilic archaea are interesting in that, potentially, their metal(loid)s resistance capabilities can be harnessed. Al-Mailem et al.

(2011) reported the capability of *Halococcus, Halobacterium* and *Haloferax* to resist and volatilize mercury (Hg). Williams et al. (2013) discussed the tolerance of *Natronobacterium gregoryi* and *Halobacterium saccharovorum* to 0.001 and 0.01 mM of cadmium (Cd) and zinc (Zn), respectively. Das et al. (2014) investigated the tolerance and intracellular accumulation of Cd by *Haloferax*, whereas Salgaonkar et al. (2015) reported the resistance of halophilic archaea to zinc oxide nanoparticles (ZnO NPs) for the first time.

Metal(loid)s resistance in halophilic archaea also make them possible candidates for the environmentally-sound synthesis of metal nanoparticles (NPs) which can be employed in various fields. For example, the selenium nanoparticles (SeNPs) synthesized by *Halococcus salifodinae* BK18 could be used as a chemotherapeutic agent against cancer as they stopped the proliferation of cancerous HeLa cell lines when studied *in vitro* (Srivastava et al. 2014). Also, silver nanoparticles (AgNPs) synthesized by *Halococcus salifodinae* BK3 are reported to have anti-bacterial activity against both Gram-positive (*Staphylococcus aureus* and *Micrococcus luteus*) and Gram-negative (*Escherichia coli* and *Pseudomonas aeruginosa*) bacteria (Srivastava et al. 2013). Since metal uptake and synthesis of NPs are intracellular, haloarchaea have an added advantage as they can be used for metal(loid)s bioremediation and NPs synthesis.

5.3.8 Other Applications

The use of halophilic archaea for exo-polysaccharide production has also a large potential with current utilisation as stabilisers, thickeners, gelling agents and emulsifiers in the pharmaceutical, paint, paper and textile industries (Litchfield 2011; Ventosa and Nieto 1995). Further examples of the wide range of applications for halophiles include such diverse areas as microbially enhanced oil recovery (MEOR) processes, use of gas vesicles for bioengineering, liposomes with increased resistance for cosmetic industry, and saline soil recovery for agriculture, among several others (Litchfield 2011; Oren 2002; Ventosa and Nieto 1995).

5.4 Screening Methodologies

5.4.1 Archaeal Pigments

When grown on agar medium containing high NaCl concentrations, most extremely halophilic archaea display bright red-orange pigmentation, imparted by carotenoids, and can therefore be easily segregated from their non-archaeal counterparts.

5.4.1.1 Haloarchaeal Pigments Extraction and Characterization

Haloarchaeal pigments can be extracted from cells by using solvents, individually or combined (Salgaonkar et al. 2015). In particular, the ultraviolet (UV)-visible spectra of the haloarchaeal C-50 bacterioruberin pigment show characteristic absorption maxima and peaks, while high-performance liquid chromatography (HPLC) analysis presents multiple elution peaks (Yatsunami et al. 2014; Bodaker et al. 2009).

5.4.2 Polyhydroxyalkanoates

Various methods are employed for the screening of intracellular accumulated PHA. The primary method relies on cell staining or the staining agent being incorporated during growth, with binding to PHA granules which fluorescence when exposed to UV light (Legat et al. 2010; Ostle and Holt 1982; Spiekermann et al. 1999). Quantitative PHA production is estimated by acidic hydrolysis, and characteristic absorption peaks (Slepecky 1961). The presence of intracellular PHA granules can also be detected with transmission electron microscopy, Fourier transform infrared spectroscopy (FTIR), or screening of target strains for the genes encoding PHA synthase (further details in e.g. Han et al. 2010; Salgaonkar and Bragança 2015).

5.4.2.1 Extraction of PHA

PHAs can be recovered by lysing cells, followed by polymer solubilization and purification (Tan et al. 2014). As halophilic archaea thrive under very high salinities, their use is associated with very low risks of contamination. Furthermore, their cells lyse in water or in low osmolarity solutions, greatly facilitating the extraction of intracellular PHA granules, and reducing production costs (Quillaguamán et al. 2010).

5.4.2.2 PHA Characterization

Characterization of PHA is very important for their application, as more than 150 monomeric units are available, which impart different properties to the polymer (Tan et al. 2014). Monomer composition is determined by techniques such as gas chromatography (GC), nuclear magnetic resonance (NMR) and spectroscopy after depolymerization (Tan et al. 2014). Furthermore gel permeation chromatography (GPC) is used to determine the polymer's average (a) molecular mass (Mw), (b) molecular mass distribution (Mn), and (c) polydispersity index (PDI; Mw/Mn) (Ashby et al. 2002).

PHA thermal properties determine the temperature conditions at which the polymer can be processed and utilized (Tan et al. 2014; Chen 2010). Thermal properties include glass transition temperature, melting temperature, and thermodegradation temperature, which are obtained using differential scanning calorimetry, differential thermal analysis, and thermogravimetric analysis. The absolute crystallinity of produced PHA polymers can be measured by X-ray diffraction (XRD) analysis (see Chanprateep 2010 and Sánchez et al. 2003 for more detailed information).

Note that PHA polymers can either be a soft elastomeric material or a hard rigid material, displaying a wide elongation at break values between 2 % and 1000 % (Chen 2010). PHA mechanical properties that are commonly evaluated include: (a) Young's modulus which provide a measure of the polymer's stiffness and ranges from the very ductile mcl-PHA to the stiffer scl-PHA (Rai et al. 2011); (b) elongation at break, which measures the extent that a material will stretch before it breaks and is expressed as a percentage of the material's original length; and (c) tensile strength, which measures the amount of force required to pull a material until it breaks (Rai et al. 2011). These assays can be performed with tensile tester instrument by standardized test methods such as the ones recommended by the American Society for Testing and Materials (ASTM) standards (Wu and Liao 2014).

5.4.3 *Enzymes*

Quantitative analysis of hydrolytic enzyme production in halophilic archaea traditionally relies on screening by plate assays wherein the substrate of the enzyme in question is provided as the sole carbon source (Kharroub et al. 2014; Kakhki et al. 2011). Any minimal halophilic medium supplemented with 20–25 % salt and having a proper nitrogen source can be used for enzymatic screening. Examples of preparation and screening methodologies are abundant and include different hydrolytic activities such as e.g. (a) amylase (Amoozegar et al. 2003), (b) cellulose and xylanse (Wejse et al. 2003), (c) pectinase (Soares et al. 1999), (d) extracellular protease (Amoozegar et al. 2008), (e) DNase (Onishi et al. 1983), and (f) chitinase (Park et al. 2000). Examples of purification procedure of enzymes obtained from halophilic archaea can be found in multiple references (e.g. Delgado-García et al. 2012; Moshfegh et al. 2013; Pérez-Pomares et al. 2003; Vidyasagar et al. 2006).

A faster alternative to plate screening is the *in silico* approach where genomic data is checked for putative enzyme genes. But the fact that the whole genome of the organism has to be known, clearly limits the use of this method.

5.4.4 *Halocins*

Halocins are commonly found in the cell-free supernatants (CFS) of halophilic archaea. Standard methodologies employ the agar well diffusion assay, in which the indicator organism is surface-spread or seeded into agar and the CFS of the producer strain is placed in wells within the same plate and allowed to diffuse. The minimum inhibitory concentration (MIC) of the halocin is assayed by serial dilution of the CFS and the activity is presented in Arbitrary Units (AU) (Atanasova et al. 2013; Salgaonkar et al. 2012).

5.4.4.1 Characterization and Purification of Halocins

After achieving significant MIC results, additional steps of characterization and purification are employed. Initial characterization plots halocin activity profiles *versus* growth phase. This provides insights on the phase of growth during which the halocin is produced. To further characterize halocin activity several parameters are tested: pH, temperature, NaCl concentration, and different solvents. It is worth noting that almost all reported halocins are hydrophobic, and reverse-phase HPLC is commonly employed for their complete purification (Meknaci et al. 2014; Price and Shand 2000).

5.4.5 Bioremediation of Metal(loid)s/Metal Nanoparticles

Resistance of haloarchaeal strains to metal(loid)s can be checked by growing strains in media with increasing concentrations of the respective metals. This will also determine the MIC, which is the minimum concentration of metal(loid)s that inhibits archaeal growth. It is worth mentioning that growth of halophilic archaea in the presence of certain metals such as silver/tellurium and selenium changes its pigmentation from red-orange to black and brick-red, respectively.

5.4.5.1 Detection of Metal(loid)s Uptake

Cells grown in the presence of metal(loid)s are hydrolysed using a solution of concentrated nitric acid: sulphuric acid (v/v), followed by complete digestion at 100° C and analysis by absorption spectrophotometry (AAS) (Das et al. 2014).

5.4.5.2 Characterization of the Nanoparticles

The cells grown in the presence of metal(loid)s are harvested, dialyzed, dried and ground using motor and pestle to fine powder (nm range). This powder is analyzed using techniques such as scanning electron microscopy-energy dispersive X-ray spectroscopy (SEM-EDX), XRD and TEM. The UV-visible spectra of silver and selenium nanoparticles show absorption maxima at 440 and 270nm, respectively.

5.5 Current and Future Trends in Mining for Applications

Intensive research efforts currently aim to unleash the full biotechnological potential of halophilic archaea. The recent introduction of genetically optimized efficient expression systems for genes from halophilic sources, has removed a major limitation for large-scale applications. The most promising systems are based on fast growing aerobic extreme halophiles, such as *Haloferax volcanii* (Allers et al. 2010) or *Halobacterium* sp. NRC-1 (Karan et al. 2013), which can even be used for high-yielding protein expression in bioreactors (Strillinger et al. submitted). Additionally, different strategies were reported to optimize *E. coli* for archaeal protein expression (e.g. Connaris et al. 1998; Cao et al. 2008). With the appropriate molecular biotechnology tools in place, developments of more efficient and reliable bioprospecting tools are underway to eliminate remaining bottlenecks. Comprehension of the full capacity of halophilic archaea will arise from understanding their biodiversity and a detailed insight into their molecular functions. Hence, since less than 1 % of the viable organisms within a particular niche are cultivable (Amann et al. 1995), accessing and harvesting genomic material of these microorganisms represents the main challenge. To some extent, introducing specialized laboratory equipment to mimic the extreme conditions of the native habitats will facilitate more efficient laboratory cultivation of halophiles from samples. However, major contributions are expected to come from metagenomic approaches as well as SAG libraries.

5.5.1 Next Generation Sequencing Methods

The advent of cheaper and faster second or next-generation sequencing (NGS) platforms, enabled a shift towards novel culture-independent genome and transcriptome analysis methods. These methods are based on direct DNA and/or RNA isolation from environmental samples and fall into the following classes: (i) metagenomics (DNA based), (ii) metatranscriptomics (RNA based) and (iii) single cell genomics (DNA based).

Metagenomics identification of microbial communities commonly relies on sequencing of the 16S rRNA; however, the same concept can be applied directly for the sequencing of metagenomic DNA samples (Von Mering et al. 2007). Introduction of metagenomics lead to the identification of thousands of novel protein families from diverse environments (Yooseph et al. 2007). Metatranscriptomics based on mRNA (Sorek and Cossart 2010) complements the DNA-based metagenomic approach and provides an understanding of the genomically active genes of microbial population at a given time point from a specific environment. This method requires the isolation of mRNA, which is translated into cDNA before sequencing. The resulting short sequences (reads) of typically a few hundred base pairs for NGS are subsequently assembled and annotated.

5.5.2 DNA Assembly, the First Milestone for Successful Data Mining

Assembling the comparatively short reads is a challenging task, since reads are derived from a myriad of organisms, which form the sampled community. Hence the bioinformatic assembly algorithms applied need to accurately resolve the correct position, and the specific biological entity (e.g. DNA from microbial genomes, viruses or plasmids) for each read (Mick and Sorek 2014). Every single DNA fragment is therefore compared, to all others, to identify overlapping sequences as merging points. Bioinformatics challenges include (a) defining the exact length of naturally occurring and quite common repeats (identical sequence repetitions), (b) differentiating between random overlaps and defined overlaps, (c) defining the correct orientation of the DNA sequence, (d) identifying sequencing errors from the real sequence, (e) correctly identifying the organism (from the pool of diverse genetic material in the environmental sample) from which the sequence originates, and (f) accounting for different sequence depths (amount of sequencing). One should note that before sequencing, the DNA is amplified using random primers, which show statistical variations in binding DNA, resulting in regions that are more or less often amplified per amplification cycle. Assembler programs therefore require about 8 copies of each piece of genome (Baker 2012).

Until recently, the assembly of genomes relied on the genetic material from a single organism or on reference genomes. These approaches led to problems when trying to separate complex metagenomic data into specific biological entities. For samples from archaeal and/or extremophilic communities, which include genomic material that commonly extend far beyond what is covered by reference databases, other assembly strategies are required to interpret metagenomic data without relying on reference sequences. Nielsen et al. (2014) established a new method and demonstrated its power on the analysis of the complex human gut microbiome. The protocol facilitates the extraction of single genomes from complex microbial samples and uses the relative abundance of an organism in the community, which fluctuates over time between different samplings of the same environment. By tracking the changes in abundance of genes between different sampling times, genes showing highly correlated abundance are clustered together. It was shown that such a correlation corresponds with a high probability of belonging to the same genome (Mick and Sorek 2014). Strain-level resolution in metagenomics can be used to identify variations in highly flexible genomic parts, which are coexisting with the relatively stable core components (Kashtan et al. 2014) and thus provides insight into genes essential for adaptation to dramatic changes in environment. Those genes may illuminate the microbial mechanisms involved in environmental adaption. Limitations of this approach include the need for access to a fairly large number of independent samplings of one niche, or related niches, which is required for statistical analysis. However, due to the amount of sequences per sample, the sequence depth can be reduced (Mick and Sorek 2014).

The advent of single-cell genomics (Lasken 2007) allowed identification of different species in an environment, while eliminating the challenge of assigning DNA reads (fragments) to different genomes. Single cell genomics is based on multiplication of the DNA of a single cell through multiple displacement amplifications. As a result, a few femtograms of DNA are enough to provide the microgram amounts of DNA necessary for library construction and sequencing (Lasken 2007). Equal and complete amplification of the minimal amount of source DNA must be achieved to obtain unbiased results, which represents the major challenge of this method.

5.5.3 *Current Representation of Archaeal Genomes in Largest Databases*

Compared to other domains of life, genomic analysis of archaea is still in its infancy, but interest is growing. Correspondingly, current genomic information of archaeal origin represents only 0.4 to 3 % of the data available from major databases listing genomic and oligonucleotide sequences (Table 5.3). The availability of reference sequences is crucial for genome annotation (see below) and therefore continuous publication of fully assembled and annotated archaeal genomes is required to facilitate genomic assembly, improve reliability and accelerate bioinformatic processing of archaeal data.

5.5.4 *Genome Annotation, the Second Milestone in Successful Information Mining*

Genome annotation connects DNA sequences to biological information. The value of a genome is determined by its annotation (Stein 2001). Inaccurate annotations lead to incorrect *in silico* identification of enzymes of interest and are particularly

Table 5.3 Representation of archaeal genomes in selected large-scale online databases

Database	Total genomes	Archaeal genomes
Gen Bank[a]	12,882	382 (3.0 %)
Ensembl[b]	23,000	297 (1.3 %)
Sequence read archive (SRA)[c]	832,167	3205 (0.4 %)
Genomes OnLine Database (GOLD)[d]	64,799	986 (1.5 %)

[a]Gen Bank (Benson et al. 2015) [*http://www.ncbi.nlm.nih.gov/genome/browse/*]
[b]Ensembl (Cunningham et al. 2015). Note: Data from their newest release (release 27) was used, as it does not differentiate between bacteria and archaea [*http://bacteria.ensembl.org/index.html*]
[c]Sequence read archive (SRA) (Leinonen et al. 2011). Note: Samples are listed, as it does not list genomes [*http://www.ncbi.nlm.nih.gov/Traces/sra/sra.cgi?view=samples*]
[d]GOLD database (Reddy et al. 2015). Projects are listed, as it does not list genomes [*https://gold.jgi-psf.org/distribution#*]

problematic when systems biology approaches are used to understand the functions of a cell at the molecular level based on a model of pathways or specific enzymes. Starting from an assembled standard genome, the annotation can be divided into three steps: First, parts of the genome that do not code for proteins are excluded (non-coding RNA); second, the prediction of protein-coding genes (open reading frames) in the genome is undertaken; and third, a biological function is assigned to the proteins. Depending on the goal of the genome annotation a further focus might be the identification of regulatory elements or non-coding RNA (e.g. tRNA, and rRNA).

The standard gene annotation approach relies on gene homology to genes already annotated and available from the common genomic databases. Unfortunately, annotation reliability is indirectly proportional to the variance of the two compared respective proteins' primary structures. Since novel genomes from uncommon habitats are expected to show a lower homology to any gene described so far, the reliability of genome annotation is, in general, decreased. The situation is complicated by error propagation. Also, experimental validation of the encoded protein's function exists only for a small and continuously diminishing fraction of gene sequences available from databases. Originally, the functions of novel genes were annotated based on gene sequences with experimentally verified function. Based on these novel determined genes, further genes were annotated and so on. While in this chain, two proteins in a row are always highly similar, a low similarity of the last annotated gene and the experimental verified source may result, depending on how many non experimentally verified genes are in-between. From an experimental and protein engineering point of view, faulty annotations are a fundamental problem.

Analysis of state-of-the-art annotation pipelines reveals a surprisingly high level of uncertainty in gene annotation. Annotations of the same *E. coli* strain by the leading annotation pipelines yielded about 5.5 % false positives and a significantly higher rate of false positives may be expected for novel genomes (Alam et al. 2013, Grötzinger et al. 2014). Hence, several bioinformatics groups work on strategies to increase annotation reliability, typically by including additional data. For example Alam et al. (2013), combined several strategies, including comparison of predicted 16S rRNA genes with the NCBI prokaryotic 16S rRNA gene database to retrieve taxonomic information and rank the obtained BLAST hits (Altschul et al. 1990). BLAST against several databases resulted in coverage of most known genes. Additionally, the analysis of gene distribution in different pathways helped to evaluate expected and annotated gene presence. Software such as the InterProScan database (Jones et al. 2014; Mitchell et al. 2015) introduced predictions of protein functions based on the number of domains or active sites. Other approaches are focusing on highly reliable annotation of a selected set of single proteins instead of a whole genome annotation, e.g. when mining genomic data for enzymes of interest to biotechnology (Grötzinger et al. 2014). The analysis of annotation metadata is particularly useful for this approach. These metadata contain information on the presence of conserved domains such as active centers or binding pockets, and can be identified during the annotation process. Presence of domains that are relevant for protein activity should increase annotation reliability. Despite the progress made in annotation of proteins with described function, the correct assignment of function and pathway location of proteins that are not described remains a major hurdle.

5.5.5 *Potential and Challenges of Upcoming Generations of DNA Sequencing*

The advent of the third-generation DNA sequencing (single molecule sequencing) not only brings a further reduction in sequencing costs, but also increases read lengths to several thousand base pairs. This not only reduces the complexity of the genome assembly process, or the assignment of specific genomes from a metagenomic DNA pool, but also increases the overall quality of genomes and therefore may even eliminate the concept of draft genomes completely (Land et al. 2015). At the moment about 10 % of all draft genomes are of too poor quality to be used (Land et al. 2014). Third generation sequencing can theoretically produce a finished genome in a few hours and simultaneously identify specific methylation sites (Land et al. 2015).

Although DNA assembly might be simplified in the future, the challenge of proper genomic annotation remains and new challenges will arise from the management of the constantly increasing stream of data. Experimentalists are in need for tools to help them make sense of their massive amount of data, while currently bioinformatics research is struggling to analyze, compare, interpret and visualize data at the pace at which sequencing throughput increase (Land et al. 2015). Bioinformatics progress heavily relies on the use of supercomputers because the amount and the complexity of genomic data are growing significantly faster than the increase in computing and storage capabilities of current systems. The development of new algorithms will require dividing the entire data processing into more manageable tasks, so that it can be addressed on smaller computer clusters, by cloud computing, or by outsourcing and accession via the web. This will assist the end-user as it does not require direct access to a supercomputer.

The need to minimize the amount of metadata included in every sequenced data (Kottmann et al. 2008) illustrates the problems that arise from handling the increasing data volume and complexity. Such metadata include, e.g. geographic location and habitat from which the sample was taken, and details of the sequencing method used which is necessary for efficient assembly, and assigning specific features such as tolerance to specific extreme environments. However, as described above, insufficient reliability of annotations for genomic material from uncommon environments mandates that annotation pipelines include a significantly enriched body of metadata. Future annotation of halophilic archaea could particularly benefit from metadata-based precise domain architecture prediction (e.g. if functionally associated domains are in close proximity such as active catalytically center and cofactor binding pocket). In detail pathway analysis can be used to evaluate how many of the other enzymes, required to provide the cofactor, or substrate, or use of a product, are represented in the organism. It may therefore provide a reliable measure for the probability of correct annotation.

Static tables or images such as charts or plots cannot illustrate accurately the highly complex information available within genomic datasets. Therefore, new approaches to analyse and visualize data are also necessary, apart from novel algorithms,. Linking integrated databases/warehouses (e.g. INDIGO (Alam et al. 2013), to visualization tools such as Krona (Ondov et al. 2014), can be used to illustrate clusters or correlations of genomic information. The integrated databases provide annotations, and direct access to metadata quickly, which can be visualized on multi-level pie charts using standard web browsers. The unprecedented rate of development of genomic sequencing methods effectively shifted the major costs of biomining from sequencing to the genome assembly and functional annotation, and data analysis and management procedures (Land et al. 2015).

The combination of novel culture independent sequencing techniques with new bioinformatics annotation and data analysis tools now permits the analysis of natural microbial communities *in situ*. Future results will therefore provide insights into microbial distribution patterns, and their individual (SAG & assigned genomes), or uniform (meta-genomics/transcriptomics) molecular functions. Hence, a powerful set of techniques is at hand to mine archaeal sources, which will harvest an increasing amount of the biotechnological potential of halophilic archaea. The appropriate utilization of these tools in combination with laboratory-based analysis, will not only increase our understanding of symbiotic and other interactions in microbial communities, but will also provide access to whole sets of enzymes from the same environments. This information can be used to establish multi-enzyme reactions in industry and consequently provide more sustainable solutions for the pharmaceutical and biotechnological industry.

5.6 Research Initiatives of Interest for Bioprospecting Archaea

It is a difficult task to list research initiatives on bioprospecting Archaea. First, there is no agreed definition of the term "bioprospecting", and although there is a general understanding that it involves research for commercial purposes (outlined in Fig. 5.3), it is usually difficult to distinguish, in practice, between basic and applied research (Arico and Salpin 2005). Additionally, large-scale research initiatives usually have a wide-scope and unsurprisingly no such program has specifically targeted Archaea. However, given their importance in extreme environments, and their newly found relevance in marine ecosystems, one can rightfully assume that research initiatives focusing on such locations include Archaea as major targets.

Research on extremophiles and their applications has boomed recently as evidenced by an increasing number of publications in high-impact journals and patents. The importance of this field is further attested by concerted funding initiatives in the USA (NSF and NASA's programs Life in Extreme Environments, Exobiology and Astrobiology), the EU (Biotechnology of Extremophiles, Extremophiles as Cell Factories, ILEE- Investigating Life in Extreme Environments, and CAREX-

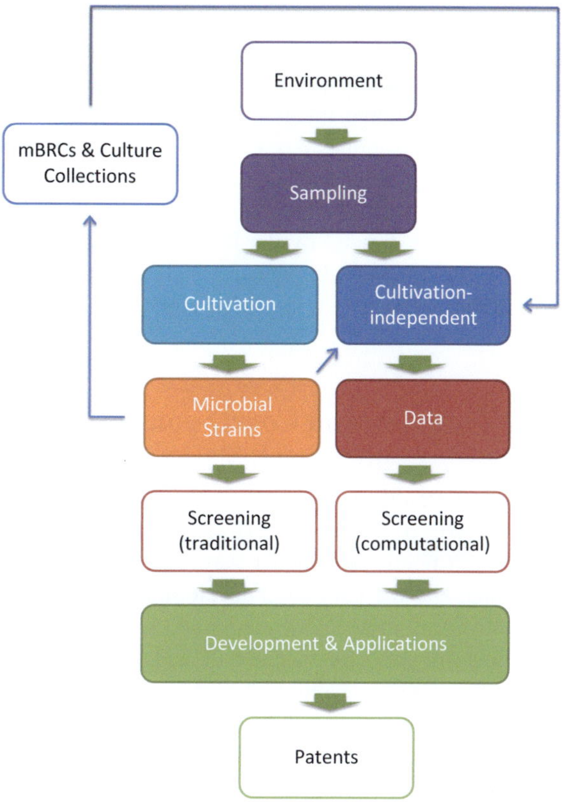

Fig. 5.3 General outline of the different possible steps involved in bioprospecting activities

Coordination Action for Research Activities on Life in Extreme Environments), and Japan (JAMSTEC Frontier Research System for Extremophiles program) (Jamieson 2015; Rothschild and Mancinelli 2001).

During this period, environmental and marine research initiatives and programs have seen an impressive increase in scope, reach, complexity, and dimension. Many of these projects have a global scale and include a very wide variety of measured parameters. A non-extensive list of more visible initiatives would include the Census of Marine Life (http://www.coml.org), Global Ocean Sampling (GOS; http://www.jcvi.org/cms/research/projects/gos/overview, MaCuMBA (http://www.macumbaproject.eu), Malaspina (http://www.expedicionmalaspina.es), MAMBA (http://mamba.bangor.ac.uk), TARA Oceans (http://www.embl.de/tara-oceans), and Micro B3 (http://www.microb3.eu).

Several other initiatives target general genomic and metagenomic data generation, frequently involved in filling current gaps in our understanding of specific environments or phylogenetic groups. Noteworthy examples include the Earth Microbiome Project (EMP; www.earthmicrobiome.org), the Genomic Encyclopedia of Bacteria and Archaea (GEBA; www.jgi.doe.gov/programs/GEBA), and the Marine Microbial Genome Sequencing Project (http://camera.calit2.net/microgenome).

Microbial Biological Resource Centers (mBRCs) and culture collections also play an important role, fueling the bio-economy as sources of microbiological resources, data, and expertise. It is worth noting the current programs that are moving towards regional integration of mBRCs, and the promotion of a more active interaction with industry (e.g. the EU-funded Microbial Research Infrastructure; www.mirri.org). Closer interactions between industrial and research institutions are further highlighted by the recent wave of clusters formed within the Bioindustrie 2021 initiative, funded by the Bundes Ministerium für Bildung und Forschung in Germany (http://www.bioindustrie2021.eu), and looking into fostering new innovations in bioproducts (e.g. biofuels, biopolymers, and biocatalysts).

5.7 Overview and Conclusions

Archaea were originally perceived as evolutionary oddities with restricted importance. However, a significant shift in our understanding of their diversity, ecology, and impact is currently under way. Increased exploration efforts in multiple environments, and the continued development, and application of new methodologies for cultivation, molecular-based studies, and *in silico* approaches will further promote this shift, and are expected to lead the way towards a wave of new discoveries. Furthermore, correct annotation of genomes still remains one of the major challenges in genomic data mining. Different strategies are evolving and improved algorithms together with experimental data established in the laboratory are poised to handle these challenges.

Halophilic archaea are a prime example of the increasing reach and range of applications and are perceived as rising stars for industrial biotechnology (e.g. biocatalysis, bioengineering, biofuel, pharmaceuticals). Further bioprospecting initiatives will foster new innovations in bioproducts, and help to fuel the bio-economy.

Acknowledgments The authors of this publication were partially supported by competitive research funding from King Abdullah University of Science and Technology (KAUST), and by KAUST baseline research funds to VBB.

References

Akolkar AV, Durai D, Desai AJ (2010) *Halobacterium* sp. SP1(1) as a starter culture for accelerating fish sauce fermentation. J Appl Microbiol 109:44–53

Al-Mailem DM, Al-Awadh H, Sorkhoh NA et al (2011) Mercury resistance and volatilization by oil utilizing haloarchaea under hypersaline conditions. Extremophiles 15:39–44

Alam I, Antunes A, Kamau AA et al (2013) INDIGO – Integrated data warehouse of microbial genomes with examples from the red sea extremophiles. PLoS One 8(12):e82210. doi:10.1371/journal.pone.0082210

Allers T, Barak S, Liddell S et al (2010) Improved strains and plasmid vectors for conditional overexpression of His-tagged proteins in *Haloferax volcanii*. Appl Environ Microbiol 76(6):1759–1769. doi:10.1128/AEM.02670-09

Altschul SF, Gish W, Miller W et al (1990) Basic local alignment search tool. J Mol Biol 215(3):403–410. doi:10.1016/S0022-2836(05)80360-2

Amann RI, Ludwig W, Schleifer KH (1995) Phylogenetic identification and in situ detection of individual microbial cells without cultivation. Microbiol Rev 59(1):143–169

Amoozegar MA, Makhdoumi-Kakhki A, Shahzadeh Fazeli SA et al (2012) *Halopenitus persicus* gen. nov., sp. nov., an archaeon from an inland salt lake. Int J Syst Evol Microbiol 62:1932–1936

Amoozegar MA, Malekzadeh F, Malik KA (2003) Production of amylase by newly isolated moderate halophile, *Halobacillus* sp. strain MA-2. J Microbiol Methods 52(3):353–359

Amoozegar MA, Salehghamari E, Khajeh K et al (2008) Production of an extracellular thermohalophilic lipase from a moderately halophilic bacterium, *Salinivibrio* sp. strain SA-2. J Basic Microbiol 48(3):160–167

Antunes A, Rainey F, Wanner G et al (2008) A new lineage of halophilic, wall-less, contractile bacteria from a brine-filled Deep of the Red Sea. J Bacteriol 190:3580–3587

Antunes A, Ngugi DK, Stingl U (2011a) Microbiology of the Red Sea (and other) deep-sea anoxic brine lakes. Environ Microbiol Rep 3:416–433

Antunes A, Alam I, Bajic VB, Stingl U (2011b) Genome sequence of *Salinisphaera shabanensis*, a gammaproteobacterium from the harsh, variable environment of the brine-seawater interface of the Shaban Deep in the Red Sea. J Bacteriol 193(17):4555–4556

Antunes A, Alam I, Simões MF et al (2015) First insights on the viral communities of the deep-sea anoxic brines of the Red Sea. Genomics Proteomics Bioinformatics (accepted)

Arico S, Salpin C (2005) Bioprospecting of genetic resources in the deep seabed: scientific, legal and policy aspects. UNU-IAS Report pp 1–72. http://www.ias.unu.edu

Ashby R, Solaiman D, Foglia T (2002) Poly(ethylene glycol)-mediated molar mass control of short-chain- and medium-chain-length poly(hydroxyalkanoates) from *Pseudomonas oleovorans*. Appl Microbiol Biotechnol 60:154–159

Atanasova NS, Pietilä MK, Oksanen HM (2013) Diverse antimicrobial interactions of halophilic archaea and bacteria extend over geographical distances and cross the domain barrier. MicrobiologyOpen 2(5):811–825. doi:10.1002/mbo3.115

Baker M (2012) De novo genome assembly: what every biologist should know. Nat Methods 9(4):333–337. doi:10.1038/nmeth.1935

Balch WE, Magrum LJ, Fox GE et al (1977) An ancient divergence among the bacteria. J Mol Evol 9(4):305–311

Bardavid RE, Mana L, Oren A (2007) *Haloplanus natans* gen. nov., sp. nov., an extremely halophilic, gas-vacuolate archaeon isolated from Dead Sea–Red Sea water mixtures in experimental outdoor ponds. Int J Syst Evol Microbiol 57:780–783

Barns SM, Delwiche CF, Palmer JD, Pace NR (1996) Perspectives on archaeal diversity, thermophily and monophyly from environmental rRNA sequences. Proc Natl Acad Sci U S A 93(17):9188–9193

Benson DA, Clark K, Karsch-Mizrachi I et al (2015) GenBank. Nucleic Acids Res, 43(Database issue):D30–D35. doi:10.1093/nar/gku1216

Bini E (2010) Archaeal transformation of metals in the environment. FEMS Microbiol Ecol 73:1–16

Bodaker I, Beja O, Sharon I et al (2009) Archaeal diversity in the Dead Sea: microbial survival under increasingly harsh conditions. Nat Resour Environ Issues 15(1):25

Bougouffa S, Yang JK, Lee OO et al (2013) Distinctive microbial community structure in highly stratified deep-sea brine water columns. Appl Environ Microbiol 79(11):3425–3437

Boutaiba S, Bhatnagar T, Hacene H et al (2006) Preliminary characterisation of a lipolytic activity from an extremely halophilic archaeon, *Natronococcus* sp. J Mol Catal B Enzym 41(1):21–26

Brochier-Armanet C, Boussau B, Gribaldo S, Forterre P (2008) Mesophilic Crenarchaeota: proposal for a third archaeal phylum, the Thaumarchaeota. Nat Rev Microbiol 6(3):245–252

Brochier C, Gribaldo S, Zivanovic Y et al (2005) Nanoarchaea: representatives of a novel archaeal phylum or a fast-evolving euryarchaeal lineage related to Thermococcales? Genome Biol 6(5):R42

Burns DG, Janssen PH, Itoh T et al (2007) *Haloquadratum walsbyi* gen. nov., sp. nov., the square haloarchaeon of Walsby, isolated from saltern crystallizers in Australia and Spain. Int J Syst Evol Microbiol 57:387–392

Burns DG, Janssen PH, Itoh T, Kamekura M et al (2010) *Halonotius pteroides* gen. nov., sp. nov., an extremely halophilic archaeon recovered from a saltern crystallizer. Int J Syst Evol Microbiol 60:1196–1199

Cao Y, Liao L, Xu XW et al (2008) Characterization of alcohol dehydrogenase from the haloalkaliphilic archaeon *Natronomonas pharaonis*. Extremophiles 12(3):471–476. doi:10.1007/s00792-007-0133-7

Castelle CJ, Wrighton KC, Thomas BC et al (2015) Genomic expansion of domain archaea highlights roles for organisms from new phyla in anaerobic carbon cycling. Curr Biol 25(6):690–701

Castillo AM, Gutiérrez MC, Kamekura M et al (2006a) *Halostagnicola larsenii* gen. nov., sp. nov., an extremely halophilic archaeon from a saline lake in Inner Mongolia, China. Int J Syst Evol Microbiol 56:1519–1524

Castillo AM, Gutiérrez MC, Kamekura M et al (2006b) *Halovivax asiaticus* gen. nov., sp. nov., a novel extremely halophilic archaeon isolated from Inner Mongolia, China. Int J Syst Evol Microbiol 56:765–770

Chaga G, Porath J, Illéni T (1993) Isolation and purification of amyloglucosidase from *Halobacterium sodomense*. Biomed Chromatogr 7(5):256–261

Chanprateep S (2010) Current trends in biodegradable polyhydroxyalkanoates. J Biosci Bioeng 110:621–632

Chen GQ (2010) Introduction of bacterial plastics PHA, PLA, PBS, PE, PTT, and PPP. In: Chen GQ (ed) Plastics from bacteria: natural functions and applications. Springer, Berlin/Heidelberg, pp. 1–16

Choi J, Lee SY (1999) Factors affecting the economics of poly- hydroxyalkanoate production by bacterial fermentation. Appl Microbiol Biotechnol 51:13–21

Connaris H, West SM, Hough DW, Danson MJ (1998) Cloning and overexpression in *Escherichia coli* of the gene encoding citrate synthase from the hyperthermophilic Archaeon *Sulfolobus solfataricus*. Extremophiles 2(2):61–66

Cui HL, Qiu XX (2014) *Salinarubrum litoreum* gen. nov., sp. nov.: a new member of the family *Halobacteriaceae* isolated from Chinese marine solar salterns. Antonie van Leeuwenhoek 105:135–141

Cui HL, Zhang WJ (2014) *Salinigranum rubrum* gen. nov., sp. nov., a member of the family *Halobacteriaceae* isolated from a marine solar saltern. Int J Syst Evol Microbiol 64:2029–2033

Cui HL, Gao X, Yang X, Xu XW (2010a) *Halorussus rarus* gen. nov., sp. nov., a new member of the family *Halobacteriaceae* isolated from a marine solar saltern. Extremophiles 14:493–499

Cui HL, Gao X, Sun FF et al (2010b) *Halogranum rubrum* gen. nov., sp. nov., a halophilic archaeon isolated from a marine solar saltern. Int J Syst Evol Microbiol 60:1366–1371

Cui HL, Li XY, Gao X et al (2010c) *Halopelagius inordinatus* gen. nov., sp. nov., a new member of the family *Halobacteriaceae* isolated from a marine solar saltern. Int J Syst Evol Microbiol 60:2089–2093

Cui HL, Yang X, Mou YZ (2011a) *Salinarchaeum laminariae* gen. nov., sp. nov.: a new member of the family *Halobacteriaceae* isolated from salted brown alga *Laminaria*. Extremophiles 15:625–631

Cui HL, Gao X, Yang X, Xu XW (2011b) *Halolamina pelagica* gen. nov., sp. nov., a new member of the family *Halobacteriaceae*. Int J Syst Evol Microbiol 61:1617–1621

Cui HL, Yang X, Gao X, Xu XW (2011c) *Halobellus clavatus* gen. nov., sp. nov. and *Halorientalis regularis* gen. nov., sp. nov., two new members of the family *Halobacteriaceae*. Int J Syst Evol Microbiol 61:2682–2689

Cui HL, Mou YZ, Yang X et al (2012) *Halorubellus salinus* gen. nov., sp. nov. and *Halorubellus litoreus* sp. nov., novel halophilic archaea isolated from a marine solar saltern. Syst Appl Microbiol 35:30–34

Cunningham F, Amode MR, Barrell D et al (2015) Ensembl 2015. Nucleic Acids Res 43(Database issue):D662–D669. doi:10.1093/nar/gku1010
Daffonchio D, Borin S, Brusa T et al (2006) Stratified prokaryote network in the oxic–anoxic transition of a deep-sea halocline. Nature 440(7081):203–207
Das D, Salgaonkar BB, Mani K, Bragança JM (2014) Cadmium resistance in extremely halophilic archaeon *Haloferax* strain BBK2. Chemosphere 112:385–392
DasSarma S, Arora P (2001) Halophiles. In: Encyclopedia of life sciences. Nature Publishing Group, London. http://www.els.net
DasSarma P, Coker JA, Huse V, DasSarma S (2010) Halophiles, industrial applications. In: Flickinger MC (ed) Encyclopedia of industrial biotechnology: bioprocess, bioseparation, and cell technology. Wiley, Hoboken, pp. 1–43
Delgado-García M, Valdivia-Urdiales B, Aguilar-González CN et al (2012) Halophilic hydrolases as a new tool for the biotechnological industries. J Sci Food Agric 92:2575–2580
de Macario EC, Macario AJ (2009) Methanogenic archaea in health and disease: a novel paradigm of microbial pathogenesis. Int J Med Microbiol 299(2):99–108
Demirjian DC, Moris-Varas F, Cassidy CS (2001) Enzymes from extremophiles. Curr Opin Chem Biol 5(2):144–151
Divya G, Achana T, Manzano RA (2013) Polyhydroxinates, a sustainable alternative to petro-based plastics. J Pet Environ Biotechnol 4(3):1000143
Echigo A, Minegishi H, Shimane Y et al (2013) *Halomicroarcula pellucida* gen. nov., sp. nov., a non-pigmented, transparent-colony-forming, halophilic archaeon isolated from solar salt. Int J Syst Evol Microbiol 63:3556–3562
Elazari-Volcani B (1957) Genus XII. *Halobacterium* Elazari-Volcani, 1940. In: Breed RS, Murray EGD, Smith NR (eds) Bergey's manual of determinative bacteriology, 7th edn. Williams & Wilkins, Baltimore, pp. 207–212
Eme L, Doolittle WF (2015) Microbial diversity: a bonanza of phyla. Curr Biol 25(6): R227–R230
Ferrera I, Takacs-Vesbach CD, Reysenbach AL (2008) Archaeal ecology. In: Encyclopedia of life sciences *(eLS)*. Wiley, Chichester
Fish SA, Shepherd TJ, McGenity TJ, Grant WD (2002) Recovery of 16S ribosomal RNA gene fragments from ancient halite. Nature 417(6887):432–436
Fox GE, Magrum LJ, Balch WE et al (1977) Classification of methanogenic bacteria by 16S ribosomal RNA characterization. Proc Natl Acad Sci U S A 74(10):4537–4541
Galinski EA, Trüper HG (1994) Microbiol behaviour in salt-stressed ecosystems. FEMS Microbiol Rev 15:95–108
Giménez MI, Studdert CA, Sánchez JJ, De Castro RE (2000) Extracellular protease of *Natrialba magadii*: purification and biochemical characterization. Extremophiles 4(3):181–188
Good WA, Hartman PA (1970) Properties of the amylase from *Halobacterium halobium*. J Bacteriol 104(1):601–603
Grötzinger SW, Alam I, Ba Alawi WB et al (2014) Mining a database of single amplified genomes from red sea brine pool extremophiles-improving reliability of gene function prediction using a profile and pattern matching algorithm (PPMA). Front Microbiol 5:134. doi:10.3389/fmicb.2014.00134
Guan Y, Hikmawan T, Antunes A et al (2015) Diversity of methanogens and sulfate-reducing bacteria in the interfaces of five deep-sea anoxic brines of the Red Sea. Res Microbiol (accepted)
Gupta RS, Naushad S, Baker S (2015) Phylogenomic analyses and molecular signatures for the class *Halobacteria* and its two major clades: a proposal for division of the class *Halobacteria* into an emended order *Halobacteriales* and two new orders, *Haloferacales* ord. nov. and *Natrialbales* ord. nov. Int J Syst Evol Microbiol 65:1050–1069
Gutiérrez MC, Castillo AM, Kamekura M et al (2007) *Halopiger xanaduensis* gen. nov., sp. nov., an extremely halophilic archaeon isolated from saline Lake Shangmatala in Inner Mongolia, China. Int J Syst Evol Microbiol 57:1402–1407
Guy L, Ettema TJ (2011) The archaeal 'TACK'superphylum and the origin of eukaryotes. Trends Microbiol 19(12):580–587

Hampp N (2000) Bacteriorhodopsin as a photochromic retinal protein for optical memories. Chem Rev 100:1755–1776

Han J, Lu Q, Zhou L et al (2007) Molecular characterization of the phaECHm genes, required for biosynthesis of poly (3-hydroxybutyrate) in the extremely halophilic archaeon *Haloarcula marismortui*. Appl Environ Microbiol 73(19):6058–6065

Han J, Hou J, Liu H et al (2010) Wide distribution among halophilic Archaea of a novel polyhydroxyalkanoate synthase subtype with homology to bacterial type III synthases. Appl Environ Microbiol 76:7811–7819

Hezayen FF, Tindall BJ, Steinbüchel A, Rehm BH (2002) Characterization of a novel halophilic archaeon, *Halobiforma haloterrestris* gen. nov., sp. nov., and transfer of *Natronobacterium nitratireducens* to *Halobiforma nitratireducens* comb. nov. Int J Syst Evol Microbiol 52:2271–2280

Holmes ML, Scopes RK, Moritz RL et al (1997) Purification and analysis of an extremely halophilic β-galactosidase from *Haloferax alicantei*. BBA-Protein Struct M 1337(2):276–286

Huang TY, Duan KJ, Huang SY, Chen CW (2006) Production of polyhydroxyalkanoates from inexpensive extruded rice bran and starch by *Haloferax mediterranei*. J Ind Microbiol Biotechnol 33:701–706

Huber H, Hohn MJ, Rachel R et al (2002) A new phylum of Archaea represented by a nanosized hyperthermophilic symbiont. Nature 417(6884):63–67

Inoue K, Itoh T, Ohkuma M, Kogure K (2011) *Halomarina oriensis* gen. nov., sp. nov., a halophilic archaeon isolated from a seawater aquarium. Int J Syst Evol Microbiol 61:942–946

Itoh T, Yamaguchi T, Zhou P, Takashina T (2005) *Natronolimnobius baerhuensis* gen. nov., sp. nov. and *Natronolimnobius innermongolicus* sp. nov., novel haloalkaliphilic archaea isolated from soda lakes in Inner Mongolia, China. Extremophiles 9:111–116

Izotova LS, Strongin AY, Chekulaeva LN et al (1983) Purification and properties of serine protease from *Halobacterium halobium*. J Bacteriol 155(2):826–830

Jamieson A (2015) The hadal zone: life in the deepest oceans. Cambridge University Press, Cambridge

Jones P, Binns D, Chang HY et al (2014) InterProScan 5: genome-scale protein function classification. Bioinformatics 30(9):1236–1240. doi:10.1093/bioinformatics/btu031

Joye SB, Samarkin VA, MacDonald IR et al (2009) Metabolic variability in seafloor brines revealed by carbon and sulphur dynamics. Nat Geosci 2(5):349–354

Kakhki AM, Amoozegar MA, Khaledi EM (2011) Diversity of hydrolytic enzymes in haloarchaeal strains isolated from salt lake. Int J Environ Sci Technol 8(4):705–714

Kamekura M, Dyall-Smith ML (1995) Taxonomy of the family *Halobacteriaceae* and the description of two new genera *Halorubrobacterium* and *Natrialba*. J Gen Appl Microbiol 41:333–350

Kamekura M, Dyall-Smith ML, Upasani V et al (1997) Diversity of alkaliphilic halobacteria: proposals for transfer of *Natronobacterium vacuolatum*, *Natronobacterium magadii*, and *Natronobacterium pharaonis* to *Halorubrum*, *Natrialba*, and *Natronomonas* gen. nov., respectively, as *Halorubrum vacuolatum* comb. nov., *Natrialba magadii* comb. nov., and *Natronomonas pharaonis* comb. nov., respectively. Int J Syst Bacteriol 47:853–857

Kamekura M, Seno Y (1990) A halophilic extracellular protease from a halophilic archaebacterium strain 172 P1. Biochem Cell Biol 68(1):352–359

Kamekura M, Seno Y, Holmes ML, Dyall-Smith ML (1992) Molecular cloning and sequencing of the gene for a halophilic alkaline serine protease (halolysin) from an unidentified halophilic archaea strain (172P1) and expression of the gene in *Haloferax volcanii*. J Bacteriol 174(3):736–742

Karan R, Capes MD, DasSarma P, DasSarma S (2013) Cloning, overexpression, purification, and characterization of a polyextremophilic beta-galactosidase from the Antarctic haloarchaeon *Halorubrum lacusprofundi*. BMC Biotechnol 13:3. doi:10.1186/1472-6750-13-3

Karan R, Capes MD, DasSarma S (2012) Function and biotechnology of extremophilic enzymes in low water activity. Aquatic Biosystems 8(4) www.aquaticbiosystems.org/content/8/1/4

Karner MB, DeLong EF, Karl DM (2001) Archaeal dominance in the mesopelagic zone of the Pacific Ocean. Nature 409(6819):507–510

Kashtan N, Roggensack SE, Rodrigue S et al (2014) Single-cell genomics reveals hundreds of coexisting subpopulations in wild *Prochlorococcus*. Science 344(6182):416–420. doi:10.1126/science.1248575

Kharroub K, Gomri MA, Aguilera M, Monteoliva-Sanchez M (2014) Diversity of hydrolytic enzymes in haloarchaea isolated from Algerian sabkhas. Afr J Microbiol Res 8(52):3992–4001

Kis-Papo T, Oren A (2000) Halocins: are they involved in the competition between halobacteria in saltern ponds? Extremophiles 4(1):35–41

Kobayashi T, Kanai H, Hayashi T et al (1992) Haloalkaliphilic maltotriose-forming alpha-amylase from the archaebacterium *Natronococcus* sp. strain Ah-36. J Bacteriol 174(11):3439–3444

Kobayashi T, Kanai H, Aono R et al (1994) Cloning, expression, and nucleotide sequence of the alpha-amylase gene from the haloalkaliphilic archaeon *Natronococcus* sp. strain Ah-36. J Bacteriol 176(16):5131–5134

Kottmann R, Gray T, Murphy S et al (2008) A standard MIGS/MIMS compliant XML schema: toward the development of the Genomic Contextual Data Markup Language (GCDML). OMICS 12(2):115–121. doi:10.1089/omi.2008.0A10

Krishnan G, Altekar W (1991) An unusual class I (Schiff base) fructose-1,6-bisphosphate aldolase from the halophilic archaebacterium *Haloarcula vallismortis*. Eur J Biochem 195(2):343–350

Land M, Hauser L, Jun SR et al (2015) Insights from 20 years of bacterial genome sequencing. Funct Integr Genomics 15(2):141–161. doi:10.1007/s10142-015-0433-4

Land ML, Hyatt D, Jun SR et al (2014) Quality scores for 32,000 genomes. Stand Genomic Sci 9:20. doi:10.1186/1944-3277-9-20

Lasken RS (2007) Single-cell genomic sequencing using multiple displacement amplification. Curr Opin Microbiol 10(5):510–516

Law JH, Slepecky RA (1961) Assay of poly-β-hydroxybutyric acid. J Bacteriol 82:33–36

Lee H-S (2013) Diversity of halophilic Archaea in fermented foods and human intestines and their application. J Microbiol Biotechnol 23(12):1645–1653

Legat A, Gruber C, Zangger K et al (2010) Identification of polyhydroxyalkanoates in *Halococcus* and other haloarchaeal species. Appl Microbiol Biotechnol 87:1119–1127

Leininger S, Urich T, Schloter M et al (2006) Archaea predominate among ammonia-oxidizing prokaryotes in soils. Nature 442(7104):806–809

Leinonen R, Sugawara H, Shumway M (2011) The sequence read archive. Nucleic Acids Res 39(Database issue):D19–D21. doi:10.1093/nar/gkq1019

Li Y, Xiang H, Liu J et al (2003) Purification and biological characterization of halocin C8, a novel peptide antibiotic from *Halobacterium* strain AS7092. Extremophiles 7(5):401–407

Litchfield CD (2011) Potential for industrial products from the halophilic Archaea. J Ind Microbiol Biotechnol 38(10):1635–1647

Liszka MJ, Clark ME, Schneider E, Clark DS (2012) Nature versus nurture: developing enzymes that function under extreme conditions. Ann Rev Chem Biol Eng 3:77–102

LPSN-List of Prokaryotic Names with Standing in Nomenclature (2015) List of prokaryotic names with standing in nomenclature. www.bacterio.net

Madern D, Ebel C, Zaccai G (2000) Halophilic adaptation of enzymes. Extremophiles 4:91–98

Makhdoumi-Kakhki A, Amoozegar MA, Bagheri M et al (2012a) *Haloarchaeobius iranensis* gen. nov., sp. nov., an extremely halophilic archaeon isolated from a saline lake. Int J Syst Evol Microbiol 62:1021–1026

Makhdoumi-Kakhki A, Amoozegar MA, Ventosa A (2012b) *Halovenus aranensis* gen. nov., sp. nov., an extremely halophilic archaeon from Aran-Bidgol salt lake. Int J Syst Evol Microbiol 62:1331–1336

Mapelli F, Borin S, Daffonchio D (2012) Microbial diversity in deep hypersaline anoxic basins. In: Stan-Lotter H, Fendrihan S (eds) Adaption of microbial life to environmental extremes. Springer, Wien/New York, pp. 21–36

Margesin R, Schinner F (2001) Potential of halotolerant and halophilic microorganisms for biotechnology. Extremophiles 5:73–83
Marhuenda-Egea F, Bonete MJ (2002) Extreme halophilic enzymes in organic solvents. Curr Opin Biotechnol 13:385–389
McGenity TJ, Grant WD (1995) Transfer of *Halobacterium saccharovorum*, *Halobacterium sodomense*, *Halobacterium trapanicum* NRC 34021 and *Halobacterium lacusprofundi* to the genus *Halorubrum* gen. nov., as *Halorubrum saccharovorum* comb. nov., *Halorubrum sodomense* comb. nov., *Halorubrum trapanicum* comb., nov., and *Halorubrum lacusprofundi* comb. nov. Syst Appl Microbiol 18:237–243
McGenity TJ, Gemmell RT, Grant WD (1998) Proposal of a new halobacterial genus *Natrinema* gen. nov., with two species *Natrinema pellirubrum* nom. nov. and *Natrinema pallidum* nom. nov. Int J Syst Bacteriol 48:1187–1196
McGenity TJ, Gemmell RT, Grant WD, Stan-Lotter H (2000) Origins of halophilic microorganisms in ancient salt deposits. Environ Microbiol 2(3):243–250
Meknaci R, Lopes P, Servy C et al (2014) Agar-supported cultivation of *Halorubrum* sp. SSR, and production of halocin C8 on the scale-up prototype platotex. Extremophiles 18(6):1049–1055
Meng J, Xu J, Qin D et al (2014) Genetic and functional properties of uncultivated MCG archaea assessed by metagenome and gene expression analyses. ISME J 8(3):650–659
Mick E, Sorek R (2014) High-resolution metagenomics. Nat Biotechnol 32(8):750–751. doi:10.1038/nbt.2962
Minegishi H, Echigo A, Nagaoka S et al (2010) *Halarchaeum acidiphilum* gen. nov., sp. nov., a moderately acidophilic haloarchaeon isolated from commercial solar salt. Int J Syst Evol Microbiol 60:2513–2516
Minegishi H, Kamekura M, Kitajima-Ihara T et al (1998) *Halogeometricum borinquense* gen. nov., sp. nov., a novel halophilic archaeon from Puerto Rico. Int J Syst Bacteriol 48:1305–1312
Mitchell A, Chang HY, Daugherty L et al (2015) The InterPro protein families database: the classification resource after 15 years. Nucleic Acids Res 43(Database issue):D213–D221. doi:10.1093/nar/gku1243
Mohamed YM, Ghazy MA, Sayed A et al (2013) Isolation and characterization of a heavy metal-resistant, thermophilic esterase from a Red Sea brine pool. Sci Rep 3:3358. doi:10.1038/srep03358
Moshfegh M, Shahverdi AR, Zarrini G, Faramarzi MA (2013) Biochemical characterization of an extracellular polyextremophilic α-amylase from the halophilic archaeon *Halorubrum xinjiangense*. Extremophiles 17(4):677–687
Mou YZ, Qiu XX, Zhao ML et al (2012) *Halohasta litorea* gen. nov. sp. nov., and *Halohasta litchfieldiae* sp. nov., isolated from the Daliang aquaculture farm, China and from Deep Lake, Antarctica, respectively. Extremophiles 16(6):895–901
Namwong S, Tanasupawat S, Visessanguan W et al (2007) *Halococcus thailandensis* sp. nov., from fish sauce in Thailand. Int J Syst Evol Microbiol 57:2199–2203
Nielsen HB, Almeida M, Juncker AS et al (2014) Identification and assembly of genomes and genetic elements in complex metagenomic samples without using reference genomes. Nat Biotechnol 32(8):822–828. doi:10.1038/nbt.2939
Nunoura T, Takaki Y, Kakuta J et al (2010) Insights into the evolution of Archaea and eukaryotic protein modifier systems revealed by the genome of a novel archaeal group. Nucleic Ac Res 1–20. doi:10.1093/nar/gkq1228
O'Connor EM, Shand RF (2002) Halocins and sulfolobicins: the emerging story of archaeal protein and peptide antibiotics. J Ind Microbiol Biotechnol 28:23–31
Ollivier B, Caumette P, Garcia J-L, Mah R (1994) Anaerobic bacteria from hypersaline environments. Microbiol Rev 58(1):27–38
Ondov B, Bergman N, Phillippy A (2014) Krona: interactive metagenomic visualization in a web browser. In: Nelson KE (ed) Encyclopedia of metagenomics. Springer, New York, pp. 1–8

Onishi H, Mori T, Takeuchi S et al (1983) Halophilic nuclease of a moderately halophilic *Bacillus* sp.: production, purification, and characterization. Appl Environ Microbiol 45(1):24–30
Oren A (1983) A thermophilic amyloglucosidase from *Halobacterium sodomense*, a halophilic bacterium from the Dead Sea. Curr Microbiol 8(4):225–230
Oren A (2000) Life at high salt concentrations. In: The prokaryotes: an evolving electronic resource for the microbiological community, 3rd edn, release 3.1, Springer, New York. http://link.springer-ny.com/link/service/books/10125
Oren A (2002) Diversity of halophilic microorganisms: environments, phylogeny, physiology, and applications. J Ind Microbiol Biotechnol 28:56–63
Oren A (2010) Industrial and environmental applications of halophilic microorganisms. Environ Technol 31:825–834
Oren A, Gurevich P, Gemmell RT, Teske A (1995) *Halobaculum gomorrense* gen. nov., sp. nov., a novel extremely halophilic archaeon from the Dead Sea. Int J Syst Bacteriol 45:747–754
Oren A, Elevi R, Watanabe S et al (2002) *Halomicrobium mukohataei* gen. nov., comb. nov., and emended description of *Halomicrobium mukohataei*. Int J Syst Evol Microbiol 52:1831–1835
Ostle AG, Holt J (1982) Nile blue A as a fluorescent stain for poly-beta-hydroxybutyrate. Appl Environ Microbiol 44:238–241
Park SH, Lee JH, Lee HK (2000) Purification and characterization of chitinase from a marine bacterium, Vibrio sp. 98CJ11027. J Microbiol 38:224–229
Paula DP, Gleny A, Martha H et al (2013) Kinetics of arsenite removal by halobacteria from a highly and Andean Chilean Salar. Aquat Biosyst 9:8. www.aquaticbiosystems.org/content/9/1/8.
Pérez-Pomares F, Bautista V, Ferrer J et al (2003) α-amylase activity from the halophilic archaeon *Haloferax mediterranei*. Extremophiles 7(4):299–306
Preston CM, Wu KY, Molinski TF, DeLong EF (1996) A psychrophilic crenarchaeon inhabits a marine sponge: cenarchaeum symbiosum gen. nov., sp. nov. Proc Natl Acad Sci U S A 93(13):6241–6246
Price LB, Shand RF (2000) Halocin S8: a 36-amino-acid microhalocin from the haloarchaeal strain S8a. J Bacteriol 182(17):4951–4958
Probst AJ, Auerbach AK, Moissl-Eichinger C (2013) Archaea on human skin. PLoS One 8(6):e65388. doi:10.1371/journal.pone.0065388
Probst AJ, Moissl-Eichinger C (2015) "Altiarchaeales": uncultivated Archaea from the subsurface. Life 5(2):1381–1395
Quillaguamán J, Guzmán H, Van-Thuoc D, Hatti-Kaul R (2010) Synthesis and production of polyhydroxyalkanoates by halophiles: current potential and future prospects. Appl Microbiol Biotechnol 85:1687–1696
Rai R, Keshavarz T, Roether JA et al (2011) Medium chain length polyhydroxyalkanoates, promising new biomedical materials for the future. Mater Sci Eng R Rep 72:29–47
Reed CJ, Lewis H, Trejo E et al (2013) Protein adaptations in archaeal extremophiles. Archaea 2013. http://dx.doi.org/10.1155/2013/373275
Reddy TB, Thomas AD, Stamatis D et al (2015) The genomes onLine database (GOLD) v.5: a metadata management system based on a four level (meta)genome project classification. Nucleic Acids Res 43(Database issue):D1099–D1106. doi:10.1093/nar/gku950
Riley MA, Wertz JE (2002) Bacteriocins: evolution, ecology, and application. Ann Rev Microbiol 56(1):117–137
Rinke C, Schwientek P, Sczyrba A et al (2013) Insights into the phylogeny and coding potential of microbial dark matter. Nature 499(7459):431–437
Rodríguez-Valera F (1988) Characteristics and microbial ecology of hypersaline environments. In: Rodríguez-Valera F (ed) Halophilic bacteria. CRC Press, Boca Raton, pp. 3–30
Rothschild LJ, Mancinelli RL (2001) Life in extreme environments. Nature 409(6823):1092–1101
Ryu K, Kim J, Dordick JS (1994) Catalytic properties and potential of an extracellular protease from an extreme halophile. Enzym Microb Technol 16(4):266–275

Sagar S, Esau L, Hikmawan T et al (2013) Cytotoxic and apoptotic evaluations of marine bacteria isolated from brine-seawater interface of the red sea. BMC Complement Altern Med 13(1):29

Salgaonkar BB, Bragança JM (2015) Biosynthesis of poly(3-hydroxybutyrate-*co*-3-hydroxyvalerate) by *Halogeometricum borinquense* strain E3. Int J Biol Macromol 78:339–346

Salgaonkar BB, Das D, Bragança JM (2015) Resistance of extremely halophilic archaea to zinc and zinc oxide nanoparticles. Appl Nanosci. doi:10.1007/s13204-015-0424-8

Salgaonkar BB, Kabilan M, Nair A et al (2012) Interspecific interactions among members of family *Halobacteriaceae* from natural solar salterns. Probiotics Antimicrob Proteins 4(2):98–107

Sánchez RJ, Schripsema J, da Silva LF et al (2003) Medium-chain-length polyhydroxyalkanoic acids (PHAmcl) produced by *Pseudomonas putida* IPT 046 from renewable sources. Eur Polym J 39:1385–1394

Savage KN, Krumholz LR, Oren A, Elshahed MS (2007) *Haladaptatus paucihalophilus* gen. nov., sp. nov., a halophilic archaeon isolated from a low-salt, sulfide-rich spring. Int J Syst Evol Microbiol 57:19–24

Savage KN, Krumholz LR, Oren A, Elshahed MS (2008) *Halosarcina pallida* gen. nov., sp. nov., a halophilic archaeon from a low-salt, sulfide-rich spring. Int J Syst Evol Microbiol 58:856–860

Sayed A, Ghazy MA, Ferreira AJ (2014) A novel mercuric reductase from the unique deep brine environment of Atlantis II in the Red Sea. J Biol Chem 289(3):1675–1687

Schoop G (1935) *Halococcus litoralis*, ein obligat halphiler Farbstoffbildner. Dtsch Tierarztl Wochenschr 43:817–820

Shimane Y, Hatada Y, Minegishi H et al (2010) *Natronoarchaeum mannanilyticum* gen. nov., sp. nov., an aerobic, extremely halophilic archaeon isolated from commercial salt. Int J Syst Evol Microbiol 60:2529–2534

Shimane Y, Hatada Y, Minegishi H et al (2011) *Salarchaeum japonicum* gen. nov., sp. nov., an aerobic, extremely halophilic member of the Archaea isolated from commercial salt. Int J Syst Evol Microbiol 61:2266–2270

Siam R, Mustafa GA, Sharaf H et al (2012) Unique prokaryotic consortia in geochemically distinct sediments from Red Sea Atlantis II and Discovery Deep brine pools. PLoS One 7(8):e42872. doi:10.1371/journal.pone.0042872

Skerman VBD, McGowan V, Sneath PHA et al (1980) Approved lists. Int J Syst Bacteriol 30:225–420

Soares MM, Silva RD, Gomes E (1999) Screening of bacterial strains for pectinolytic activity: characterization of the polygalacturonase produced by *Bacillus* sp. Rev Microbiol 30(4):299–303

Spang A, Saw JH, Jørgensen SL et al (2015) Complex archaea that bridge the gap between prokaryotes and eukaryotes. Nature 521(7551):173–179

Spiekermann P, Rehm BH, Kalscheuer R (1999) A sensitive, viable-colony staining method using Nile red for direct screening of bacteria that accumulate polyhydroxyalkanoic acids and other lipid storage compounds. Arch Microbiol 171:73–80

Srivastava P, Bragança J, Ramanan SR, Kowshik M (2013) Synthesis of silver nanoparticle synthesis using haloarchaeal isolate *Halococcus salifodinae* BK3. Extremophiles 17:821–831

Srivastava P, Bragança J, Kowshik M (2014) *In vivo* synthesis of selenium nanoparticles by *Halococcus salifodinae* BK18 and their anti-proliferative properties against HeLa cell line. Biotechnol Prog 30:1480–1487

Song HS, Cha IT, Yim KJ et al (2014) *Halapricum salinum* gen. nov., sp. nov., an extremely halophilic archaeon isolated from non-purified solar salt. Antonie Van Leeuwenhoek 105:979–986

Sorek R, Cossart P (2010) Prokaryotic transcriptomics: a new view on regulation, physiology and pathogenicity. Nat Rev Genet 11(1):9–16. doi:10.1038/nrg2695

Stein L (2001) Genome annotation: from sequence to biology. Nat Rev Genet 2(7):493–503. doi:10.1038/35080529

Stepanov VM, Rudenskaya GN, Revina LP et al (1992) A serine proteinase of an archaebacterium, *Halobacterium mediterranei*. A homologue of eubacterial subtilisins. Biochem J 285:281–286

Studdert CA, De Castro RE, Seitz KH, Sánchez JJ (1997) Detection and preliminary characterization of extracellular proteolytic activities of the haloalkaliphilic archaeon *Natronococcus occultus*. Arch Microbiol 168(6):532–535

Tan GY, Chen CL, Li L et al (2014) Start a research on biopolymer polyhydroxyalkanoate (PHA): a review. Polymers 6:706–754

Tapingkae W, Tanasupawat S, Itoh T et al (2008) *Natrinema gari* sp. nov., a halophilic archaeon isolated from fish sauce in Thailand. Int J Syst Evol Microbiol 58:2378–2383

Tindall BJ, Ross HNM, Grant WD (1984) *Natronobacterium* gen. nov. and *Natronococcus* gen. nov., two new genera of haloalkaliphilic archaebacteria. Syst Appl Microbiol 5:41–57

Torreblanca M, Rodriguez-Valera F, Juez G et al (1986) Classification of non-alkaliphilic halobacteria based on numerical taxonomy and polar lipid composition, and description of *Haloarcula* gen. nov. and *Haloferax* gen. nov. Syst Appl Microbiol 8:89–99

Torreblanca M, Meseguer I, Ventosa A (1994) Production of halocin is a practically universal feature of archaeal halophilic rods. Lett Appl Microbiol 19(4):201–205

Trivedi S, Choudhary OP, Gharu J (2011) Different proposed applications of bacteriorhodopsin. Recent Pat DNA Gen Seq 5:35–40

van den Burg B (2003) Extremophiles as a source for novel enzymes. Curr Opin Microbiol 6:213–218

van der Wielen PW, Bolhuis H, Borin S et al (2005) The enigma of prokaryotic life in deep hypersaline anoxic basins. Nature 307:121–123

Ventosa A, Gutiérrez MC, Kamekura M, Dyall-Smith ML (1999) Proposal to transfer *Halococcus turkmenicus*, *Halobacterium trapanicum* JCM 9743 and strain GSL-11 to *Haloterrigena turkmenica* gen. nov., comb. nov. Int J Syst Bacteriol 49:131–136

Ventosa A, Nieto JJ (1995) Biotechnological applications and potentialities of halophilic microorganisms. World J Microbiol Biotechnol 11:85–94

Ventosa A, Nieto JJ, Oren A (1998) Biology of moderately halophilic aerobic bacteria. Microbiol Mol Biol Rev 62:504–544

Ventosa A, Sánchez-Porro C, Martín S, Mellado E (2005) Halophilic archaea and bacteria as a source of extracellular hydrolytic enzymes. In: Adaptation to life at high salt concentrations in Archaea, Bacteria, and Eukarya. Springer, Dordrecht, pp. 337–354

Vidyasagar M, Prakash S, Litchfield C, Sreeramulu K (2006) Purification and characterization of a thermostable, haloalkaliphilic extracellular serine protease from the extreme halophilic archaeon *Halogeometricum borinquense* strain TSS101. Archaea 2(1):51–57

von Mering C, Hugenholtz P, Raes J et al (2007) Quantitative phylogenetic assessment of microbial communities in diverse environments. Science 315(5815):1126–1130. doi:10.1126/science.1133420

Vreeland RH, Rosenzweig WD, Powers DW (2000) Isolation of a 250 million-year-old halotolerant bacterium from a primary salt crystal. Nature 407(6806):897–900

Vreeland RH, Straight S, Krammes J et al (2002) *Halosimplex carlsbadense* gen. nov., sp. nov., a unique halophilic archaeon, with three 16S rRNA genes, that grows only in defined medium with glycerol and acetate or pyruvate. Extremophiles 6:445–452

Wainø M, Tindall BJ, Ingvorsen K (2000) *Halorhabdus utahensis* gen. nov., sp. nov., an aerobic, extremely halophilic member of the Archaea from Great Salt Lake, Utah. Int J Syst Evol Microbiol 50:183–190

Wainø M, Ingvorsen K (2003) Production of β-xylanase and β-xylosidase by the extremely halophilic archaeon *Halorhabdus utahensis*. Extremophiles 7(2):87–93

Wang Y, Yang J, Lee OO et al (2011) Hydrothermally generated aromatic compounds are consumed by bacteria colonizing in Atlantis II Deep of the Red Sea. ISME J 5(10):1652–1659

Waters E, Hohn MJ, Ahel I et al (2003) The genome of *Nanoarchaeum equitans*: insights into early archaeal evolution and derived parasitism. Proc Natl Acad Sci U S A 100(22):12984–12988

Wejse PL, Ingvorsen K, Mortensen KK (2003) Purification and characterisation of two extremely halotolerant xylanases from a novel halophilic bacterium. Extremophiles 7(5):423–431

Williams GP, Gnanadesigan M, Ravikumar S (2013) Biosorption and bio-kinetic properties of solar saltern halobacterial strains for managing Zn^{2+}, As^{2+} and Cd^{2+} metals. Geomicrobiol J 30:497–500

Woese CR, Fox GE (1977) Phylogenetic structure of the prokaryotic domain: the primary kingdoms. Proc Natl Acad Sci U S A 74(11):5088–5090

Woese CR, Kandler O, Wheelis ML (1990) Towards a natural system of organisms: proposal for the domains Archaea, Bacteria, and Eucarya. Proc Natl Acad Sci U S A 87(12):4576–4579

Wu CS, Liao HT (2014) The mechanical properties, biocompatibility and biodegradability of chestnut shell fibre and polyhydroxyalkanoate composites. Polym Degrad Stab 99:274–282

Xu Y, Zhou P, Tian X (1999) Characterization of two novel haloalkaliphilic archaea *Natronorubrum bangense* gen. nov., sp. nov. and *Natronorubrum tibetense* gen. nov., sp. nov. Int J Syst Bacteriol 49:261–266

Xue Y, Fan H, Ventosa A et al (2005) *Halalkalicoccus tibetensis* gen. nov., sp. nov., representing a novel genus of haloalkaliphilic archaea. Int J Syst Evol Microbiol 55:2501–2505

Yatsunami R, Ando A, Yang Y et al (2014) Identification of carotenoids from the extremely halophilic archaeon *Haloarcula japonica*. Front Microbiol 5:100. doi:10.3389/fmicb.2014.00100

Yooseph S, Sutton G, Rusch DB et al (2007) The sorcerer II global ocean sampling expedition: expanding the universe of protein families. PLoS Biol 5(3):e16. doi:10.1371/journal.pbio.0050016

Yu TX (1991) Protease of haloalkaliphiles. In: Horikoshi K, Grant WD (eds) Superbugs: microorganisms in extreme environments. Springer, New York, pp. 76–83

Chapter 6
Bioprospecting Soil Metagenomes for Antibiotics

Chiara Borsetto and Elizabeth M.H. Wellington

Abstract The recent emergence of antibiotic resistant pathogens, such as methicillin resistant *Staphylococcus aureus* (MRSA) and strains of multidrug resistant *Mycobacterium tuberculosis*, has raised awareness of a new class of global health threats. Unless alternative solutions are found, we can expect to see a considerable rise in incurable infection and fatality in developed and developing regions. Less than 10 % of the microbial natural product diversity has been explored and these uncultured communities have remained unexploited for pharmaceutical and biocatalytic applications. Soil contains the greatest diversity of bacteria compared to all other habitats. Thus the aim of this review is to explore the ways in which the uncultured majority of bacteria in soil has been explored and attempts made to exploit functional diversity using the extensive metagenomes of bacterial communities in soil. We focus on the capture of genes with metabolic potential to produce bioactive metabolites by a combination of approaches. These have involved (a) making metagenomic libraries in suitable expression hosts and examination of metabolic diversity by phenotypic screening, (b) co-culture, (c) sequence target PCR screening, (d) specialised metabolite analysis and (e) sequence analysis of the total assembled metagenome.

6.1 Introduction

> *"We know more about the movement of celestial bodies than about the soil underfoot."*
> *(Leonardo da Vinci)*

Soil still remains one of the most challenging environments for ecological understanding of how community structure relates to function (Fierer et al. 2012; Bardgett and van der Putten 2014). Many studies have reported impressive bacterial diversity within soils; the most significant early breakthroughs were achieved using DNA re-association techniques, microscopy and molecular analysis targeting 16S rRNA

C. Borsetto • E.M.H. Wellington (✉)
School of Life Sciences, University of Warwick, Coventry, UK
e-mail: E.M.H.Wellington@warwick.ac.uk

R. Paterson, N. Lima (eds.), *Bioprospecting*, Topics in Biodiversity and Conservation 16, DOI 10.1007/978-3-319-47935-4_6

phylogenetic marker genes for bacteria (Torsvik et al. 1998; Tiedje et al. 2001; Derakshani et al. 2001). Below ground diversity will be affected by physicochemical properties of the soil, but most importantly the plant root systems will exert a significant impact on the rhizosphere microbial communities (Bardgett et al. 2014). In addition, soil microfauna present a further habitat for colonisation and bacteria have been found growing in close association with Collembola, which can act as vectors for bacteria (Hoffmann et al. 1999). Soil is a complex environment that greatly changes in texture and composition, not only in relation to the stratification (topsoil, bulk soil, permafrost, etc.), but also depending on the climate (Wall et al. 2010).

Early attempts to exploit the microbial diversity in soil focused on isolation techniques which proved to be highly successful for the discovery of novel antibiotics in the 1960's (Drews 2000). However, it soon became apparent that different approaches were required to explore greater diversity without rediscovering the same molecules. A good example of this is streptomycin, which was discovered as a product of the soil *Streptomyces griseus* firstly isolated by Waksman (Schatz and Waksman 1944) and then rediscovered numerous times in screens of strains closely related to *S. griseus*, but frequently incorrectly elevated to species status, for example, *S. bikiniensis* (Johnstone and Waksman 1948), *S. griseinus* and *S. globisporus*. In contrast to this, many other distinct species have also been identified as producers of streptomycin as *S. cinnamoneus*, *S. mashuensis*, *S. glaucescens* and *S. galbus* (Laskaris 2009). The rapid advances in molecular ecology focused on phylogenetic markers, and revealed the uncultured majority in soil providing estimates of more than 95 % of bacterial diversity had yet to be cultured (Torsvik et al. 1998).

In the past, the prospecting for genetic biodiversity and the exploitation of flora and fauna for discovery and commercialisation of products has been greatly criticised. The reasons for this relate to the unequal sharing of the benefits of patents and resulting commercial benefits with the source countries. The original explorers collecting novel plant material and other sources of medicinal benefits relating to local know-how and folklore, failed to attribute any rights to local people and often this process was referred to as biopiracy (Sheridan 2005; Mackey and Liang 2012). Particular concern was focused on pharmaceutical companies' exploitation of genetic and chemical biodiversity for commercial purposes. Agreements and international laws were eventually set in motion with the Convention on Biological Diversity (CBD) (Anon. 1992) signed by 168 countries in 1992 at the Rio Earth Summit (the United Nations Conference on Environment and Development). Three major objectives were set and regulations determined on:

- access to the genetic resources and technologies;
- fair and equal benefit-sharing within the parties;
- sustainable use of biological diversity.

The CDB therefore established the rights and responsibilities amongst the Contracting Parties (sovereign states) to achieve these major goals. Private organisations such as companies or academic institutions were automatically bound to the CBD depending on national laws. The number of countries that have signed the CDB is 198. Further regulations on the fair and equal sharing of benefits arising from the use of genetic biodiversity, were recently set in the Nagoya Protocol (entered into force from 2014) implementing those listed in the CDB.

In the antibiotic discovery field, the biological diversity present in different soil types globally was a source for detection and commercialisation of natural products with antimicrobial activity from the beginning of drug discovery era (Demain and Sanchez 2009). The environment (in particular soil) has been explored for novel genes of interest related to the production of bioactive molecules since microorganisms have been identified as the main producers of compounds with activity for pharmaceutical, industrial and agricultural purposes (Keller and Zengler 2004).

In particular, awareness of the urgent need for discovering molecules with new antimicrobial activity against resistant pathogens has arisen with the continued emergence of bacteria resistant to multiple antibiotics (Levy and Marshall 2004). The likely misuse of antibiotics for non-exclusive treatment of infections has increased the selection pressure on antimicrobial resistance genes; as early as the 1960s reports of resistant nosocomial pathogens such as *Staphylococcus aureus* were recorded (Jevons 1961). During the last two decades multi-resistant pathogens such as *Mycobacterium tuberculosis* appeared, bringing a serious threat to the future of bacterial infection treatment (Velayati et al. 2009). The fact that some of the common pathogens such as the ESKAPE (*Enterococcus faecium*, *Staphylococcus aureus*, *Klebsiella pneumonia*, *Acinetobacter baumanii*, *Pseudomonas aeruginosa* and *Enterobacter*) group are gaining resistance to last-resort treatments (e.g. *K. pneumonia* resistant to 3rd generation cephalosporins and carbapenems (Srinivasan et al. 2014)) or are treatable only with second-line drugs (WHO 2014), has induced governments to address antibiotic resistance as a serious and urgent problem and pharmaceutical companies to reinvest in drug discovery programmes (Frearson and Wyatt 2010; Harbarth et al. 2015). The absence of novel antibiotics and effective interventions to prevent infectious diseases will lead to a post-antibiotic era where we will return to a level of fatality experienced in the 1930–1940s (WHO 2014).

6.2 Antibiotics Discovery from Soil

6.2.1 Strain Isolation

Since the discovery of penicillin from the fungus *Penicillium chrysogenum* by Alexander Fleming in 1928 (Fleming 1929), the era of antibiotic discovery started and developed rapidly. Soil has been one of the major sources of exploitation for strain isolation for the identification of novel compounds with antimicrobial activity (Keller and Zengler 2004) as mentioned previously. Different soils have been explored using various isolation conditions with particular interest for the isolation of groups of bacteria that were able to produce compounds with antimicrobial activity (Table 6.1). Actinobacteria and in particular the genus *Streptomyces*, represented one of the major groups producing inhibitory bioactive compounds effective against other bacteria since the beginning of antibiotic discovery programs (Berdy 2012). A famous example is streptomycin produced by *S. griseus*, discovered by Waksman's group which was notable for the broad spectrum of activity of this aminoglycoside (Schatz and Waksman 1944). Soils from all over the world have been collected and tested over the last century, leading to the isolation of different actinobacteria (Table 6.1), some of which produce useful natural products (Table 6.2).

A great effort has been made to develop new methods and techniques to access the biological diversity present in the environment with improved media for isolation of particular groups of bacteria (Stevenson et al. 2004; George et al. 2011), in addition to new cultivation techniques (Zengler et al. 2002; Ling et al. 2015; Vester et al. 2015) of recalcitrant novel bacterial groups. A recent success was the β-proteobacteria *Eleftheria terrae* producing the novel non-ribosomal peptide teixobactin (Ling et al. 2015) isolated using an isolation chip (iChip) system (Nichols et al. 2010). In this system, miniature diffusion chambers are loaded with liquid-

Table 6.1 Examples of species isolated from soil samples collected all over the world

Species	Soil location for isolation	Reference
Streptomyces venezuelae	Caracas, Venezuela	Ehrlich et al. (1948)
Streptomyces bikiniensis	Bikini atoll, Pacific Island	Johnstone (1947)
Streptomyces costaricanus	Costa Rica	Esnard et al. (1995)
Streptomyces beijiangensis	Beijing, China	Li et al. (2002)
Streptomyces africanus	Cape Town, South Africa	Meyers et al. (2004)
Amycolatopsis taiwanensis	Yilan county, Taiwan	Tseng et al. (2006)
Amycolatopsis australiensis	Australia	Tan et al. (2006)

Table 6.2 Examples of known antibiotics produced by single species isolated from soil samples

Antibiotic	Class	Species	Soil location for isolation	Reference
Streptomycin	Aminoglycoside	*Streptomyces griseus*	Russia	Waksman et al. (1948)
Aureomycin	Tetracycline	*Streptomyces aureofaciens*	Missouri, USA	Duggar (1948)
Erythromycin	Macrolide	*Saccharopolyspora erythraea*	Philippines	Anon (1952)
Vancomycin	Glycopeptide	*Amycolatopsis orientalis*	Orient	Brigham and Pittenger (1956)
Daptomycin	Lipopeptide	*Streptomyces roseosporus*		Allen et al., (1987)
Platensimycin	Diterpenoid - FASII inhibitor	*Streptomyces platensis*	South Africa	Wang et al. (2006)
Teixobactin	Glycopeptide	*Eleftheria terrae*	Maine, USA	Ling et al. (2015)

agar based media inoculated with diluted soil samples and incubated *in situ* (in direct contact with the soil from where the sample was taken) leading to a recovery of a single isolate per channel. Data showed that the system, in comparison to classical isolation in petri dishes, recovers a significantly higher diversity of bacteria with a possibility to cultivate *in vitro* of 10–15 % (Nichols et al. 2010). This technique, which preserved the natural environmental conditions, promoted the isolation of 'unculturable' bacteria but was still labour intensive; therefore it was not considered suitable for routine screening efforts in drug discovery. Other available strategies for isolation of novel strains with interesting phenotypes for antimicrobial discovery rely on resistance to particular classes of antibiotics. For instance, Thaker et al. (2014) used a double selection in the classical Waksman's soil bacterial isolation approach; this included selection for a particular group of soil bacteria, the Actinomycetes, combined with phenotypic selection for glycopepetide antibiotic resistance. The strains isolated were able to grow on glycopeptide containing plates and contained glycopeptide resistance genes and were producers of related glycopeptides. This screening/isolation technique allowed the identification of derivatives of glycopeptide structures that could have been used to improve activity compared to the existing products. In general, however, strain isolation techniques still suffer from the limited knowledge of microbial communities in the soil and their physiology (Keller and Zengler 2004).

6.2.2 Metagenomics

Metagenomics overcomes the problem of isolation and allows more in depth study on the structure and function of bacteria within the microbial community (Handelsman 2004; Allen and Banfield 2005). Baker et al. (2006) pioneered the extraction of genes and partial genomes from environmental DNA to construct a metagenome of the biofilm community from acid mine drainage forming underground at the Richmound Mine at Iron Mountain, California. However, the very first studies on characterisation of uncultured communities using a metagenomic approach were carried out in marine communities where extraction of larger fragments of DNA was possible as illustrated by Stein et al. (1996). Community genomics in the oceans provided a paradigm shift in comprehension of uncultured bacterial groups, presenting an improved understanding and knowledge base of marine communities, their metabolisms, biochemical pathways and genomic variability (DeLong, 2005). This culture-independent approach is now a key component of the efforts to exploit bacterial communities in soil. Previous studies reported that only 1 % of the total bacterial community in soil has been accessed with culture techniques (Torsvik et al. 1996), with a remaining 99 % still unexplored and available for further discovery of strains and molecules of interest. Metagenomics allows accessing the 'unculturable' fraction of microbial communities without the need to isolate and cultivate microorganisms in the laboratory. Since the introduction of the concept of metagenomics (Handelsman et al. 1998; Rondon et al. 2000), the technologies involved in this process have developed and improved. Next-Generation Sequencing (NGS) technologies, which allow exploring the DNA diversity present in the environments on a molecular level, are constantly evolving (van Dijk et al. 2014). Since the advent of the first techniques, multiple platforms and systems have been introduced into the market rapidly, to respond to the ever-increasing demand of deeper sequencing with lower error rates and costs (Metzker 2010; Loman et al. 2012).

The 454 pyrosequencing approach was a pioneer in the field of NGS technologies, followed by Illumina (HiSeq and MiSeq), SOLiD, Ion Torrent, Pacific Biosciences and more recently Oxford Nanopore (Branton et al. 2008; Glenn 2011; Luo et al. 2012; Quail et al. 2012; Wang et al. 2014). The possibility to sequence amplicons of different lengths, or shotgun sequence DNA samples at different depths, provided new horizons in the exploration of microbial diversity at a molecular level. Various environments have been used to provide metagenomes which were subsequently annotated and are available to interrogate community diversity at the functional level.

The Global Ocean Sampling Expedition programmes (Rusch et al. 2007) is an example of a large, deep and annotated metagenome. Water samples from various oceans were collected, deep sequenced and analysed creating a 'reference' metagenome for further studies. A similar project was launched in 2009 for an international soil project termed the "TerraGenome", which combined the efforts of the global scientific community for the construction (sequencing and annotation) of a 'reference' soil metagenome using Park Grass soil samples, a well characterised soil from Rothamsted Research station (UK) (Vogel et al. 2009). Metagenomic approaches aim to capture community diversity using libraries which are available for future industrial applications by analyzing the functional genomic diversity. Large metagenomic libraries can constitute an invaluable source of unexplored and unexploited diversity, inaccessible with classical methods that could be of great interest for green, white and red biotechnology industries. In particular, metagenomic libraries from soil could provide a resource for the discovery of new chemical structures with antimicrobial activity against resistant pathogens such as methicillin resistant *Staphylococcus aureus* (Kallifidas et al. 2012). Some examples of heterologously expressed antibiotic gene clusters recovered from metagenomic libraries created from soil samples are reported in Table 6.3.

There are still a number of bottlenecks in the process of exploiting of metagenomic libraries including bias in the extraction efficiency, difficulties in DNA extraction to provide large fragments, computationally demanding bioinformatics and heterologous expression of the captured DNA.

Table 6.3 Examples of antibiotics isolated from metagenomic libraries from soil and expressed in heterologous hosts

Antibiotic	Class	Soil location for library	Reference
Utahmycin A and B	2-azaanthraquinone	Utah	Bauer et al. (2010)*
Fluostatins F, G and H		Anzo Borrego desert, California	Feng et al. (2010)*
Tetarimycin A	Tetracycline	Arizona desert	Kallifidas et al. (2012)*
Fasamycin A and B	FASII inhibitor	Arizona desert	Feng et al. (2012)*
Arixanthomycins A–C	pentangular polyphenol	Arizona desert	Kang and Brady (2014)*

**Streptomyces albus* was the heterologous host

6.3 Metagenomic Tools: Problems and Successes for Antibiotic Discovery

The following summarises the main bottlenecks and illustrates the advantages of metagenomics which can be presented as a generalised flow diagram (Fig. 6.1) for the exploitation of biodiversity. The focus is on antibiotic discovery although the term "antibiotic" is used in a general sense to encompass natural product diversity.

6.3.1 Sample Selection

A wide variety of strategies have been used to select soil type for sampling: an initial understanding of the community structure is a useful precursor for site selection. Examples in the past showed that exploring extreme environments allowed the isolation of enzymes with improved characteristics for industrial application such as detergents used at low temperature. In the drug discovery field, soil and more recently marine environments, have been identified as rich sources of diversity (Gerwick and Moore 2012). Soil is a physically more complex environment and requires sampling strategies enabling characterisation of the vertical and horizontal structure at the macrosite and microsite levels. Differences in the microbial community can depend on the physical factors such as sample depth, due to the presence of different biotic and abiotic stress factors (Fierer et al. 2003).

Soil contains a great variety of habitats with many physicochemical gradients, for example plant roots release exudates providing a nutritional gradient for bacteria to exploit. These gradients will depend on the type of plant and to some extent, on the soil structure and climatic effects (Marschner et al. 2001; Fierer et al. 2003). Therefore this strong interaction between plants roots and their microbiomes results

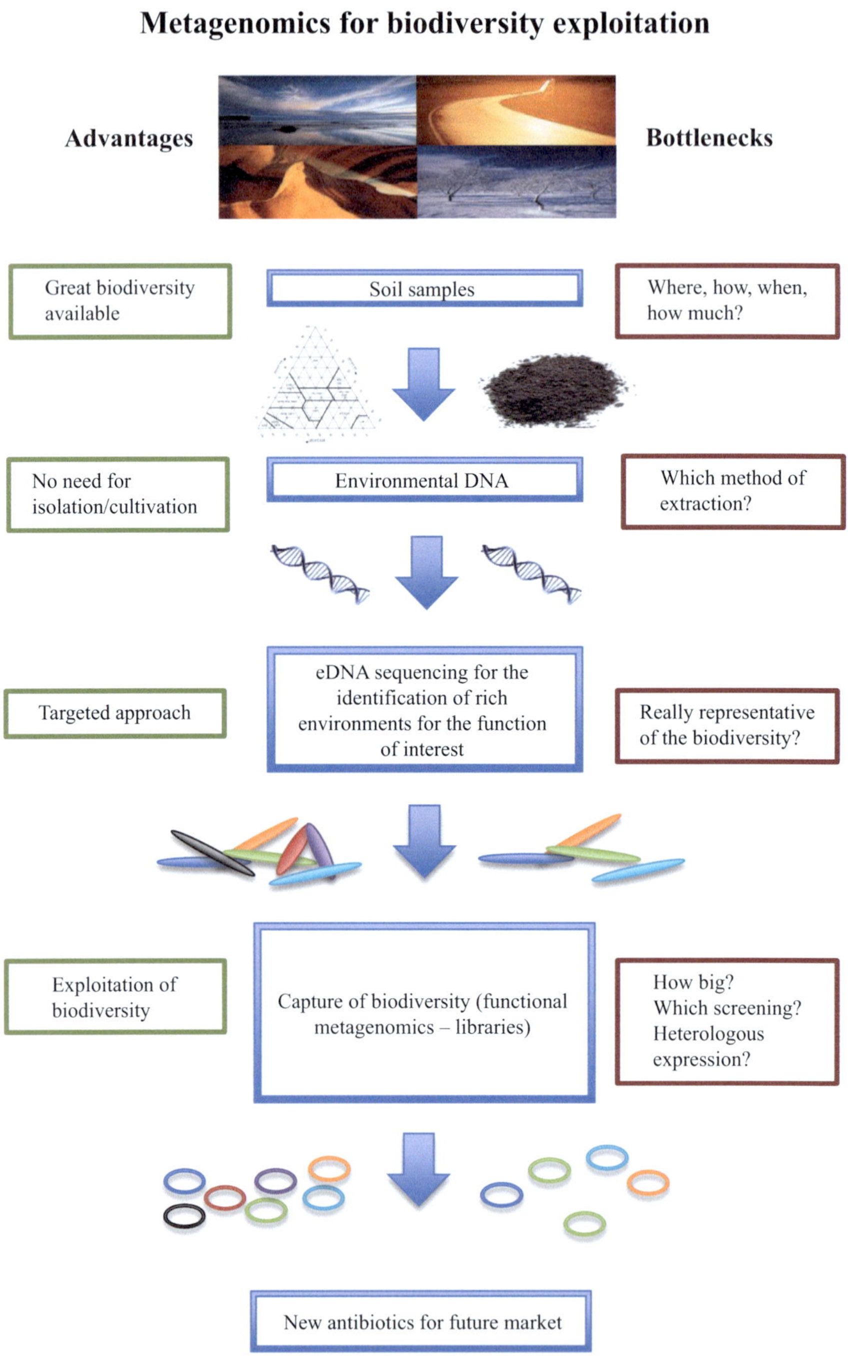

Fig. 6.1 Metagenomics for biodiversity exploitation. Representation of the metagenomic approach for functional biodiversity exploitation with advantages (*left*) and bottlenecks (*right*)

in a different community than that present in the surrounding bulk soil (Berg and Smalla 2009; Berendsen et al. 2012; Bulgarelli et al. 2012). Sampling strategies are rarely discussed in detail and have received little attention in terms of exploring microbial diversity in soil.

One gram of soil contains roughly ten thousand particles. Each particle may have one hundred thousand sites for microbial colonisation and with this level of complexity it is a considerable challenge to devise an appropriate sampling strategy. Considering the heterogeneity, mixing original samples to reduce the variability and making it more homogeneous for subsampling, could be a valuable option. When subsampling soil for analysis, the quantity of soil extracted will inevitably result in the prevalence of different groups of microbes due to biases in community structure (Kang and Mills 2006). Replicate soil samples of sizes from 0.1 g up to 10 g were used for DNA extraction and these extracts were subsequently analysed for 16S rRNA bacterial diversity. Discriminant analysis of the molecular data was done to determine the optimal size for sampling the bacterial population diversity. The authors recommend a size specific study of comparable analysis to obtain the optimal sample size. For the meadow sampled this represented 0.25 g, the soil was stored frozen at –80° C and subsequently thawed, mixed, broken up by hand, and then sieved (Kang and Mills 2006).

6.3.2 DNA Extraction

DNA extraction is one of the key steps in metagenomics. The diversity and coverage of the metagenome will depend on the efficiency of DNA extraction, although this frequently relies on aggressive physical procedures which can fragment the DNA. This in turn compromises insert size and thus we have the paradox of metagenomic library construction; methods that are very efficient for DNA extraction are not optimal for isolation of large DNA fragments necessary for library construction.

There are multiple DNA extraction protocols available in the literature to treat soil samples that can be mainly classified into direct or indirect methods. The direct method requires lysis in situ within the soil during the extraction process, whereas the indirect method involves recovery of cells and their lysis as a separate step. Both methods produce bias, although the direct method has been favoured for metagenomics possibly because the efficiency of cell extraction from soil is difficult to evaluate. However, it is probably low and will not work for mycelium or spores of filamentous bacteria. For example, many of the actinobacteria previously favoured for high diversity of antibiotic gene clusters, produce mycelia and spores, which often means they are poorly represented in metagenomic libraries because gentle lysis procedures do not lysis spores efficiently and direct methods do not recover mycelia (Morris et al. 1999).

The average size of DNA extracted using different approaches varies and for PCR amplicon-based diversity analysis, high yield of good quality DNA is acceptable, coupled with a fragment size in the distribution 10–30 kb. Therefore, direct or indirect methods that allow the extraction of inhibitor-free DNA are preferred, such as the commercially available kits FastDNA™ Spin kit for soil (MP Biomedical), or PowerSoil® DNA Isolation Kit (MO BIO). High molecular weight DNA is required for metagenomic library preparation with a size range depending on the vector used to clone the DNA in relation to the size of the genes of interest (from 5 kb for single enzymes up to 200 kb for antibiotic clusters). Gentle methods of extraction are fundamental to obtain high quality DNA to create libraries (e.g. Brady (2007)).

The indirect method involves the collection of cellular biomass by separation from soil using various centrifugation steps and a density separation protocol, involving the use of the Nycodenz polymer which results in the recovery of a layer of cells at the interface between the polymer and the upper aqueous layer (Bakken and Lindahl 1995; Berry et al. 2003). Other methods have involved the dispersal of soil using Chelex 100 prior to centrifugation (Herron and Wellington 1990). The recovered biomass is then lysed, giving clean DNA of a greater size rage than that recovered with direct lysis (Berry et al. 2003). The direct method enables a lysis in situ in the total soil sample with or without dispersants such as sodium dodecyl sulfate (SDS) or Chelex 100, but with the addition of an enzymatic lytic mixture. This second method can be applied to the soil sample: in addition, agarose can be used to support the soil during the lysis procedure and the gel plug is then subjected to gel electrophoresis. The direct chemical lysis approach allows the recovery of high molecular weight (HMW) DNA but with inhibitors such as humic acids depending on the soil type. Humic compounds can be co-purified with DNA in the extraction process and they can constitute a problem in the following steps of PCR amplification, sequencing and cloning (Tebbe and Vahjen 1993). Prevention of inhibitor build up can be achieved by pre-treatment of the sample with the addition of $AlNH_4(SO_4)_2$ to remove the inhibitors through chemical flocculation (Braid et al. 2003), or the addition of 1-Hexadecyltrimethylammonium bromide (CTAB) and polyvinylpolypyrrolidone (PVPP) to precipitate humic compounds. However, PVPP has been shown to complex with DNA giving a reduced recovery of DNA than using CTAB (Zhou et al. 1996). Another option to purify DNA of inhibitors is the use of cesium chloride (CsCl) density centrifugation. However, this is labour-intense and might not remove the inhibitors completely. The use of the pre-treatments described can help in reducing the amount of inhibitors prior to agarose gel electrophoresis for quality control and/or size fractionation of the DNA.

Depending on the efficacy of lysis method all bacterial groups may not be represented equally with DNA from some groups being recovered more easily (Bürgmann et al. 2001). Texture and composition of the soil could play a crucial role in the performance of the lysis methods; the content of clay, for example, can decrease the efficiency of DNA purification coupled with co-extraction of inhibitors (Braid et al. 2003).

6.3.3 Bioinformatic Pipelines for Sequence Analysis

High-throughput sequencing technologies developed rapidly since their introduction to the market, allowing the study of microbial communities and ecological relationships amongst microorganisms in more depth. International projects have been established to examine specific habitats such as the TerraGenome project with the focus on soil (Vogel et al. 2009) and the Global Ocean Sampling Expedition (Vogel et al. 2009; Rusch et al. 2007) to sequence and collect metagenomes as references for future study. Amplicon sequencing has been used to study environments for specific genes of interest such as bacterial 16S rRNA for phylogenetic diversity (Logares et al. 2013), secondary metabolites biodiversity potential (Charlop-Powers et al. 2014) or antibiotic resistome phylogeny and distribution (Allen et al. 2008; Wright and Poinar 2012).

Different bioinformatics tools were developed at the same time to enable deep processing of large volumes of sequence data generated from NGS platforms. Some of the main bioinformatics pipelines available for metagenome data analysis and annotation are listed below, the:

- MetaGenomics Rapid Annotation using Subsystems Technology (MG-RAST) server is an open-source analysis pipeline for phylogenetic and functional annotation of assembled or non-assembled prokaryotic metagenomic sequences (Meyer et al. 2008).
- EBI metagenomics portal is an automated pipeline to archive and analyse metagenomic data following the European Nucleotide Archive (ENA) data requisitions and the Genomic Standards Consortium (GSC) guidelines providing phylogenetic and functional annotation (Hunter et al. 2014).
- Integrated Microbial Genomes/Metagenomic analysis (IMG/M) platform allows a comparative analysis of metagenomes against a comprehensive set of reference genomes of bacteria, archaea, eukarya, plasmids, viruses and genome fragments directly sequence by the Department of Energy – Joint Genome Institute (DOE-JGI), or is publicly available (Markowitz et al. 2014).

Common databases used for phylogenetic analysis and taxonomy assignment are, for example, Greengenes (DeSantis et al. 2006) and the Ribosomal Database Project (RDP) (Cole et al. 2014). However improvements in the analysis of data are continuously being made and reflect the extensive application of methods for classification of 16S rRNA sequences (Vinje et al. 2015).

Metagenome sequencing must be performed with consideration of the sample size in relation to the final aim of the experiment (e.g. few samples sequenced in depth for the identification of rare microorganisms or many samples sequenced with less depth to determine microbial community dynamics). The choice depends on the experimental aim and will influence the bioinformatics approach, for example in a diverse community a draft genome of a prevalent group will be more readily recovered if the sample is deeply sequenced (Knight et al. 2012). Special purpose analysis platforms have been developed for identification and characterisation of key functional groups such as the antibiotic gene clusters and include Antibiotics & Secondary Metabolite Analysis SHell (AntiSMASH) (Medema et al. 2011), Natural Product Domain Seeker (NaPDoS) (Ziemert et al. 2012) and Environmental Surveyor of Natural Product Diversity (eSNaP) (Reddy et al. 2014).

6.3.4 Metagenomic Library Construction

6.3.4.1 Creation and Vectors Available

To explore environmental biodiversity one of the key tools is metagenomic libraries. The creation of these allows analysis and capture of genes for industrial or pharmaceutical use. Different types of libraries can be created in a variety of vectors depending on the genes of interest: plasmid libraries allow inserts of up to 10 kb so HMW DNA is unnecessary, therefore it is useful to capture single genes such as those coding for enzymes with application in the food and detergent industry. Gene clusters require larger inserts for use with fosmid or cosmid vectors (up to 40 kb). These two types of vector are similar but they differ in the origin of replications and stability of the insert inside the host (usually *E. coli*). Cosmids contain *E. coli* bacteriophage λ *cos* sites which allow the recognition of the vector from the lambda bacteriophage head particles during *in vitro* reactions. This vector has been used for classical cloning with great success (Collins and Hohn 1978) and has been more recently applied in single strain genomic and metagenomic library preparation. Lambda bacteriophage headful packaging can incorporate approximately 50 kb, therefore metagenomic libraries with insert size of up to 40–45 kb per clone can be created. One of the recent examples of cosmid vector use is the cosmid pWEB (Epicentre) which was used to screen a metagenomic library for novel antimicrobial activity (Brady 2007; Owen et al. 2013). Fosmid vectors were developed by engineering a cosmid vector in order to overcome instability problems showed by rearrangements (Kim et al. 1992). Fosmids contain the bacterial F-plasmid replicon system that maintains a single copy of the vector per cell; therefore fosmid vectors are more stable than cosmids. Optimization of the original pFOS1 fosmid vector (Kim et al. 1992) was performed in subsequent years and engineered fosmids are currently available as commercial kits for the creation of metagenomic libraries. Following the ligation of the inserted DNA, packaging can occur inside λ bacteriophage particles to create coliphages *in vitro* and the phages used to infect *E. coli* to create the library. Selection of positive clones is performed using selectable markers such as antibiotic resistance genes. Usually the transduction efficiency is high allowing the recovery of thousands of clones containing inserted DNA from the sample tested.

Bacterial Artificial Chromosomes (BACs) are vectors that can be used to insert up to 300 kb of DNA and were first engineered in the early 1990s (Shizuya et al. 1992). These vectors can be transferred through transformation, instead of transduction, into *E. coli* but there is a lower efficacy which causes a bottleneck in the preparation of large libraries. Nevertheless, the BAC vectors can insert up to 300–350 kb (even though the average size is ~120kb) and offer an alternative for the capture of larger gene clusters beyond 50 kb. Other possible vectors similar to BACs that can be transformed into *E. coli* to capture the isolated eDNA are the phage P1-derived vectors such as P1-derivate Artificial Chromosome (PAC) which can allow insertion up to 300 kb with an average insert size of 130–150 kb (Ioannou et al. 1994). An example of an engineered BAC vector is the *E. coli – Streptomyces* Artificial Chromosome (ESAC) vectors such as pPAC-S1 (Sosio et al., 2000) which have been used for the creation of genomic libraries that can be easily replicated and manipulated in *E. coli* and stably maintained in a different host such as *Streptomyces coelicolor* for the expression of the inserted DNA. They have proved useful for the manipulation of antibiotic gene clusters and had been used to create metagenomic libraries for this purpose (Berry et al. 2003). These vectors can autonomously replicate in *E. coli* and integrate stably into *Streptomyces* chromosomes in a specific site of recombination recognising phage ϕC31 *attB* sites. This ability makes these vectors useful for library construction as the positive clones containing the genes of interest can be identified in *E. coli* by PCR screening and then transferred using intergeneric conjugation to different *Streptomyces* spp. for heterologous expression and characterization.

In addition to the vector choice other aspects to be considered are the coverage and storage. The size of the library is the reflection of the depth of analysis and success in recovery of targeted genes or gene clusters. The size of the library is one of the fundamental problems: to provide a prospective of the enormous diversity present within the soil, one would need to produce approximately 10^{11} BAC clones to achieve complete coverage of the bacterial genomes present within a gram of soil (Daniel 2005). Clearly a compromise has to be reached and this will depend on method of targeting habitats, sample types and the use of high-throughput clone handling procedures. In addition, it may be possible to recover target DNA selectively but this requires a priori selection thus imposing an extreme bias (Pivetal et al. 2014).

Libraries can contain between thousands to hundreds of thousands of clones (depending on the insert size captured by the vector chosen) that need to be stored and maintained in suitable systems for future high-throughput screening. Innovative systems relying on semi-liquid media, where clones grow individually even though they are pooled, have been introduced with great success (Elsaesser and Paysan 2004, Hrvatin and Piel 2007) in addition to the classical arrayed clone pools in 96 or 384-wells plates.

6.3.4.2 Screening and Heterologous Expression

The size of metagenomic library will dictate the screening approach and high-throughput methods will be required for the majority of targets. Screening methods can be divided into two main categories: target directed screening or whole cell screening. Usually with the first method, a specific target for inhibition is employed which is often an enzyme assay, while growth inhibitory biological activity is detected with an unspecific target in the case of the second method. An example of the targeted screening method was applied to isolates which resulted in the identification of platensimycin, an antibiotic that inhibits fatty acid biosynthesis. This compound is produced by *Streptomyces platensis*, which was isolated from a South African soil sample (Wang et al. 2006). Platensimycin is a diterpenoid that inhibits the Fatty Acid Synthase type II FabF enzyme by blocking the binding of the substrate. The activity was detected by reducing the level of FabF in the cell by antisense DNA thus providing a *Staphylococcus aureus* strain which was sensitive to FabF. This approach compensates for the very low level of natural product produced by the wild-type isolates which is also likely to be true in cloned antibiotic gene clusters. Fatty acid biosynthesis has already been identified as a possible new target for antibiotic discovery and has been introduced in previous phenotypic screening by companies such as GlaxoSmithKlein (Heath and Rock 2004).

Phenotypic and activity screening for drug discovery usually relies on the modification of a target that can be detected as inhibition of sensitive bacterial cells, for example growth inhibition or a change in the morphology of the expressing clone. Heterologous hosts used for expression of metagenomic library DNA provide an ideal background for detection of novel antibiotics allowing any new bioactive product could be detected by de novo activity against whole cell bioassay of a known pathogen (Brady 2007). These types of screenings usually involve considerable efforts for processing large number of clones, so automated assays have been developed for high-throughput screening programmes (Moy et al. 2009; Nybond et al. 2013; Rajamuthiah et al. 2014).

Another available approach based on molecular similarity is PCR screening for genes of interest. This is often performed in tandem with phenotypic and activity screening methods, where host expression system incompatibility occurs (Gabor et al. 2004). For instance, *E. coli*, which is the usual host for genomic and metagenomic libraries, is not an optimal heterologous host, as it is able to express only 40 % of the total eDNA captured in the library due to the absence of primary substrates or differences in the translation system (Gabor et al. 2004). Therefore, a PCR screening approach is usually performed for the identification of novel antibiotic clusters. There are a few examples of degenerate primers that can be used for metagenomic library screening to discover antibiotic gene clusters. These include primers targeting genes encoding for polyketide synthase (PKS) and non-ribosomal peptide synthetase (NRPS) (Ayuso-Sacido and Genilloud 2005; Schirmer et al. 2005; Laskaris 2009; Parsley et al. 2011; Owen et al. 2013) targeting specific regions. These two enzymes are the main systems involved in biosynthesis of a large proportion of antibiotics and their genes are responsible for encoding multi-complex enzymes that are organized in modules. Within each module, in NRPS and PKS enzymes systems respectively, there are conserved domains involved in the (a) recognition and activation of specific substrate units (Adenylation – A and Acyltransferase – AT), (b) binding of the extended activated units (PCP and ACP) and (c) elongation of the forming non-ribosomal peptide or polyketide adding the activated units (condensation – C and ketosynthase – KS). In both described systems, additional domains could be present to modify the forming compound (e.g. amino acid epimerization, methylation, reduction in NRPS and β-ketoreduction, dehydration and enoylreduction in PKS). The thioesterase domain (TE) performs the final release of the compound (Donadio et al. 2007; Strieker et al. 2010; Challis et al. 2000). The presence of conserved domains that can be targeted with degenerate primers is the basis of the PCR screening assay for metagenomic libraries aiming to identify gene clusters that could potentially lead to novel bioactive structures. Another possible approach is PCR screening for conserved enzyme coding genes present in a particular class of antibiotics of interest, such as the OxyC-like enzyme, strictly involved in a specific C-C bond formation in the aglycone of vancomycin- and teicoplanin-like glycopeptides (Banik and Brady 2008).

After the identification of the positive clones for the genes of interest, expression is the next challenge in metagenomics. Heterologous hosts have been used for expression, recovery and characterization of enzymes with a commercial application, and examples include cellulases from *Zymomonas mobilis* for ethanol production (Linger et al. 2010) and gene clusters coding for antimicrobial metabolites isolated from a single isolate (Zhang et al. 2010). The choice of the heterologous host for expression of eDNA can be difficult; few heterologous super-hosts have been engineered over previous decades, especially for secondary metabolite expression. However, one of the most widely used heterologous hosts is *Streptomyces coelicolor* with derivate strains such as M1152 and M1154 (Gomez-Escribano and Bibb 2011). The main characteristic of these strains is that, being a secondary metabolite producer, *S. coelicolor* has all the necessary machinery for antibiotic production. This host was further modified by removing known antibiotic gene clusters from the chromosome in order to obtain a lower secondary metabolite background production and to increase the availability of substrates for production of metabolites from introduced gene clusters. Other valuable heterologous hosts belonging to the same genus are *S. lividans* and *S. albus* J1074 (Gullon et al. 2006; Lombo et al. 2006). These have been employed to express silent gene clusters and to move gene clusters from genetically intractable producers into a more readily controlled host background (Baltz 2010). These engineered *Streptomyces* hosts are usually the first choice when expressing antibiotic clusters. However the expression of eDNA (clusters or single genes) will demand a much wider range of hosts due to the high diversity of soil bacteria with very different physiologies. Craig et al. (2010) showed that a wide diversity of heterologous hosts from some of the main bacterial phyla populating the soil microbial communities (i.e. *Alphaproteobacteria*, *Beta proteobacteria* and *Gammaproteobacteria*) (Fierer et al. 2007) increased the detection of novel phenotypes. The tested heterologous hosts (*Agrobacterium tumefaciens, Burkholderia graminis, Caulobacter vibrioides, E. coli, Pseudomonas putida* and *Ralstonia metallidurans*) belonged to the superphylum *Proteobacteria* (Craig et al. 2010).

6.4 Exploitation of Soil Biodiversity

The soil bacterial community has been the source of a large number of useful enzymes which have been applied in biotransformation, catalysis of specific reactions, degradation of pollutants, production of vitamins and pharmaceuticals and many other useful molecules (Adrio and Demain 2014). The advent of metagenomics has enabled the exploitation of this bacterial diversity via culture-independent roots essential for the large number of, as yet uncultured, soil bacteria (Lee and Lee 2013). A recent example is the identification of a novel salt-tolerant chitobiosidase from a metagenomic library created from a chitin-amended disease suppressive soil sample, which could potentially be applied in biodegradation of chitin (Cretoiu et al. 2015). The amendment of soil with different nitrogen sources such as chitin has been studied and results showed a change in the bacterial community by enrichment (Jacquiod et al. 2013; Johnson-Rollings et al. 2014). This method could be applied in functional metagenomics for drug discovery to increase the relative abundance of specific groups such as actinobacteria, one of the main sources of natural products diversity. Moreover, the application of different enrichment treatments might be helpful in the recovery of spore-forming bacteria via germination of spores and easier extraction of the DNA for large insert libraries. Variations in the abiotic conditions could lead to the germination of spores, giving access to higher amount of DNA and increasing the chances to discover novel enzymes and compounds with interesting activities. However, a counter argument would be the loss of some groups due to restricted nutrient availability. Enrichment or partial selection of specific bacterial groups could still result in the recovery of DNA from novel bacteria that were not previously isolated and were recalcitrant to laboratory domestication, considering the extensive diversity of soil bacteria. Soil amendments can be conducted in laboratory based microcosms and a variety of incubation conditions and substrates amendments can be made. Manipulation of soil bacterial communities has led to the recovery of novel bacterial genomes using metagenomic analysis (Delmont et al. 2015). This study proves the principle of enrichment as the majority of bacteria in soil will be novel. The interaction between plants roots with microbial communities in the rhizosphere can change in relation to the plant type because of differences in the plants exudates as mentioned previously in this chapter. The study of these hotspots in soil could be of particular interest for the recovery of bacteria with varying mechanisms of competition involving antagonism.

The biogeography of microbial diversity has been exploited in the past and continues to be of interest in the development of metagenomics. Finlay (2002) argued that the small size of prokaryotes would result in their ubiquitous distribution and follow the well-known paradigm 'everything is everywhere, but the environment selects', therefore the only difference amongst environments is in the relative abundance of each species in relation to the environmental conditions (De Wit and Bouvier 2006). However, poorly represented groups will not be detected by metagenomics and therefore biogeography will provide a vital clue for the exploitation of different groups (Keller and Zengler 2004). An initial study conducted by Charlop-Powers et al. (2014) on 96 soil samples revealed that there is a strong correlation between soil type characteristics and secondary metabolite biosynthetic diversity based on PKS ketosynthase domain and NRPS adenylation domain amplicon sequencing. An extended study on a more global scale was performed by the same research group on 185 soil samples representing different environments such as coastal sediments, deserts, rainforests and temperate forests collected through the project "drugs from dirt" (www.drugsfromdirt.org). The analysis of the same biosynthetic domains (KS and A) showed that there was a stronger relationship between soil samples collected at a closer physical distance than from similar biome environments with different geographic locations (Charlop-Powers et al. 2015). Moreover, interesting "hotspots" for biomedically relevant compounds and specific gene clusters have been identified using eSNaPD analysis (Reddy et al. 2014). Biogeographical studies using metagenomic techniques are likely to become an essential tool in the race to discover novel antibiotics for combating the alarming problem of antibiotic resistance.

6.5 Conclusions

Drug discovery has always focussed on exploiting natural biodiversity using the latest technologies available to achieve this. One problem that still needs to be addressed is whether the newer DNA based technologies will lead to an extensive catalogue of clinically important antibiotics. The relationships between microbes and the roles of many metabolites in the soil is still unclear or unknown and the soil remains an environment with many challenges for detailed studies of community function in relation to the extensive diversity discussed in this chapter. Natural product antibiotics are an invaluable resource for medical treatments but the role they play in nature is still uncertain. The awakening of cryptic or silent antibiotic gene clusters by co-culture with other specific strains suggests that antibiotics do play a role in microbial community interactions (Seyedsayamdost et al. 2012). Understanding the physiological role of these compounds would promote a better design of high-throughput screening programmes facilitating discovery. In tandem with microbial diversity exploitation the problem of microbial community extinction needs to be addressed. Pristine environments such as the Antarctic could be a source of novel genes because of the extreme conditions that shaped these different microbial communities (Pearce et al. 2012; Cowan et al. 2014). Global climate change and increasing human activities could result in extinction of part of this precious microbial biodiversity. Finally, the production of extensive metagenomic libraries will go someway towards preserving genetic diversity for future generations and will provide a blueprint of the community structure and function.

References

Adrio JL, Demain AL (2014) Microbial enzymes: tools for biotechnological processes. Biomolecules 4:117–139

Allen EE, Banfield JF (2005) Community genomics in microbial ecology and evolution. Nat Rev Microbiol 3:489–498

Allen NE, Hobbs JN, Alborn WE Jr (1987) Inhibition of peptidoglycan biosynthesis in gram-positive bacteria by LY146032. Antimicrob Agents Chemother 31:1093–1099

Allen HK, Moe LA, Rodbumrer J, Gaarder A, Handelsman J (2008) Functional metagenomics reveals diverse [beta]-lactamases in a remote alaskan soil. *ISME J* 3:243–251

ANON (1952) Erythromycin. Br Med J 2:1085–1086

ANON. 1992. Convention on biological diversity

Ayuso-Sacido A, Genilloud O (2005) New PCR primers for the screening of NRPS and PKS-I systems in actinomycetes: detection and distribution of these biosynthetic gene sequences in major taxonomic groups. Microb Ecol 49:10–24

Baker BJ, Tyson GW, Webb RI, Flanagan J, Hugenholtz P, Allen EE, Banfield JF (2006) Lineages of acidophilic archaea revealed by community genomic analysis. Science 314:1933–1935

Bakken, L. R. & Lindahl, V. 1995. Recovery of bacterial cells from soil. *In:* JD Van Elsas, JT Trevors (eds.) Nucleic acids in the environment: methods and applications. Heidelberg: Springer.

Baltz RH (2010) *Streptomyces* and *Saccharopolyspora* hosts for heterologous expression of secondary metabolite gene clusters. J Ind Microbiol Biotechnol 37:759–772

Banik JJ, Brady SF (2008) Cloning and characterization of new glycopeptide gene clusters found in an environmental DNA megalibrary. Proc Natl Acad Sci U S A 105:17273–17277

Bardgett RD, van der Putten WH (2014) Below ground biodiversity and ecosystem functioning. Nature 515:505–511

Bardgett RD, Mommer L, de Vries FT (2014) Going underground: root traits as drivers of ecosystem processes. Trends Ecol Evol 29:692–699

Bauer JD, King RW, Brady SF (2010) Utahmycins A and B, azaquinones produced by an environmental DNA clone. J Nat Prod 73:976–979

Berdy J (2012) Thoughts and facts about antibiotics: where we are now and where we are heading. J Antibiot 65:385–395

Berendsen RL, Pieterse CMJ, Bakker PAHM (2012) The rhizosphere microbiome and plant health. Trends Plant Sci 17:478–486

Berg G, Smalla K (2009) Plant species and soil type cooperatively shape the structure and function of microbial communities in the rhizosphere. FEMS Microbiol Ecol 68:1–13

Berry AE, Chiocchini C, Selby T, Sosio M, Wellington EM (2003) Isolation of high molecular weight DNA from soil for cloning into BAC vectors. FEMS Microbiol Lett 223:15–20

Brady SF (2007) Construction of soil environmental DNA cosmid libraries and screening for clones that produce biologically active small molecules. Nat Protoc 2:1297–1305

Braid MD, Daniels LM, Kitts CL (2003) Removal of PCR inhibitors from soil DNA by chemical flocculation. J Microbiol Methods 52:389–393

Branton D, Deamer DW, Marziali A, Bayley H, Benner SA, Butler T, di Ventra M, Garaj S, Hibbs A, Huang X, Jovanovich SB, Krstic PS, Lindsay S, Ling XS, Mastrangelo CH, Meller A, Oliver JS, Pershin YV, Ramsey JM, Riehn R, Soni GV, Tabard-Cossa V, Wanunu M, Wiggin M, Schloss JA (2008) The potential and challenges of nanopore sequencing. Nat Biotechnol 26:1146–1153

Brigham RB, Pittenger RC (1956) *Streptomyces orientalis*, n. Sp., the source of vancomycin. Antibiot Chemother (Northfield) 6:642–647

Bulgarelli D, Rott M, Schlaeppi K, van Themaat EVL, Ahmadinejad N, Assenza F, Rauf P, Huettel B, Reinhardt R, Schmelzer E, Peplies J, Gloeckner FO, Amann R, Eickhorst T, Schulze-Lefert P (2012) Revealing structure and assembly cues for Arabidopsis root-inhabiting bacterial microbiota. Nature 488:91–95

Bürgmann H, Pesaro M, Widmer F, Zeyer J (2001) A strategy for optimizing quality and quantity of DNA extracted from soil. J Microbiol Methods 45:7–20
Challis GL, Ravel J, Townsend CA (2000) Predictive, structure-based model of amino acid recognition by nonribosomal peptide synthetase adenylation domains. Chem Biol 7:211–224
Charlop-Powers Z, Owen JG, Reddy BV, Ternei MA, Brady SF (2014) Chemical-biogeographic survey of secondary metabolism in soil. Proc Natl Acad Sci U S A 111:3757–3762
Charlop-Powers Z, Owen JG, Reddy BVB, Ternei MA, Guimarães DO, de Frias UA, Pupo MT, Seepe P, Feng Z, Brady SF (2015) Global biogeographic sampling of bacterial secondary metabolism. Elife 4:e05048
Cole JR, Wang Q, Fish JA, Chai B, Mcgarrell DM, Sun Y, Brown CT, Porras-Alfaro A, Kuske CR, Tiedje JM (2014) Ribosomal database project: data and tools for high throughput rrna analysis. Nucleic Acids Res 42:D633–D642
Collins J, Hohn B (1978) Cosmids: a type of plasmid gene-cloning vector that is packageable in vitro in bacteriophage lambda heads. Proc Natl Acad Sci U S A 75:4242–4246
Cowan DA, Makhalanyane TP, Dennis PG, Hopkins DW (2014) Microbial ecology and biogeochemistry of continental Antarctic soils. Front Microbiol 5:154
Craig JW, Chang FY, Kim JH, Obiajulu SC, Brady SF (2010) Expanding small-molecule functional metagenomics through parallel screening of broad-host-range cosmid environmental DNA libraries in diverse proteobacteria. Appl Environ Microbiol 76:1633–1641
Cretoiu, M., Berini, F., Kielak, A., Marinelli, F. & van Elsas, J. 2015. A novel salt-tolerant chitobiosidase discovered by genetic screening of a metagenomic library derived from chitin-amended agricultural soil. Appl Microbiol Biotechnol, 1–17
Daniel R (2005) The metagenomics of soil. Nat Rev Microbiol 3:470–478
de Wit R, Bouvier T (2006) 'Everything is everywhere, but, the environment selects'; what did Baas Becking and Beijerinck really say? Environ Microbiol 8:755–758
Delmont TO, Eren AM, Maccario L, Prestat E, Esen OC, Pelletier E, Le Paslier D, Simonet P, Vogel TM (2015) Reconstructing rare soil microbial genomes using in situ enrichments and metagenomics. Front Microbiol 6:358
Delong EF (2005) Microbial community genomics in the ocean. Nat Rev Microbiol 3:459–469
Demain AL, Sanchez S (2009) Microbial drug discovery: 80 years of progress. J Antibiot 62:5–16
Derakshani M, Lukow T, Liesack W (2001) Novel bacterial lineages at the (sub)division level as detected by signature nucleotide-targeted recovery of 16S rrna genes from bulk soil and rice roots of flooded rice microcosms. Appl Environ Microbiol 67:623–631
Desantis TZ, Hugenholtz P, Larsen N, Rojas M, Brodie EL, Keller K, Huber T, Dalevi D, Hu P, Andersen GL (2006) Greengenes, a chimera-checked 16S rrna gene database and workbench compatible with ARB. Appl Environ Microbiol 72:5069–5072
Donadio S, Monciardini P, Sosio M (2007) Polyketide synthases and nonribosomal peptide synthetases: the emerging view from bacterial genomics. Nat Prod Rep 24:1073–1109
Drews J (2000) Drug discovery: a historical perspective. Science 287:1960–1964
Duggar BM (1948) Aureomycin: a product of the continuing search for new antibiotics. Ann N Y Acad Sci 1241:163–169
Ehrlich J, Gottlieb D et al (1948) *Streptomyces venezuelae*, n. Sp., the source of chloromycetin. J Bacteriol 56:467–477
Elsaesser R, Paysan J (2004) Liquid gel amplification of complex plasmid libraries. Biotechniques 37(200):202
Esnard J, Potter TL, Zuckerman BM (1995) *Streptomyces costaricanus* sp. Nov., isolated from nematode-suppressive soil. Int J Syst Bacteriol 45:775–779
Feng Z, Chakraborty D, Dewell SB, Reddy BV, Brady SF (2012) Environmental DNA-encoded antibiotics fasamycins A and B inhibit fabf in type II fatty acid biosynthesis. J Am Chem Soc 134:2981–2987
Feng Z, Kim JH, Brady SF (2010) Fluostatins produced by the heterologous expression of a TAR reassembled environmental DNA derived type II PKS gene cluster. J Am Chem Soc 132:11902–11903

Fierer N, Bradford MA, Jackson RB (2007) Toward an ecological classification of soil bacteria. Ecology 88:1354–1364
Fierer N, Leff JW, Adams BJ, Nielsen UN, Bates ST, Lauber CL, Owens S, Gilbert JA, Wall DH, Caporaso JG (2012) Cross-biome metagenomic analyses of soil microbial communities and their functional attributes. Proc Natl Acad Sci U S A 109:21390–21395
FIERER N, SCHIMEL JP, HOLDEN PA (2003) Variations in microbial community composition through two soil depth profiles. Soil Biol Biochem 35:167–176
Finlay BJ (2002) Global dispersal of free-living microbial eukaryote species. Science 296:1061–1063
Fleming A (1929) On the antibacterial action of cultures of a Penicillium, with special reference to their use in the isolation of *B. Influenzae*. 1929. Bull World Health Organ 79:780–790
Frearson J, Wyatt P (2010) Drug discovery in academia- the third way? Expert Opin Drug Discovery 5:909–919
Gabor EM, Alkema WB, Janssen DB (2004) Quantifying the accessibility of the metagenome by random expression cloning techniques. Environ Microbiol 6:879–886
George IF, Hartmann M, Liles MR, Agathos SN (2011) Recovery of as-yet-uncultured soil acidobacteria on dilute solid media. Appl Environ Microbiol 77:8184–8188
Gerwick WH, Moore BS (2012) Lessons from the past and charting the future of marine natural products drug discovery and chemical biology. Chem Biol 19:85–98
Glenn TC (2011) Field guide to next-generation DNA sequencers. Mol Ecol Resour 11:759–769
Gomez-Escribano JP, Bibb MJ (2011) Engineering *Streptomyces coelicolor* for heterologous expression of secondary metabolite gene clusters. Microb Biotechnol 4:207–215
Gullon S, Olano C, Abdelfattah MS, Brana AF, Rohr J, Mendez C, Salas JA (2006) Isolation, characterization, and heterologous expression of the biosynthesis gene cluster for the antitumor anthracycline steffimycin. Appl Environ Microbiol 72:4172–4183
Handelsman J (2004) Metagenomics: application of genomics to uncultured microorganisms. Microbiol Mol Biol Rev 68:669–685
Handelsman J, Rondon MR, Brady SF, Clardy J, Goodman RM (1998) Molecular biological access to the chemistry of unknown soil microbes: a new frontier for natural products. Chem Biol 5:R245–R249
Harbarth S, Theuretzbacher U, Hackett J, Consortium D-A (2015) Antibiotic research and development: business as usual? J Antimicrob Chemother 70:1604–1607
Heath RJ, Rock CO (2004) Fatty acid biosynthesis as a target for novel antibacterials. Curr Opin Investig Drugs 5:146–153
Herron PR, Wellington EM (1990) New method for extraction of streptomycete spores from soil and application to the study of lysogeny in sterile amended and nonsterile soil. Appl Environ Microbiol 56:1406–1412
Hoffmann A, Thimm T, Tebbe CC (1999) Fate of plasmid-bearing, luciferase marker gene tagged bacteria after feeding to the soil microarthropod *Onychiurus fimatus* (Collembola). FEMS Microbiol Ecol 30:125–135
Hrvatin S, Piel J (2007) Rapid isolation of rare clones from highly complex DNA libraries by PCR analysis of liquid gel pools. J Microbiol Methods 68:434–436
Hunter S, Corbett M, Denise H, Fraser M, Gonzalez-Beltran A, Hunter C, Jones P, Leinonen R, Mcanulla C, Maguire E, Maslen J, Mitchell A, Nuka G, Oisel A, Pesseat S, Radhakrishnan R, Rocca-Serra P, Scheremetjew M, Sterk P, Vaughan D, Cochrane G, Field D, Sansone SA (2014) EBI metagenomics – a new resource for the analysis and archiving of metagenomic data. Nucleic Acids Res 42:D600–D606
Ioannou PA, Amemiya CT, Garnes J, Kroisel PM, Shizuya H, Chen C, Batzer MA, De Jong PJ (1994) A new bacteriophage P1-derived vector for the propagation of large human DNA fragments. Nat Genet 6:84–89
Jacquiod S, Franqueville L, Cécillon S, Vogel TM, Simonet P (2013) Soil bacterial community shifts after chitin enrichment: an integrative metagenomic approach. PLoS One 8:e79699
Jevons MP (1961) "Celbenin" – resistant Staphylococci. BMJ 1:124–125

Johnson-Rollings AS, Wright H, Masciandaro G, Macci C, Doni S, Calvo-Bado LA, Slade SE, Vallin Plou C, Wellington EMH (2014) Exploring the functional soil-microbe interface and exoenzymes through soil metaexoproteomics. ISME J 8:2148–2150

Johnstone DB (1947) Soil actinomycetes of Bikini atoll, with special reference to an antibiotic-producing organism. J Bacteriol 54:25

Johnstone DB, Waksman SA (1948) The production of streptomycin by *Streptomyces bikiniensis*. J Bacteriol 55:317–326

Kallifidas D, Kang HS, Brady SF (2012) Tetarimycin A, an MRSA-active antibiotic identified through induced expression of environmental DNA gene clusters. J Am Chem Soc 134:19552–19555

Kang HS, Brady SF (2014) Arixanthomycins A-C: phylogeny-guided discovery of biologically active edna-derived pentangular polyphenols. ACS Chem Biol 9:1267–1272

Kang S, Mills AL (2006) The effect of sample size in studies of soil microbial community structure. J Microbiol Methods 66:242–250

Keller M, Zengler K (2004) Tapping into microbial diversity. Nat Rev Microbiol 2:141–150

Kim UJ, Shizuya H, De Jong PJ, Birren B, Simon MI (1992) Stable propagation of cosmid sized human DNA inserts in an F factor based vector. Nucleic Acids Res 20:1083–1085

Knight R, Jansson J, Field D, Fierer N, Desai N, Fuhrman JA, Hugenholtz P, Van Der Lelie D, Meyer F, Stevens R, Bailey MJ, Gordon JI, Kowalchuk GA, Gilbert JA (2012) Unlocking the potential of metagenomics through replicated experimental design. Nat Biotechnol 30:513–520

Laskaris, P. 2009. *Evolution of the streptomycin and viomycin biosynthetic clusters and resistance genes*. Thesis (Ph.D.) – University of Warwick, 2009

Lee MH, Lee SW (2013) Bioprospecting potential of the soil metagenome: novel enzymes and bioactivities. Genome Inform 11:114–120

Levy SB, Marshall B (2004) Antibacterial resistance worldwide: causes, challenges and responses. Nat Med 10:S122–S129

Li WJ, Zhang LP, Xu P, Cui XL, Lu ZT, Xu LH, Jiang CL (2002) *Streptomyces beijiangensis* sp. Nov., a psychrotolerant actinomycete isolated from soil in China. Int J Syst Evol Microbiol 52:1695–1699

Ling LL, Schneider T, Peoples AJ, Spoering AL, Engels I, Conlon BP, Mueller A, Schaberle TF, Hughes DE, Epstein S, Jones M, Lazarides L, Steadman VA, Cohen DR, Felix CR, Fetterman KA, Millett WP, Nitti AG, Zullo AM, Chen C, Lewis K (2015) A new antibiotic kills pathogens without detectable resistance. Nature 517:455–459

Linger JG, Adney WS, Darzins A (2010) Heterologous expression and extracellular secretion of cellulolytic enzymes by *Zymomonas mobilis*. Appl Environ Microbiol 76:6360–6369

Logares, R., Sunagawa, S., Salazar, G., Cornejo-Castillo, F. M., Ferrera, I., Sarmento, H., Hingamp, P., Ogata, H., De Vargas, C., Lima-Mendez, G., Raes, J., Poulain, J., Jaillon, O., Wincker, P., Kandels-Lewis, S., Karsenti, E., Bork, P. & Acinas, S. G. 2013. Metagenomic 16S rdna Illumina tags are a powerful alternative to amplicon sequencing to explore diversity and structure of microbial communities. Environ Microbiol

Loman NJ, Misra RV, Dallman TJ, Constantinidou C, Gharbia SE, Wain J, Pallen MJ (2012) Performance comparison of benchtop high-throughput sequencing platforms. Nat Biotechnol 30:434–439

Lombo F, Velasco A, Castro A, De La Calle F, Brana AF, Sanchez-Puelles JM, Mendez C, Salas JA (2006) Deciphering the biosynthesis pathway of the antitumor thiocoraline from a marine actinomycete and its expression in two streptomyces species. Chembiochem 7:366–376

Luo C, Tsementzi D, Kyrpides N, Read T, Konstantinidis KT (2012) Direct comparisons of Illumina vs. Roche 454 sequencing technologies on the same microbial community DNA sample. PLoS One 7:e30087

Mackey TK, Liang BA (2012) Integrating biodiversity management and indigenous biopiracy protection to promote environmental justice and global health. Am J Public Health 102:1091–1095

Markowitz VM, Chen IM, Chu K, Szeto E, Palaniappan K, Pillay M, Ratner A, Huang J, Pagani I, Tringe S, Huntemann M, Billis K, Varghese N, Tennessen K, Mavromatis K, Pati A, Ivanova NN, Kyrpides NC (2014) IMG/M 4 version of the integrated metagenome comparative analysis system. Nucleic Acids Res 42:D568–D573

Marschner P, Yang CH, Lieberei R, Crowley DE (2001) Soil and plant specific effects on bacterial community composition in the rhizosphere. Soil Biol Biochem 33:1437–1445

Medema MH, Blin K, Cimermancic P, De Jager V, Zakrzewski P, Fischbach MA, Weber T, Takano E, Breitling R (2011) Antismash: rapid identification, annotation and analysis of secondary metabolite biosynthesis gene clusters in bacterial and fungal genome sequences. Nucleic Acids Res 39:W339–W346

Metzker ML (2010) Sequencing technologies - the next generation. Nat Rev Genet 11:31–46

Meyer F, Paarmann D, D'souza M, Olson R, Glass EM, Kubal M, Paczian T, Rodriguez A, Stevens R, Wilke A, Wilkening J, Edwards RA (2008) The metagenomics RAST server – a public resource for the automatic phylogenetic and functional analysis of metagenomes. BMC Bioinf 9:386

Meyers PR, Goodwin CM, Bennett JA, Aken BL, Price CE, Van Rooyen JM (2004) *Streptomyces africanus* sp. Nov., a novel streptomycete with blue aerial mycelium. Int J Syst Evol Microbiol 54:1531–1535

Morris NZ, Standage S, Wellington EMH (1999) Extraction of DNA from spores in soil. In: Van Elsas JD, Akkermans A, De Bruijn FJ (eds) Molecular microbial ecology manual. Kluwer, Dordrecht

Moy TI, Conery AL, Larkins-Ford J, Wu G, Mazitschek R, Casadei G, Lewis K, Carpenter AE, Ausubel FM (2009) High-throughput screen for novel antimicrobials using a whole animal infection model. ACS Chem Biol 4:527–533

Nichols D, Cahoon N, Trakhtenberg EM, Pham L, Mehta A, Belanger A, Kanigan T, Lewis K, Epstein SS (2010) Use of ichip for high-throughput in situ cultivation of "uncultivable" microbial species. Appl Environ Microbiol 76:2445–2450

Nybond S, Karp M, Tammela P (2013) Antimicrobial assay optimization and validation for HTS in 384-well format using a bioluminescent *E. Coli* K-12 strain. Eur J Pharm Sci 49:782–789

Owen JG, Reddy BV, Ternei MA, Charlop-Powers Z, Calle PY, Kim JH, Brady SF (2013) Mapping gene clusters within arrayed metagenomic libraries to expand the structural diversity of biomedically relevant natural products. Proc Natl Acad Sci U S A 110:11797–11802

Parsley LC, Linneman J, Goode AM, Becklund K, George I, Goodman RM, Lopanik NB, Liles MR (2011) Polyketide synthase pathways identified from a metagenomic library are derived from soil *Acidobacteria*. FEMS Microbiol Ecol 78:176–187

Pearce DA, Newsham KK, Thorne MA, Calvo-Bado L, Krsek M, Laskaris P, Hodson A, Wellington EM (2012) Metagenomic analysis of a southern maritime antarctic soil. Front Microbiol 3:403

Pivetal J, Toru S, Frenea-Robin M, Haddour N, Cecillon S, Dempsey NM, Dumas-Bouchiat F, Simonet P (2014) Selective isolation of bacterial cells within a microfluidic device using magnetic probe-based cell fishing. Sensors Actuators B Chem 195:581–589

Quail MA, Smith M, Coupland P, Otto TD, Harris SR, Connor TR, Bertoni A, Swerdlow HP, Gu Y (2012) A tale of three next generation sequencing platforms: comparison of ion torrent, Pacific biosciences and illumina miseq sequencers. BMC Genomics 13:341

Rajamuthiah R, Fuchs BB, Jayamani E, Kim Y, Larkins-Ford J, Conery A, Ausubel FM, Mylonakis E (2014) Whole animal automated platform for drug discovery against multi-drug resistant *Staphylococcus aureus*. PLoS One 9:e89189

Reddy BVB, Milshteyn A, Charlop-Powers Z, Brady SF (2014) Esnapd: a versatile, web-based bioinformatics platform for surveying and mining natural product biosynthetic diversity from metagenomes. Chem Biol 21:1023–1033

Rondon MR, August PR, Bettermann AD, Brady SF, Grossman TH, Liles MR, Loiacono KA, Lynch BA, Macneil IA, Minor C, Tiong CL, Gilman M, Osburne MS, Clardy J, Handelsman J, Goodman RM (2000) Cloning the soil metagenome: a strategy for accessing the genetic and functional diversity of uncultured microorganisms. Appl Environ Microbiol 66:2541–2547

Rusch DB, Halpern AL, Sutton G, Heidelberg KB, Williamson S, Yooseph S, wu D, Eisen JA, Hoffman JM, Remington K, Beeson K, Tran B, Smith H, Baden-Tillson H, Stewart C, Thorpe J, Freeman J, Andrews-Pfannkoch C, Venter JE, Li K, Kravitz S, Heidelberg JF, Utterback T, Rogers Y-H, Falcón LI, Souza V, Bonilla-Rosso G, Eguiarte LE, Karl DM, Sathyendranath S, Platt T, Bermingham E, Gallardo V, Tamayo-Castillo G, Ferrari MR, Strausberg RL, Nealson K, Friedman R, Frazier M, Venter JC (2007) The Sorcerer II global ocean sampling expedition: northwest Atlantic through eastern tropical Pacific. PLoS Biol 5:e77

Schatz A, Waksman SA (1944) Effect of streptomycin and other antibiotic substances upon *Mycobacterium tuberculosis* and related organisms. Exp Biol Med 57:244–248

Schirmer A, Gadkari R, Reeves CD, Ibrahim F, Delong EF, Hutchinson CR (2005) Metagenomic analysis reveals diverse polyketide synthase gene clusters in microorganisms associated with the marine sponge *Discodermia dissoluta*. Appl Environ Microbiol 71:4840–4849

Seyedsayamdost MR, Traxler MF, Clardy J, Kolter R (2012) Old meets new: using interspecies interactions to detect secondary metabolite production in actinomycetes. Methods Enzymol 517:89–109

Sheridan C (2005) EPO neem patent revocation revives biopiracy debate. Nat Biotechnol 23:511–512

Shizuya H, Birren B, Kim UJ, Mancino V, Slepak T, Tachiiri Y, Simon M (1992) Cloning and stable maintenance of 300-kilobase-pair fragments of human DNA in *Escherichia coli* using an F-factor-based vector. Proc Natl Acad Sci U S A 89:8794–8797

Sosio M, Giusino F, Cappellano C, Bossi E, Puglia AM, Donadio S (2000) Artificial chromosomes for antibiotic-producing actinomycetes. Nat Biotechnol 18:343–345

Srinivasan VB, Singh BB, Priyadarshi N, Chauhan NK, Rajamohan G (2014) Role of novel multidrug efflux pump involved in drug resistance in *Klebsiella pneumoniae*. PLoS One 9:e96288

Stein JL, Marsh TL, Wu KY, Shizuya H, Delong EF (1996) Characterization of uncultivated prokaryotes: isolation and analysis of a 40-kilobase-pair genome fragment from a planktonic marine archaeon. J Bacteriol 178:591–599

Stevenson BS, Eichorst SA, Wertz JT, Schmidt TM, Breznak JA (2004) New strategies for cultivation and detection of previously uncultured microbes. Appl Environ Microbiol 70:4748–4755

Strieker M, Tanović A, Marahiel MA (2010) Nonribosomal peptide synthetases: structures and dynamics. Curr Opin Struct Biol 20:234–240

Tan GY, Robinson S, Lacey E, Goodfellow M (2006) *Amycolatopsis australiensis* sp. Nov., an actinomycete isolated from arid soils. Int J Syst Evol Microbiol 56:2297–2301

Tebbe CC, Vahjen W (1993) Interference of humic acids and DNA extracted directly from soil in detection and transformation of recombinant DNA from bacteria and a yeast. Appl Environ Microbiol 59:2657–2665

Thaker MN, Waglechner N, Wright GD (2014) Antibiotic resistance–mediated isolation of scaffold-specific natural product producers. Nat Protoc 9:1469–1479

Tiedje, J. M. C., J.C. Cho; Murray, A.; Treves, D.; Xia, B.; Zhou, J.; 2001. Soil teeming with life: new frontiers for soil science. In*:* Rees, RM, B Ball, C. Watson, CD Campbell (ed.) Sustainable management of soil organic matter. CAB International

Torsvik V, Daae FL, Sandaa RA, Ovreas L (1998) Novel techniques for analysing microbial diversity in natural and perturbed environments. J Biotechnol 64:53–62

Torsvik V, Sørheim R, Goksøyr J (1996) Total bacterial diversity in soil and sediment communities—a review. J Ind Microbiol 17:170–178

Tseng M, Yang SF, Li WJ, Jiang CL (2006) *Amycolatopsis taiwanensis* sp. Nov., from soil. Int J Syst Evol Microbiol 56:1811–1815

van Dijk EL, Auger H, Jaszczyszyn Y, Thermes C (2014) Ten years of next-generation sequencing technology. Trends Genet 30:418–426

Velayati AA, Masjedi MR, Farnia P, Tabarsi P, Ghanavi J, Ziazarifi AH, Hoffner SE (2009) Emergence of new forms of totally drug-resistant tuberculosis bacilli: super extensively drug-resistant tuberculosis or totally drug-resistant strains in iran. Chest 136:420–425

Vester JK, Glaring MA, Stougaard P (2015) Improved cultivation and metagenomics as new tools for bioprospecting in cold environments. Extremophiles 19:17–29

Vinje H, Liland KH, Almoy T, Snipen L (2015) Comparing K-mer based methods for improved classification of 16S sequences. BMC Bioinf 16:205
Vogel TM, Simonet P, Jansson JK, Hirsch PR, Tiedje JM, van Elsas JD, Bailey MJ, Nalin R, Philippot L (2009) Terragenome: a consortium for the sequencing of a soil metagenome. Nat Rev Microbiol 7:252–252
Waksman SA, Reilly HC, Harris DA (1948) *Streptomyces griseus* (Krainsky) Waksman and Henrici. *J Bacteriol* 56:259–269
Wall DH, Bardgett RD, Kelly E (2010) Biodiversity in the dark. Nat Geosci 3:297–298
Wang J, Soisson SM, Young K, Shoop W, Kodali S, Galgoci A, Painter R, Parthasarathy G, Tang YS, Cummings R, Ha S, Dorso K, Motyl M, Jayasuriya H, Ondeyka J, Herath K, Zhang C, Hernandez L, Allocco J, Basilio A, Tormo JR, Genilloud O, Vicente F, Pelaez F, Colwell L, Lee SH, Michael B, Felcetto T, Gill C, Silver LL, Hermes JD, Bartizal K, Barrett J, Schmatz D, Becker JW, Cully D, Singh SB (2006) Platensimycin is a selective fabf inhibitor with potent antibiotic properties. Nature 441:358–361
Wang Y, Yang Q, Wang Z (2014) The evolution of nanopore sequencing. Front Genet 5:449
WHO 2014. *Antimicrobial resistance: global report on surveillance 2014*.
Wright GD, Poinar H (2012) Antibiotic resistance is ancient: implications for drug discovery. Trends Microbiol 20:157–159
Zengler K, Toledo G, Rappe M, Elkins J, Mathur EJ, Short JM, Keller M (2002) Cultivating the uncultured. Proc Natl Acad Sci U S A 99:15681–15686
Zhang H, Wang Y, Wu J, Skalina K, Pfeifer BA (2010) Complete biosynthesis of erythromycin A and designed analogs using *E Coli* as a heterologous host. Chem Biol 17:1232–1240
Zhou J, Bruns MA, Tiedje JM (1996) DNA recovery from soils of diverse composition. Appl Environ Microbiol 62:316–322
Ziemert N, Podell S, Penn K, Badger JH, Allen E, Jensen PR (2012) The natural product domain seeker napdos: a phylogeny based bioinformatic tool to classify secondary metabolite gene diversity. PLoS One 7:e34064

Chapter 7
Biotechnological Applications of the *Roseobacter* Clade

Mikkel Bentzon-Tilia and Lone Gram

Abstract The multitude of distinct niches that prevail in the marine environment has facilitated the development of very diverse marine microbiomes. This diversity is, naturally, reflected in their biochemistry and secondary metabolites and, hence, marine microbes represent a virtually untapped source of new bioactive compounds. The *Roseobacter* clade of marine α-proteobacteria represents some of the most abundant organisms in the marine environment and they may constitute as much as 20–30 % of the prokaryotic community during algal blooms. Often, they exhibit traits suggestive of a lifestyle in close association with phytoplankton; including traits related to surface colonization, iron scavenging, and the production of bioactive secondary metabolites. Despite the fact that relatively few bioactive compounds have been identified in the α-proteobacteria, the roseobacters are known to produce compounds capable of stimulating algae growth, i.e. auxins, and algaecidal compounds, i.e. the roseobacticides. In addition, the roseobacters can produce a range of antibacterial products, such as the small tropolone compound tropodithietic acid (TDA) and the nonribosomal peptide indigoidine. TDA targets a broad spectrum of Gram-positive and Gram-negative bacteria in which resistance towards the compound does not arise easily. Mining the genomes of roseobacters also reveal that they are likely capable of producing other compounds than hitherto discovered by classical bio-assay guided fractionation, since the genomes contain genes/gene clusters probably encoding unknown bioactive secondary metabolites. Therefore, bacteria of the *Roseobacter* clade may serve as potential sources of novel bioactive compounds, including novel antibiotics, which is of paramount importance in the battle against antibiotic resistant pathogenic bacteria.

The discovery of new antibiotic compounds is not the only means by which we can counter the spread of antibiotic resistance. Development of sustainable alternatives to the application of antibiotics in agri- and aquaculture may be equally important. Attributable to their inherent properties, the roseobacters may be such an alternative in the aquaculture industry. Especially at the younger stages in larviculture, disease outbreaks caused by fish pathogenic microorganisms may lead to

M. Bentzon-Tilia (✉) • L. Gram
Department of Biotechnology and Biomedicine, Technical University of Denmark, Matematiktorvet Bldg. 301, DK-2800 Kgs, Lyngby, Denmark
e-mail: mibti@bio.dtu.dk

R. Paterson, N. Lima (eds.), *Bioprospecting*, Topics in Biodiversity and Conservation 16, DOI 10.1007/978-3-319-47935-4_7

mortality rates of 100 % when antibiotic treatment is not initiated. Adding roseobacters as probiotics is promising as fish larvae challenged with fish pathogens of the genus *Vibrio* exhibit survival rates similar to, or better than, unchallenged larvae when roseobacter probionts are added. Thus, the *Roseobacter* clade is a promising source of new bioactive compounds and a possible sustainable alternative to the prophylactic administration of antibiotics in fish rearing.

Abbreviations

AHL	Acyl-homoserine lactone
c-di-GMP	Cyclic dimeric guanosinmonophosphate
DGGE	Denaturing gradient gel electrophoresis
EPS	Extracellular polymeric substance
FAO	Food and Agriculture Organization of the United Nations
GI	Gastro intestinal
MRSA	Methicillin-resistant Staphylococcus aureus
MIC	Minimal inhibitory concentration
MA	Marine agar
NRPS	Nonribosomal peptide synthetase
PSP	Paralytic shellfish poisoning
PKS	Polyketide synthase
QS	Quorum sensing
ROD	Roseovarius oyster disease
TDA	Tropodithietic acid

7.1 Bioactive Compounds in Marine Bacteria

Marine environments represent the most extensive habitats on Earth, covering 70 % of the surface and, hence, they arguably comprise the majority of Earth's biosphere. Microbes inhabiting these environments are not only the most abundant organisms on the planet with approximately 10^{30} cells (Whitman et al. 1998), but they are also exhibiting a multitude of different metabolic pathways and trophic strategies. This diversity is largely fueled by the multitude of environmental niches that prevail in marine systems. From hydrothermal vent environments to arctic sea ice brine pockets, microbes have adapted and flourish. Naturally, these adaptations manifest themselves in the diversity of the bioactive compounds produced, making the marine microbiome a promising source of enzymes and other compounds with novel pharmaceutical and biotechnological properties.

7.1.1 Marine Microbial Enzymes

Not only are microbial enzymes stable and easy to produce compared to enzymes derived from macro-organisms, but heterotrophic marine bacteria are specialized in converting complex organic matter to energy and biomass through a host of different intra- and extracellular enzymes. Hence, marine microbial enzymes involved in the degradation of biological polymers are promising candidates for bioprospecting efforts.

An example of such a biopolymer is chitin (β-1,4-linked-*N*-acetyl-glucosamine), which is the primary constituent of arthropod exoskeletons and fungal cell walls. This polysaccharide is the second most abundant biopolymer on Earth after cellulose (Gooday 1990) and it is extremely abundant in the sea with an estimated annual production in the order of 10^{10}–10^{11} metric tons (Gooday 1990; Keyhani and Roseman 1999; Zhang and Kim 2010). Hydrolytic degradation of chitin is catalyzed by a subclass of glycosyl hydrolases known as chitinases. These enzymes are likely ancient and are present in all kingdoms of life (Funkhouser and Aronson 2007), but the organisms primarily responsible for the degradation of chitin in marine environments are most likely heterotrophic bacteria (Keyhani and Roseman 1999; Zobell and Rittenberg 1938). Among others, marine bacteria of the genera *Vibrio*, *Aeromonas*, *Listonella*, *Alteromonas*, and *Pseudoalteromonas* are chitinolytic (Machado et al. 2015; Osama and Koga 1995; Soto-Gil and Zyskind 1984; Techkarnjanaruk and Goodman 1999; Tsujibo et al. 1991; Wortman et al. 1986), and several bacterial derived chitinase genes, including chitinase genes of marine origin, have been expressed heterologously in *E. coli* (Tsujibo et al. 1991), in species of the genus *Pseudomonas* (Suslow and Jones 1988), and in transgenic tobacco plants as a means of avoiding infection by plant pathogenic fungi (Lund et al. 1989).

Biological control of fungal pathogens is, however, only one of the areas of application of chitinases. Various hydrolyzed derivatives of chitin and chitosan, which is the *N*-deacytelated form of chitin, may in fact have properties relating to the elimination of cellular oxidative stress (Ngo et al. 2008, 2009, 2012), lipopolysaccharide inflammatory responses (Vo et al. 2012), and to antimicrobial activity (Je and Kim 2006). Similar to the chitinases, enzymes catalyzing different steps in the degradation of polymers produced by micro- and macro algae, such as alginate, carrageenan, agar, pectin, cellulose and lignin are also widely distributed in marine microbes. These enzymes and their substrates are highly relevant to (a) the food industry (De Ruiter and Rudolph 1997; Wang et al. 2006), (b) the fabric industry (Solbak et al. 2005), (c) biofuel production (Kim et al. 2011), (d) agriculture (Cao et al. 2007; Hien et al. 2000; Iwasaki and Matsubara 2000; Mollah et al. 2009; Yonemoto et al. 1993) and (e) clinical settings (Alkawash et al. 2006; Cotton et al. 2009; Courtois 2009; Iwamoto et al. 2005; Kurachi et al. 2005; Ramsey and Wozniak 2005).

Marine microbial enzymes are adapted to carrying out these processes under unique environmental conditions not necessarily paralleled in terrestrial environments. For instance, organisms that live in permanently cold environments produce enzymes that work effectively in the cold (D'Amico et al. 2002; Feller and Gerday 2003; Georlette et al. 2004; Smalås et al. 2000; Somero 2004), and cold-adaptation is an obvious trait to look for among marine microbial enzymes with 90 % of the ocean being below 5° C. These have the advantage that they remain active at lower temperatures than conventional enzymes from mesophilic microorganisms which has proven to be a preferential trait in a wide range of industries, including the food and feed industry, where enzymatic reactions at lower temperatures help prevent spoilage and unwanted changes in heat-sensitive products (Cavicchioli et al. 2002, 2011; Gerday et al. 2000; Russell 1998; Tutino et al. 2009). Cold-adapted proteases, lipases, and amylases offer a sustainable alternative in the detergent and laundry industries, to harsher chemicals while reducing energy consumption through the decrease in washing temperature. A reduction from 40 to 30° C in household laundry may reduce energy consumption by 30 % (Nielsen 2005).

Marine microbial enzymes may also exhibit properties such as thermostability and hyper-thermostability, or pressure and salinity tolerance - yet others may tolerate high or low pH conditions or combinations of several of these properties (Fernandez 2014), suggesting that the more extreme environments of the deep oceans may be promising hotspots for bioprospecting efforts.

7.1.2 Secondary Metabolites of Marine Microorganisms

The diversity of the marine microbiome is also reflected in the secondary metabolites produced by marine microbes. Secondary metabolites exhibiting antibacterial activity have gained increased attention, in part, due to the fact that the occurrence of antibiotic resistance in pathogenic bacteria is increasing and outpacing the discovery and development of novel antibiotics, which has stagnated significantly over recent decades (Taubes 2008). Spurred by the discovery of penicillin from the fungus *Penicillium chrysogenum* more than eight decades ago, bioprospecting for microbial secondary metabolites has mainly been focused on other filamentous soil microorganisms, which are responsible for the majority of the antibiotics that are currently in clinical use (Murphy et al. 2012). However, the conditions that prevail in the oceans select for traits specific to marine microorganisms and secondary metabolites unique to the marine microbiome may be plentiful.

One of the traits that are characteristic for the small organic molecules produced by marine microbes is the high degree of halogenation compared to terrestrially produced secondary metabolites. In the marine environment, chlorine and bromide are abundant and bromide is, in particular, often incorporated into organic compounds produced in the marine environment. Marinopyrrole B (1; Fig. 7.1) and streptochlorin (2) are examples of such halogenated secondary metabolites produced by marine filamentous bacteria of the genus *Streptomyces* (Hughes et al. 2008; Shin et al. 2007). Compound 1 was shown to exhibit significant inhibitory activity against methicillin-resistant *Staphylococcus aureus* (MRSA) (Hughes et al. 2008), whereas 2 induced apoptosis in human leukemic U937 (Park et al. 2008) and

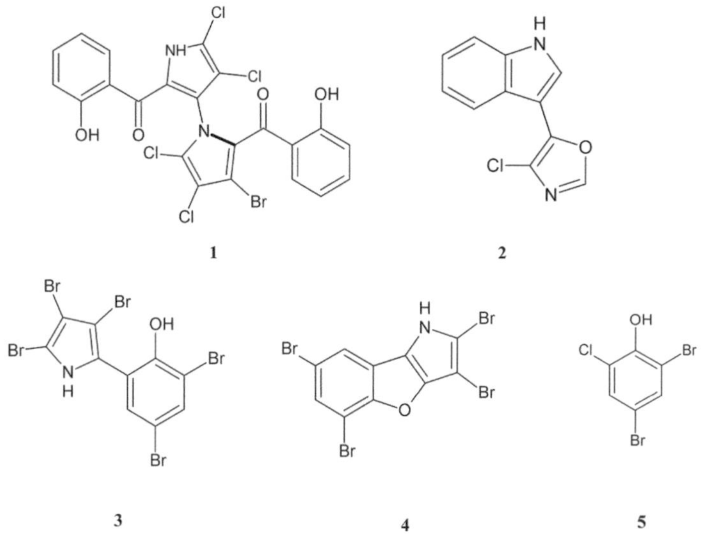

Fig. 7.1 Structures of a selection of halogenated secondary metabolites produced by marine bacteria. *1* Marinopyrrole B, *2* Streptochlorin, *3* Pentabromopseudilin, *4* 2,3,5,7-Tetrabromobenzofuro[3,2-b]pyrrole, *5* 2,4-Dibromo-6-chlorophenol

human hepatocarcinoma cells (Shin et al. 2008), suggesting it could be applied as an anti-cancer drug. In addition, the high degree of halogenation is often seen in metabolites from marine Gram-negative bacteria, for example, the antibacterial compound pentabromopseudilin (3), which was isolated from a marine pseudomonad in the 1960s (Burkholder et al. 1966). The bacterium was subsequently given the name *Pseudomonas bromoutilis* due to the highly brominated nature of this metabolite. In the genus *Pseudoalteromonas*, several halogenated compounds have been isolated (Speitling et al. 2007; Wietz et al. 2013; Yoshikawa et al. 2003), including 2,3,5,7-Tetrabromobenzofuro[3,2-b]pyrrole (4) and 2,4-Dibromo-6-chlorophenol (5), both of which exhibit inhibitory activity against different bacteria including MRSA (Fehér et al. 2010; Jiang et al. 2000).

Halogenation is but one feature of some marine natural products and a detailed description of the immense diversity observed in these small molecules is beyond the scope of this text. For a more extensive presentation of the general diversity of secondary metabolites from marine bacteria, see Murphy et al. (2012) and Wietz et al. (2013).

7.2 Bioactive Compounds in the *Roseobacter* Clade

The marine *Roseobacter* clade is a phylogenetically coherent group of bacteria in the α-proteobacterial family *Rhodobacteraceae*. The clade shares > 89 % sequence similarity in the 16S rRNA gene. It is very diverse in terms of metabolism and

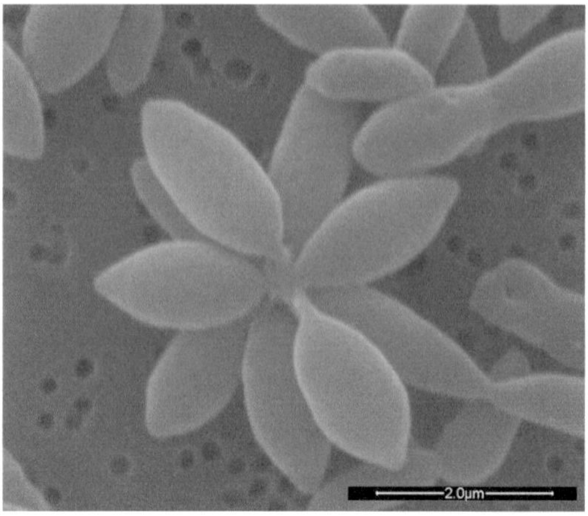

Fig. 7.2 Scanning electron micrograph of the *Roseobacter* clade member *Phaeobacter* sp. strain 27-4. Approximately 10 cells are terminally joined forming rosettes. Photograph: Michael Hansen

ecophysiology, and few specific features are shared among all the roseobacters (Buchan et al. 2005). However, most roseobacters are aerobic (except for a few facultative anaerobic denitrifyers), non-sporulating bacteria. The roseobacters are generally rod shaped, but may exhibit extended morphological features such as cell polarity and holdfast formation (Gonzalez et al. 1997) as well as formation of star-shaped cell aggregates, or rosettes (Fig. 7.2; Bruhn et al. 2005; Labrenz et al. 1998; Pukall et al. 1999; Rüger and Höfle 1992).

The members of the *Roseobacter* clade initially described were capable of performing aerobic anoxygenic photosynthesis through the production of bacteriochlorophyll *a*, and the first genus of the clade, and subsequently the clade itself, was named after the pink color accompanying production of this pigment (Shiba 1991). These members of the clade are, hence, photoheterotrophic whereas other members of the clade exhibit different trophic strategies including the transformation of sulfur-containing inorganic compounds (Ivanova et al. 2004; Lenk et al. 2012; Moran et al. 2004; Muthusamy et al. 2014; Sorokin 1994), methylotrophy (Holmes et al. 1997; Schaefer et al. 2002), and CO oxidation (Cunliffe 2011; King 2003; Moran et al. 2004). Most roseobacters, however, are heterotrophic generalists.

This high degree of metabolic diversity may, in part, be the reason why the roseobacters represent a successful, cosmopolitan clade of marine bacteria. 16S rRNA gene sequences, or cultivated members of the clade have been recovered from most marine environments including open ocean pelagic waters (e.g. Acinas et al. 1999; Giebel et al. 2009; Mullins et al. 1995; Schmidt et al. 1991; West et al. 2008), marine sediments (e.g. Lenk et al. 2012; Li et al. 1999; Lopez-Garcia et al. 2003; Mills et al. 2003), coastal environments (e.g. Allgaier et al. 2003; Dang and Lovell 2002; Gonzalez and Moran 1997; Powell et al. 2003), polar sea ice (Brinkmeyer et al. 2003; Junge et al. 2002), hyper saline marine environments (Choi and Cho 2006; Jonkers and Abed 2003; Van Trappen et al. 2004), and estuarine harbor environments (Gram et al. 2015). However, *Roseobacter* clade abundances seem highest in the upper mixed layer of the water column in temperate coastal environments and polar oceans, especially if the density of phytoplankton is high. *Roseobacter* clade numbers are positively correlated with chlorophyll *a* concentrations on a global scale with an average abundance of 3–5 % based on molecular *in situ* analyses (Wietz et al. 2010) and reports of roseobacters representing as much as 20–30 % of the total microbial community during algal blooms are common (Gonzalez et al. 2000; Gonzalez and Moran 1997; West et al. 2008; Zubkov et al. 2001). This association with eukaryotes, extends to interactions with macroalgae, marine sponges, corals, bryozoans, and especially, the accessory nidamental gland and egg capsules from squids, which host a high abundance of roseobacters (Barbieri et al. 2001; Collins et al. 2012, 2015; Grigioni et al. 2000). The nature of these interactions are not fully understood, but it can be speculated that the symbiotic association between squid eggs and roseobacters is of a mutualistic character where the bacterium may prevent predation and colonization of the newly laid eggs by potential pathogens (Barbieri et al. 2001; Biggs and Epel 1991) in return for a consistently nutrient rich environment. Similarly, in associations with phytoplankton hosts, roseobacters are presented with an environment rich in organic carbon compounds produced by the alga and in return the alga may benefit from regenerated macronutrients, sequestration of low-concentration iron, and a supply of vitamins (Geng and Belas 2010b). Ashen et al. (1999) reported that roseobacters were involved in the formation of galls, areas of abnormal excessive growth, on the thallus of a rhodophyte. This was accompanied by increased concentrations of plant growth hormones, or auxins, likely produced by the bacteria. Furthermore, the natural auxin phenylacetic acid (6; Table 7.1; Wightman and Lighty 1982) is known to be produced by members of the *Roseobacter* clade and it has been suggested that 6 is important in the symbiosis between members of the *Roseobacter* clade's *Phaeobacter* genus and the coccolithophore *Emiliania huxleyi* (Seyedsayamdost et al. 2011b). Several secondary metabolites seem to be involved in this interaction including broad spectrum antibiotic compounds and algaecides, some of which will be described below. Thus, although the α-proteobacteria are not perceived as producers of numerous different bioactive compounds, compared to members of the actinobacteria and some γ-proteobacteria (Murphy et al. 2012), members of the *Roseobacter* clade exhibits an elaborate secondary metabolism, likely owing to their adaptation to life as surface colonizers and their close association with eukaryotic organisms.

One sign of a more active secondary metabolism is the ability to orchestrate coordinated gene-expression in a density-dependent manner by quorum sensing (QS). In many proteobacterial species this is accomplished through the production and recognition of small signaling molecules such as acyl-homoserine lactones (AHLs). Initially, in the *Roseobacter* clade, potential QS molecules were found in three of five isolates and it was hypothesized to be important for the colonization of marine particulate material by these organisms (Gram et al. 2002). Subsequently, AHL production has been detected in many isolates from the clade (Berger et al. 2011; Machado et al. 2015; Martens et al. 2007; Mohamed et al. 2008; Thiel et al. 2009; Wagner-Döbler et al. 2005; Zan et al. 2012) and Cude and Buchan (2013) reported the presence of *luxI*-like AHL synthase-encoding genes in 38 of 43 publicly available roseobacter genomes, suggesting that AHL-based quorum sensing is

Table 7.1 Bioactive secondary metabolites produced by members of the *Roseobacter* clade

Compound	Structure	Bioactivity/ application	References
Phenylacetic acid (**6**)	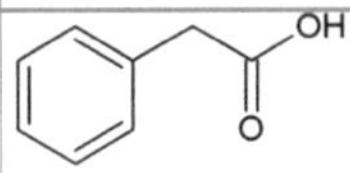	Auxin, modulation of plant and algal growth	Newton et al. (2010) Seyedsayamdost et al. (2011a); Wightman and Lighty (1982)
Indigoidine (**7**)	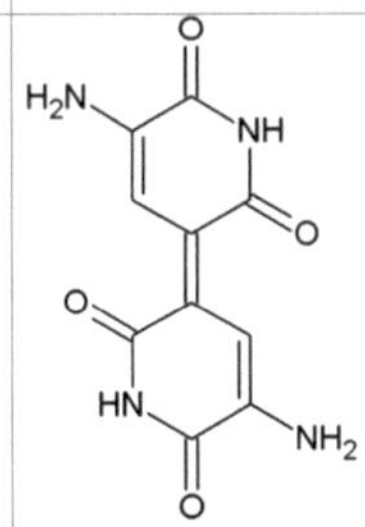	Redox indicator, industrial dye, antimicrobial[a]	Cude et al. (2012) Elazari-Volcani (1939) Takahashi et al. (2007)
Leucoindigoidine (**8**)[b]	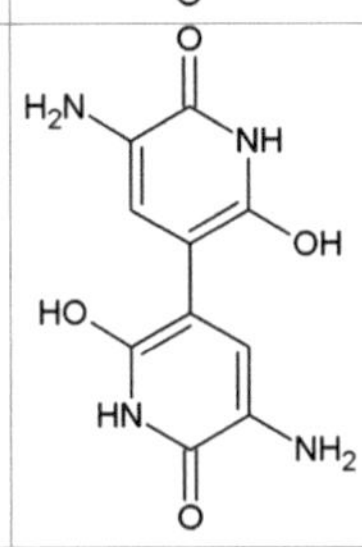	Redox indicator, antimicrobial	Cude et al. (2012) Takahashi et al. (2007)
Tropodithietic acid (**9**)	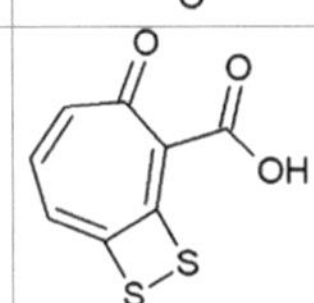	Antimicrobial	Brinkhoff et al. (2004) Kintaka et al. (1984) Tsubotani et al. (1984)
Thiotropocin (**10**)[c]	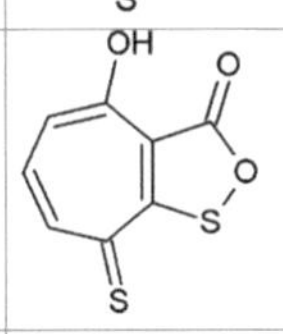	Antimicrobial	Brinkhoff et al. (2004) Kintaka et al. (1984) Tsubotani et al. (1984)
Tryptanthrin (**11**)	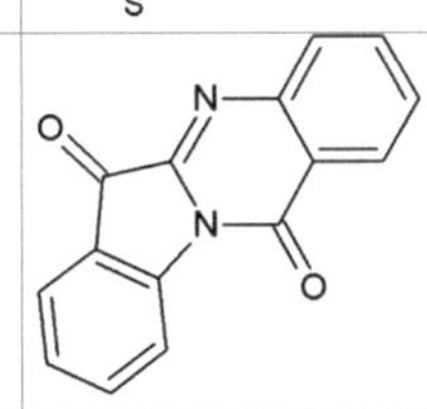	Antimicrobial, antiparasitic, anti-inflammatory, antitumor	Honda et al. (1979) Wagner-Döbler et al. (2004)
Cyclo-(Phe-Pro) (**12**)	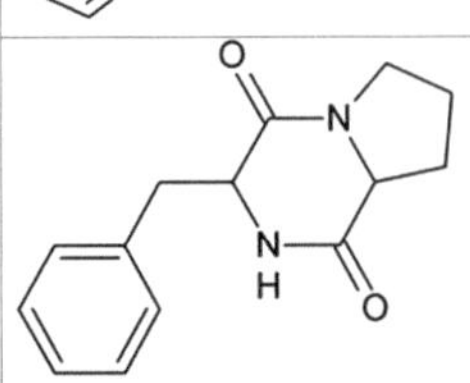	Antimicrobial	Fdhila et al. (2003) Wagner-Döbler et al. (2004)

(continued)

Table 7.1 (continued)

Compound	Structure	Bioactivity/ application	References
Cyclo-(Tyr-Pro) (**13**)	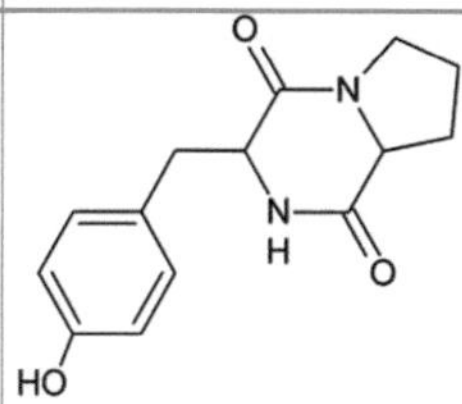	Antimicrobial	Wagner-Döbler et al. (2004)
Cyclo-(Leu-Pro) (**14**)	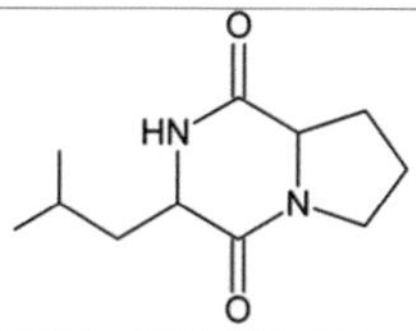	Antimicrobial	Fdhila et al. (2003) Wagner-Döbler et al. (2004)
Cyclo-(Pro-Val) (**15**)	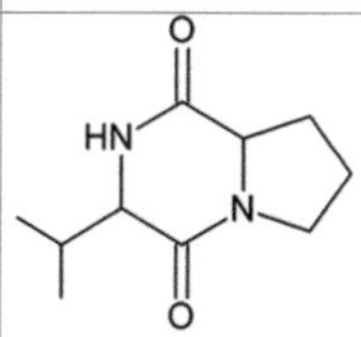	Antimicrobial	Fdhila et al. (2003)
Cyclo-(Pro-Ile) (**16**)	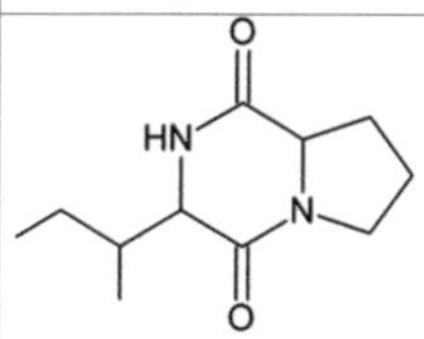	Antimicrobial	Fdhila et al. (2003)
Cyclo-*trans*-4-OH-(Pro-Phe) (**17**)	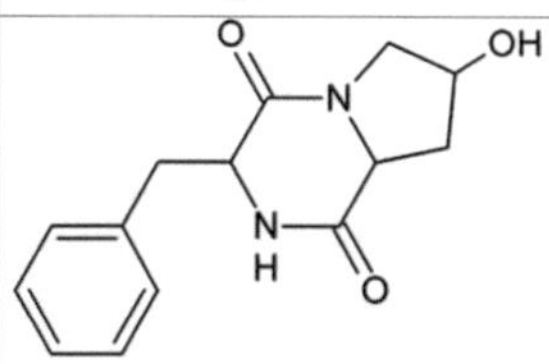	Antimicrobial	Fdhila et al. (2003)
Roseobacticide A (**18**)	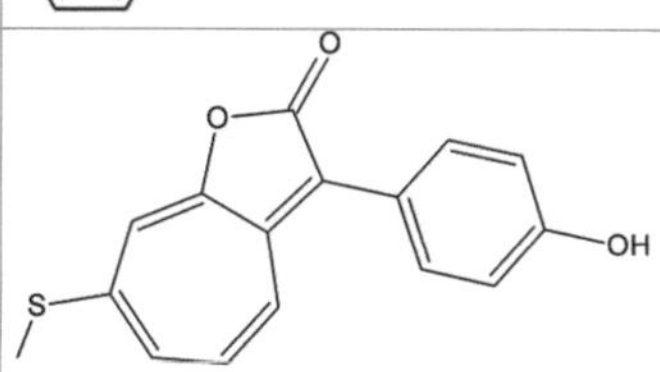	Algaecide	Seyedsayamdost et al. (2011a) Seyedsayamdost et al. (2011b) Sonnenschein, Phippen, Bentzon-Tilia, Rasmussen, Nielsen, Gram. *Roseobacter* gardening of microalgae: Phylogenetic distribution of roseobacticides and their effect on microalgae, *unpublished*

[a]It is likely the reduced form, **8**, that is responsible for the majority of the antimicrobial activity
[b]Reduced form of **7**
[c]Valence tautomer of **9**

a conserved feature in the *Roseobacter* clade. In addition, some roseobacters have been shown to utilize other signaling molecules apart from AHLs in quorum sensing, i.e. tropodithietic acid (TDA) (Geng and Belas 2010a).

Similarly, screening for the genetic potential to produce other bioactive secondary metabolites using degenerate primers targeting conserved regions of the ketosynthase domain of polyketide synthases (PKSs) and the adenylation domain of nonribosomal peptide synthetases (NRPSs) have shown that a substantial fraction of roseobacters likely harbor PKS- and NRPS-encoding gene clusters. In fact, it has been proposed that a PKS/NRPS hybrid cluster is conserved and widely distributed within the clade (Martens et al. 2007). In congruence with these findings, genome mining of 12 genome sequenced roseobacters isolated from the accessory nidamental gland of female Hawaiian bobtail squid (*Euprymna scolopes*), and subsequent comparison with 57 known *Roseobacter* genomes, showed that 28 of the 69 genomes (including the 12 isolates from the accessory nidamental gland) harbored this hybrid cluster (Collins et al. 2015). Similar observations were seen for *Roseobacter* clade bacteria obtained on the Danish ship-based expedition, Galathea 3. Here, 46 roseobacters were isolated from surface water samples taken along the full expedition. Of these roseobacters, 42 exhibited inhibitory activity against *Vibrio anguillarum* and subsequent genome mining of a few of these roseobacters have revealed the presence of the hybrid gene clusters (Machado et al. 2015). Hence, this PKS/NRPS is most likely highly conserved within roseobacters, although the product of this hybrid remains to be characterized.

7.2.1 Indigoidine

One characterized bioactive compound synthesized through an NRPS by a member of the *Roseobacter* clade, is the blue pigment indigoidine (7; Table 7.1). 7 is a bicyclic 3,3'-Bipyridyl molecule that has previously been isolated from species of the genera *Pseudomonas* (Elazari-Volcani 1939), *Erwinia* (Chatterjee and Brown 1981; Starr et al. 1966), *Arthrobacter* (Heumann et al. 1968), *Streptomyces* (Takahashi et al. 2007), and *Photorhabdus* (Brachmann et al. 2012), among others. The NRPS responsible for synthesizing 7 is comprised of a single module that acts without association with other NRPSs or PKSs and hence 7 is synthesized from a single amino acid substrate, glutamine (Takahashi et al. 2007).

Despite the fact that 7 has been known for decades (Elazari-Volcani 1939), interest in its properties has mainly focused on the conspicuous blue color, whereas the antibacterial activity of the compound was only recently realized when it was first described in a *Roseobacter* strain of the *Phaeobacter* genus (Cude et al. 2012). The ability to antagonize other bacteria had been observed previously with this strain (Slightom and Buchan 2009). Cude et al. (2012) showed that the inhibitory activity of the strain against *Vibrio fischeri* was correlated with the production of indigoidine, or rather its water soluble form leucoindigoidine (8). Interestingly, expression of the synthetase gene, *igiD*, is upregulated when the strain was grown on a solid surface or agar plates, compared to when the bacterium is growing in liquid broth cultures (Cude et al. 2012). Similarly, the expression of genes (*tdaC*) involved in production of tropodithietic acid (TDA), the antibacterial compound of *Phaeobacter inhibens* and *Ruegeria mobilis*, is induced upon attachment (D'Alvise et al. 2014). This corroborates the notion that the roseobacters may use these antibacterial compounds as a means of attaining an advantage in colonization and in their association with eukaryotes (Long and Azam 2001; Seyedsayamdost et al. 2011b; Slightom and Buchan 2009). Also, the antagonistic effect of the *Phaeobacter* strain on *Vibrio fischeri* was only observed on solid agar plates or in biofilms and not in broth cultures (Cude et al. 2012).

7.2.2 Tropodithietic Acid

The most conspicuous and well-described antimicrobial compound produced by members of the *Roseobacter* clade is not a polyketide or peptide, but the small tropolone derivative TDA (9; Table 7.1). TDA consists of a seven-membered aromatic ring structure with a hydroxyl and a carbonyl group. In addition, it has a unique disulfide moiety also known in a few other microbial secondary metabolites such as the fungal mycotoxin, gliotoxin (Gardiner et al. 2005) and holomycin from *Streptomyces* (Kenig and Reading 1979). TDA was originally described as its valence tautomer, tropothiocin (10; Table 7.1) from a soil-derived *Pseudomonas* species (Kintaka et al. 1984; Tsubotani et al. 1984), but has since been isolated from several marine strains of the *Roseobacter* clade and the *Pseudovibrio* genus. In the *Pseudovibrio* strains, which are closely related to the *Roseobacter* clade and belong to the same α-proteobacterial family, *Rhodobacteraceae*, production of TDA has been reported in one strain isolated from the rhodophyte *Delisea pulchra* (Penesyan et al. 2011) and in several strains isolated from marine sponges (Enticknap et al. 2006; Harrington et al. 2014). In addition, a marine bacterium classified as a *Caulobacter* species was isolated from phytoplankton and it was shown to be able to produce tropothiocin (Kawano et al. 1997). Production of TDA seems to be a trait, which is characteristic of several species of *Phaeobacter* and *Ruegeria* of *Roseobacter* sub-clade 1 (as defined by Newton et al. 2010). Interestingly, a common feature shared by the bacteria producing TDA and tropothiocin is that they are known to live in association with eukaryotes, suggesting that TDA is an integral part of these associations. In congruence with this notion, it seems that roseobacters mostly produce TDA under circumstances somewhat similar to those prevailing under host-associated conditions e.g. during biofilm growth. However, in contrast to production of 7 in *Phaeobacter*, which exclusively takes place during growth on solid substrates (Cude et al. 2012, 2015), TDA-producing roseobacters such as *Phaeobacter gallaeciensis*, *P. inhibens*, and *Ruegeria mobilis* do so during growth on solid surfaces and in broth cultures (Bruhn et al. 2007; Porsby et al. 2008). There is, however, one general distinction between the *Ruegeria* and the *Phaeobacter* strains with respect to how physical conditions influence production: The *Ruegeria* only produces TDA during static growth in broth cultures, whereas *Phaeobacter* strains usually produce TDA under static and shaken conditions (Porsby et al. 2008). Nevertheless, the highest yields of TDA are achieved also under static conditions for *Phaeobacter* (Porsby et al. 2008). One exception to this distinction between *Phaeobacter* and *Ruegeria* strains is a *Phaeobacter* strain, which was isolated from a Spanish turbot larval rearing unit (Hjelm et al. 2004). This strain is by phylogenetic affiliation a *Phaeobacter*, but phenotypically it resembles the *Ruegeria* in the sense that it only produces TDA under static conditions (Bruhn et al. 2005).

In *Ruegeria* and *Phaeobacter* species, growth under static culture conditions is accompanied by the formation of dense extracellular polymeric substance (EPS)-containing biofilms at the air-liquid interface. Cells growing here are aggregated; forming star-shaped rosettes of approximately five to ten terminally joined cells (Fig. 7.2; Bruhn et al. 2005, 2007; Porsby et al. 2008). It was recently shown that this mode of growth, and production of TDA, is under the influence of the intracellular signaling molecule, cyclic dimeric guanosinmonophosphate (c-di-GMP) in *Ruegeria mobilis* (D'Alvise et al. 2014). c-di-GMP usually acts as a second messenger, modulating gene expression in favor of sessile growth, which corroborates that production of TDA is part of the surface-attached stages of life of roseobacters. Furthermore, TDA-production seems to be affected by cell densities through QS. For *R. mobilis*, it has been shown that expression of the *tdaC* gene, and hence production of TDA, is under the control of TDA itself (Geng and Belas 2010a) and Berger et al. (2011) reported that production of TDA was abolished in insertion mutants of a *Phaeobacter inhibens* strain, which were deficient in expression of the *luxI* and *luxR* homologs, *pgaI* and *pgaR*. Subsequently, it was shown that this deficiency postponed the onset of production of TDA under aerated conditions, but the production was not dependent on PgaIR-mediated QS as such and concomitantly, the QS deficient *Phaeobacter inhibens* mutants were equally efficient in antagonizing the fish pathogen *Vibrio anguillarum* strain NB10 in algae cultures (Garcia et al. 2013).

In addition to an increase in rosette and biofilm formation, static *Phaeobacter* and *Ruegeria* cultures rich in iron (i.e. Marine Broth cultures) is accompanied by the formation of a distinctive brown pigment, which has given the name to the *Phaeobacter* genus (from the Greek *Phaeos*, meaning dark). Production of this pigment and TDA is positively correlated, but the two are not identical as they are separable by fractionation and bioactivity remains associated with the fraction containing TDA (Bruhn et al. 2005). The brown pigment is a precipitate composed of a complex between TDA and free iron. Furthermore, high iron concentrations are required for the maturation of TDA (D'Alvise et al. 2015), which raises the question; at what quantities are TDA synthesized *in situ*? However, the exact mechanisms behind the TDA-iron interaction have yet to be elucidated.

Many of the troponoid compounds, including TDA, exhibit antibacterial, antifungal, and/or antiviral activities (Budihas et al. 2005; Kintaka et al. 1984; Lu et al. 2015; Morita et al. 2002; Trust 1975) . TDA has been shown to be the primary active compound in the antagonism of fish pathogenic vibrios (D'Alvise et al. 2010, 2012; Garcia et al. 2013), and the compound is known to target various marine bacteria and several human pathogens (Brinkhoff et al. 2004; Kintaka et al. 1984; Porsby et al. 2011; Tsubotani et al. 1984). The spectrum of target organisms and mechanisms of action have been investigated to assess the potential of TDA as an antibiotic. Wilson et al. (2016) showed that TDA acts at the cell membrane as a proton antiporter, disrupting the proton motive force, and is hence affecting both Gram-negative and Gram-positive bacteria, corroborating previous results suggesting that TDA is a broad-spectrum antibiotic (Porsby et al. 2011). In addition to its role as a proton antiporter, the iron metabolism of the targeted bacterium is affected by TDA as well (Porsby, Knudsen, Mateiu, Gram. Technical University of Denmark, *unpublished*), suggesting that the compound may have multiple target sites within the cell. Development of resistance against TDA and similar compounds seem to be rare. Porsby et al. (2011) was unsuccessful in selecting resistant mutants by applying four different single exposure techniques and a recent adaptive laboratory evolution experiment, in which *V. anguillarum* was exposed to TDA for hundreds of generations, showed that neither tolerance nor resistance arose (Rasmussen et al. 2016). However, Harrington et al. (2014) suggested that resistance might not be as rare in indigenous marine bacteria as initially thought. Nonetheless, TDA is a promising broad-spectrum antibiotic candidate.

7.2.3 Additional Bioactive Compounds and Enzymes Produced by Roseobacters

Other secondary metabolites exhibiting antimicrobial activity have been isolated from members of the *Roseobacter* clade. The quinazolinone alkaloid tryptanthrin (11; Table 7.1), was identified in *Oceanibulbus indolifex* (Wagner-Döbler et al. 2004). Tryptanthrin does not only target bacteria and fungi (Bandekar et al. 2010; Honda et al. 1979; Kataoka et al. 2001; Mitscher and Baker 1998), but also the eukaryotic parasite *Toxoplasma* (Krivogorsky et al. 2008), and furthermore, tryptanthrin has been shown to have anti-inflammatory (e.g. Micallef et al. 2002; Recio et al. 2006) and antitumor activities (e.g. Kimoto et al. 2001; Yu et al. 2007). Three diketopiperazine cyclic dipeptides were isolated from *O. indolifex* (Wagner-Döbler et al. 2004): cyclo-(Phe-Pro) (12), cyclo-(Tyr-Pro) (13), and cyclo-(Leu-Pro) (14). In addition to cyclo-(Pro-Val) (15), cyclo-(Pro-Ile) (16), and cyclo-*trans*-4-OH-(Pro-Phe) (17), 12 and 14 were also recovered from *Roseobacter* clade members isolated from Spanish scallop (*Pecten maximus*) larvae cultures (Fdhila et al. 2003). These cyclic dipeptides proved to have significant antimicrobial activities with minimal inhibitory concentration (MIC) values between 0.03 and 0.07 μg/ml against *V. anguillarum* (Fdhila et al. 2003).

Interactions between roseobacters and eukaryotes are not exclusively of a mutualistic nature as briefly mentioned and some roseobacters may be detrimental to the host organism. This is the case with the eastern oyster (*Crassostrea virginica*), which is susceptible to colonization by *Roseovarius crassostreae*, which causes

Roseovarius Oyster Disease (ROD), previously Juvenile Oyster Disease (Boettcher et al. 2005; Maloy et al. 2007). Similarly, roseobacters living in association with the dinoflagellate *Alexandrium* are involved in production of the sodium-channel blocking paralytic shellfish toxins responsible for paralytic shell fish poisoning (PSP) (Gallacher et al. 1997).

The most recent example of the multifaceted nature of roseobacter-eukaryote interactions is the production of a family of algaecides termed roseobacticides, by members of the *Phaeobacter* genus (Seyedsayamdost et al. 2011a, b). It is hypothesized that the *Phaeobacter* and the alga *E. huxleyi* live in close association in the open oceans and as a phytoplankton bloom is initiated the bacterium colonizes the alga, initiating a mutualistic symbiosis where the bacterium receives organic carbon and supplies auxins such as 6. As the bloom deteriorates, the algal lignin breakdown product, *p*-coumaric acid, accumulates. This compound acts as an elicitor for roseobacticide A (18; Table 7.1) and B production, changing the association between the roseobacter and the algae to a parasitic one where the excretion of roseobacticides accelerates algal lysis and provides the bacterium with additional nutrients (Seyedsayamdost et al. 2011a, b).

Data on the occurrence of biotechnological relevant enzymes in the *Roseobacter* clade are currently scarce and whether roseobacters may serve as a source of potentially important enzymes is hence unknown. Nevertheless, chitinase group I genes have been amplified from roseobacters and chitinolytic phenotypes have been described (Cottrell et al. 2000). There was however, a discrepancy between the presence of chitiniase genes and the chitinolytic phenotype. Recently, genome mining showed that the chitin catabolic cascade sensor histidine kinase gene *chiS* was present in members of the *Ruegeria* genus, yet these bacteria did not (a) harbor conventional chitinase genes and (b) degrade chitin under standard laboratory conditions (Machado et al. 2015), supporting the discrepancies observed by Cottrell et al. (2000). Hence, there are likely important aspects of chitin degradation by roseobacters that we currently do not know.

Thus, the *Roseobacter* clade produces an array of different bioactive compounds; from algal growth promoters to antimicrobial compounds, toxins, and algaecidal compounds. Many of these show potential for pharmaceutical, or other industrial applications, yet mining of roseobacter genomes suggests that the known compounds only represent a fraction of the bioactive natural products produced. Hence, the *Roseobacter* clade likely represents a significant source of hitherto undescribed natural products.

7.3 Application of Roseobacter Probionts in Larviculture

The concept of bioprospecting is typically associated with the identification, extraction and use of compounds with biotechnological activities. However, the use of whole organisms in, e.g., bioremediation or as biocontrol agents is also considered here as bioprospecting. As the roseobacters have emerged as a very promising fish-larval probiotic, due to their antagonism against fish pathogenic microorganisms, we describe here the background for this use and the current status of development.

An increasing proportion of fish produced for human consumption is supplied from farmed fish. The aquaculture industry is the fastest growing sector of the food-producing industries, exhibiting an average growth rate of approximately 8 % from 1983 to 2013 according to the Food and Agriculture Organization of the United Nations (FAO; FAO – Fishery Statistical Collections 2015). Concomitantly, the amount of fish supplied through wild catches has stagnated during the same period, in part due to overfishing.

7.3.1 Disease Control in Aquaculture

Typically, fish are reared from eggs, through the larval stage to juvenile fish and finally the adult stage of the lifecycle in finfish farming. Towards the end of the larval stage as the yolk sac is depleted, the mouth opens and the larva starts to drink the surrounding water (Mangor-Jensen and Adoff 1987). Subsequently it begins foraging on prey. In the wild, larvae will feed on phytoplankton or, more commonly, zooplankton and hence live feed such as rotifers and *Artemia* are routinely supplied in the rearing of fish larvae of species such as turbot, cod, halibut, and sea bass, in order to ensure an optimal nutritional profile of the feed and to stimulate natural feeding behavior in the larvae. This practice presents some complications as the addition of zooplankton to larviculture favors the proliferation of generally opportunistic bacteria, which might prove pathogenic to the fragile larvae. Disease caused by pathogenic bacteria is believed to be one of the most serious challenges faced by the aquaculture industry (Meyer 1991) and therefore extensive measures are taken in order to limit the impact of such pathogenic bacteria on production. Procedures such as chemical disinfection of surfaces are common practice in aquaculture but does not target pathogenic bacteria efficiently in the microbiota associated with live items such as eggs and live feed. The prophylactic use of antibiotics has also been applied extensively, especially in smaller aquaculture plants and in developing countries (Cabello 2006). Misuse of antibiotics poses a series of problems relating to the spread of (a) antibiotic resistant bacteria in the environment and (b) resistance determinants between aquatic bacteria and human pathogens. Worryingly, Rhodes et al. (2000) showed that plasmids conferring oxytetracycline resistance have circulated between aquaculture and hospital environments. Similarly, antibiotic resistance determinants of the multi-resistant *Salmonella enterica* serotype Typhimurium phagetype DT104, which has been involved in extensive outbreaks (Hancock et al. 2000), may likely be derived from Asian aquaculture facilities using florfenicol in their production (Angulo and Griffin 2000; Cabello 2006). Hence, to ensure continued efficiency in the treatment of bacterial infections in animals and humans, it is recognized that sustainable alternatives to the prophylactic use of antibiotics needs to be implemented in the aqua- and agriculture industries.

7.3.1.1 Probiotics in Larviculture

Fish larvae do not have a developed adaptive immune system and vaccination, which is a widespread sustainable alternative to the prophylactic use of antibiotics in aquaculture in general, is not suitable in larviculture. Here, one approach is to manipulate the bacterial community in the larval rearing units, by (a) microbial maturation of the water, a strategy where non-opportunistic bacteria are selected for by keeping a low substrate concentration in pre-filtered water (Skjermo et al. 1997; Skjermo and Vadstein 1999), and/or (b) adding specific beneficial bacteria (i.e. probiotics) to the rearing units or to the feed. Probiotics have been defined as "live microorganisms which, when administered in adequate amounts, confer a health benefit to the host" (FAO 2001). Early studies investigating the administration of probiotics to shrimp and crab larviculture settings showed that unidentified bacteria readily isolated from soil extracts and crab larvae pond water were able to improve moult rates and larval survival when added to larval rearing units in large quantities (Maeda and Liao 1992; Nogami and Maeda 1992). Analogous observations were subsequently reported for oyster and finfish larvae. Douillet and Langdon (1993, 1994) showed that by submerging protein microcapsules intended for supplying dietary protein to filter feeders, they could obtain marine bacteria related to *Aeromonas*, which consistently increased growth rates and larval survival by approximately 20 % when supplied to cultures of growing Pacific oyster larvae. These data are, however, from monoxenic systems and consequently the probiotic effect was likely of a nutritional nature rather than due to control of pathogens from antibiotic production. Similarly, in turbot larviculture, adding probiotic bacteria to live feed (rotifers) was reported to significantly increase the weight of the fish after 20 days (Gatesoupe 1991) and subsequently, numerous reports of the positive effects of probiotics in larvi- and aquaculture have emerged.

Roseobacters seem to be indigenous to the larvi- and aquaculture setting (Table 7.2). This is important since the potential probiont is demonstrated to establish itself among, or in association with, the larvae. Furthermore, it illustrates that larvae may already be exposed to the organism, albeit at low abundances, and therefore it is unlikely that the potential probiont represents a hazard to the larvae. Accordingly, *Phaeobacter* sp. strain 27-4 (previously, *Roseobacter* sp. strain 27-4 (Martens et al. 2006), a roseobacter isolated from a Spanish turbot larval rearing facility, proved to be harmless to turbot larvae (Hjelm et al. 2004). Furthermore, *Phaeobacter* sp. strain 27-4 could antagonize the fish pathogenic bacteria *Vibrio anguillarum* and *V. splendidus* when growing in co-cultures (Hjelm et al. 2004). This antagonistic effect was observed also in challenge trials where the accumulated mortalities of challenged turbot larvae supplied with the *Phaeobacter* strain probiont through rotifers were similar to those of the unchallenged controls in some trials. Immunohistochemistry suggested that the probiont did not colonize the gut epithelium and, hence, these roseobacter probionts are likely exerting their effect in other loci to the GI tract (Planas et al. 2006). D'Alvise et al. (2013) demonstrated that cod larval mortalities consistently decreased when the *Phaeobacter* strain was added directly to the water 48 h prior to the fish pathogenic *V. anguillarum* strain (Fig. 7.3b).

Table 7.2 Overview of roseobacters isolated from larviculture facilities

Location	Aquaculture facility type	Isolation site	Genus	Antagonism of pathogens *in vitro* (+/−)	Probiotic effect *in vivo* (+/−)	References
Galicia, Spain	Scallop larval rearing unit	Larval rearing water	*Phaeobacter*	+	(-)[a]	Ruiz-ponte et al. (1998)
Galicia, Spain	Scallop larval rearing unit	Larval rearing water	*Phaeobacter*	+	ND	Fdhila et al. (2003)
Galicia, Spain	Turbot larval rearing units	Rotifers	*Phaeobacter*	+	+	Hjelm et al. (2004)
		Rearing tank wall	*Ruegeria*			
Denmark	Turbot larval rearing unit	Rearing tank wall	*Phaeobacter*	+	+	Porsby et al. (2008)
		Larval rearing water	*Ruegeria*			
		Phytoplankton				
Galicia, Spain	Mollusk hatcheries	Oyster larvae	*Phaeobacter*	+	ND	Prado et al. (2009)
		Oyster spat				
		Oyster broodstock Rearing tank wall				
		Larval rearing water				
Greece	Sea bass larval rearing unit	Larval rearing water Rearing tank wall	*Phaeobacter*	+	+	Grotkjær et al. (2016)
		Outlets				
		Outlet water				
		Live feed				

ND not determined

[a]Organic extracts had a positive effect on scallop survival, whereas challenge trails using live bacteria had limited effect

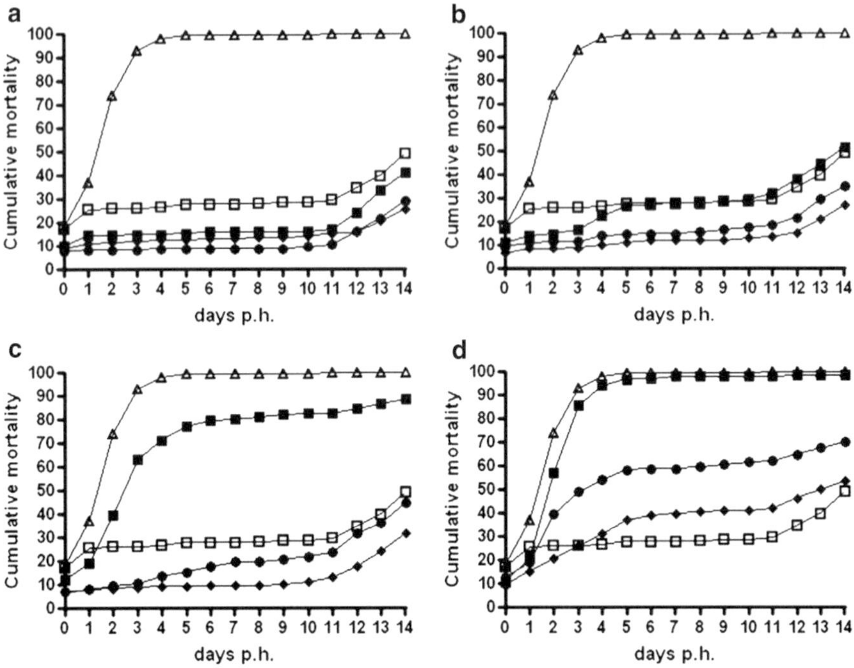

Fig. 7.3 Mortality of cod larvae in four different challenge trials. In one trial (**a**) the mortality of cod larvae was monitored for 14 days post hatch (days p.h.) when the larvae were exposed to either the fish pathogenic *V. anguillarum* strain HI610 only (Δ Positive control), *Phaeobacter* sp. strain 27–4 only (■ 27–4 alone), *Phaeobacter* sp. strain M23-3.1 only (◆ M23 alone), *Phaeobacter inhibens* strain DSM 17395 only (● DSM 17395 alone), or none of the bacterial strains (□ Negative control). In the other three trials the larvae was exposed to either *V. anguillarum* strain HI610 only (Δ Positive control), none of the bacterial strains (□ Negative control), or a combination of the pathogen and one of the three roseobacter strains: 27–4 (■), M23 (◆), DSM 17395 (●). In (**b**) the roseobacters were added 48 h prior to the pathogen, in (**c**) the roseobacter and the pathogen were added simultaneously and in (**d**) the roseobacters were added 48 h after the pathogen (Reproduced from D'Alvise et al. (D'Alvise et al. 2013) with permission from the original publisher)

Also, the probiont remained active when it was introduced at later time points (Fig. 7.3c, d).

Another roseobacter strain of the species *Phaeobacter gallaeciensis* (DSM 26640^T; previously *Roseobacter gallaeciensis* BS107 (Martens et al. 2006), was isolated from water collected in a Spanish scallop larva rearing system (Ruiz-Ponte et al. 1998). This isolate was also able to inhibit the growth of an array of different bacteria *in vitro*. However, it was not able to increase the survival of scallop larvae in challenge trials with *V. pectenicida* (Table 7.2; Ruiz-Ponte et al. 1999).

Prado et al. (2009) found that of the 52 isolates that exhibited an antagonistic effect against one or more of three target *Vibrio* strains, the four most potent isolates were from the genus *Phaeobacter*, by investigating more than 500 isolates obtained from different sites in Spanish bivalve mollusk hatcheries. One of these isolates was tested at different growth conditions and against a wide range of vibrios and other bacteria, and it was found to be efficient in inhibiting the growth of most of the target organisms. Growth on iron-rich medium, i.e. Marine agar (MA), was accompanied by formation of brown pigment and the production of TDA is likely responsible for the observed antagonism. It remains to be determined, however, whether these four *Phaeobacter* isolates exert any probiotic effects *in vivo* (Table 7.2).

The potential roseobacter probionts mentioned so far have all been isolated from systems in the mesophilic temperature range (Table 7.2). Inhibitory activity was only observed at temperatures in the range of 15–25° C for the *Phaeobacter* strain mentioned previously and not at 5 or 10° C (Bruhn et al. 2005). This indicates that these isolates would be optimal as probionts in rearing of fish larvae at temperatures above 15° C and possibly excluding their applicability in rearing of cold-water fish such as Atlantic cod, which has a growth maximum below 8° C (Jordaan and Kling, Jordaan and Kliing 2003). Roseobacters are, however, also abundant in polar and cooler temperate environments. Accordingly, roseobacters of the *Phaeobacter* and *Ruegeria* genera have been isolated from cooler larviculture environments (Table 7.2). Of 100 *Vibrio*-antagonizing isolates obtained from various sites within a Danish turbot rearing facility, 51 were members of the *Roseobacter* clade (Porsby et al. 2008). Interestingly, there seemed to be a differentiation between where in the facility the two different genera were recovered. Isolates affiliated with the *Ruegeria* genus were consistently isolated from phytoplankton, whereas the *Phaeobacter* strains were isolated from other sites in the facility. Hence, these organisms occupy different niches, which should be considered in the potential administration of these roseobacters as probionts, and is also interesting from a bioprospecting point of view in terms of finding interesting organisms in different niches. One of the obtained *Phaeobacter*-affiliated strains, was subsequently tested in cod larvae challenge trials and compared to the *Phaeobacter* and *P. inhibens* strains mentioned previously (Fig. 7.3; D'Alvise et al. 2013), which at that time was incorrectly believed to be identical to a *P. gallaeciensis* strain (Buddruhs et al. 2013). In these trials, performed at 7° C, *Phaeobacter* sp. strain M23-3.1 and *P. inhibens* strain DSM 17395 performed better than *Phaeobacter* sp. strain 27-4 when the probiont was introduced simultaneously with the pathogen or later (Fig. 7.3c, d). This could be attributable to differences in temperature preferences, but as already touched upon, the *Phaeobacter* sp. strain 27-4 also exhibits a *Ruegeria*-like physiology when it comes to production of TDA (Porsby et al. 2008) and it generally displays less inhibitory activity *in vitro* than the other *Phaeobacter* species. This likely also influences the observed differences between the three strains in the cod larva challenge trials.

The most recent illustration of the autochthony of roseobacters in marine larviculture is the isolation of thirteen TDA-producing roseobacters of the genus *Phaeobacter* from Greek Sea bass larval rearing units (Table 7.2; Grotkjær et al. 2016). Two of these strains were capable of establishing themselves in live feed cultures and they proved to antagonize the two fish pathogenic bacteria *Vibrio harveyi* and *V. anguillarum* in a similar system, suggesting that supplying roseobacter probionts through the live feed is a useful approach. Furthermore, the two tested strains had a significant probiotic effect on the *Artemia nauplii* larvae, increasing their survival significantly.

Hence it is likely that potential probiotic roseobacters are readily isolated from the larviculture environment in general and consistent with this notion, cultivation-independent studies have reported high relative abundances of roseobacteres in larviculture. For instance, one study investigated the microbial community associated with haddock larvae using 16S rRNA/DGGE analyses (Griffiths et al. 2001). In the DGGE, one band became increasingly predominant 27–37 days post hatch in 15 out of 16 tanks, and 40 days post hatch, and this band was the singularly dominant band. It proved to be from a *Sulfitobacter*-related member of the *Roseobacter* clade and, interestingly, in the one tank that did not produce this band in the DGGE, the larval population crashed on day 40. Similar observations have been reported from cod larviculture (Brunvold et al. 2007).

Thus, it is evident that roseobacters are closely associated with the live feed and the larviculture environment and, hence, they fulfill one of the important criteria for being successful probionts in the rearing of larvae, namely that they can establish themselves in the environment and that they do not pose a threat to larvae health themselves. Furthermore, they are able to antagonize fish pathogenic bacteria often associated with larval mortality, *in vitro* and *in vivo* (Table 7.2). It remains to be determined if increased roseobacter abundances, and the accompanying increased production of TDA, will affect the indigenous microbial community in larval rearing systems. Whether the production of other secondary metabolites than TDA, such as the roseobacticides, poses a problem is unknown. However, the particular *P. inhibens* strain mentioned before does not seem to impose any detrimental effects on any of the trophic levels involved in the rearing of larvae. In contrast, it is able to antagonize pathogens in all trophic levels from algae to rotifers and in the larvae (D'Alvise et al. 2012). Roseobacters could be administered through the live feed, which often acts as vectors for fish pathogenic bacteria. Alternatively, the administration of a potential *Phaeobacter* probiont could rely on the dispersal of the probiont throughout the system, utilizing the fact that the *Phaeobacter* species inhabit a diverse set of niches in the larviculture environment.

7.4 Conclusions

The *Roseobacter* clade of marine α-proteobacteria represents some of the most abundant and versatile microorganisms in the marine environment. They apply a multitude of different trophic strategies, many of which rely on some form of interaction with other organisms; eukaryotic or prokaryotic. The roseobacters produce an array of bioactive secondary metabolites involved in these interactions. Some are involved in signal transduction, others in the modulation of eukaryotic growth, yet others in the antagonism of competing microbes. These metabolites represent a largely unexploited reservoir of potential natural products and precursors for pharmaceutical and biotechnological relevant compounds including potential new antibiotics. Furthermore, some of the roseobacters seem to thrive in the rather special environment that is the larviculture setting. As a consequence of this success, potential roseobacter probionts can readily be isolated from the exact environment in which they exert their probiotic effect. The combination of these aspects of roseobacter biology makes members of this clade ideal candidates as novel probiotics in larviculture and a source of novel antimicrobial compounds. Finally, the research indicates clearly how maintaining particular niches is crucial in finding novel microorganisms relevant to bioprospecting.

Acknowledgements This work was supported by grant VKR023285 from the Villum Kahn-Rasmussen foundation and by the Innovation Fund Denmark, grant number 12-132390 (ProAqua project).

References

Acinas SG, Anton J, Rodriguez-Valera F (1999) Diversity of free-living and attached bacteria in offshore western Mediterranean waters as depicted by analysis of genes encoding 16S rRNA. Appl Environ Microbiol 65:514–522

Alkawash MA, Soothill JS, Schiller NL (2006) Alginate lyase enhances antibiotic killing of mucoid *Pseudomonas aeruginosa* in biofilms. APMIS 114:131–138

Allgaier M, Uphoff H, Felske A, Wagner-Döbler I (2003) Aerobic anoxygenic photosynthesis in *Roseobacter* clade bacteria from diverse marine habitats. Appl Environ Microbiol 69:5051–5059

Angulo FJ, Griffin PM (2000) Changes in antimicrobial resistance in *Salmonella enterica* serovar Typhimurium. Emerg Infect Dis 6:436–437

Ashen JB, Cohen JD, Goff LJ (1999) GC-SIM-MS detection and quantification of free indole-3-acetic acid in bacterial galls on the marine alga *Prionitis lanceolata* (Rhodophyta). J Phycol 35:493–500

Bandekar PP, Roopnarine KA, Parekh VJ, Mitchell TR, Novak MJ, Sinden RR (2010) Antimicrobial activity of tryptanthrins in *Escherichia coli*. J Med Chem 53:3558–3565

Barbieri E, Paster BJ, Hughes D, Zurek L, Moser DP, Teske A, Sogin ML (2001) Phylogenetic characterization of epibiotic bacteria in the accessory nidamental gland and egg capsules of the squid *Loligo pealei* (Cephalopoda: Loliginidae). Environ Microbiol 3:151–167

Berger M, Neumann A, Schulz S, Simon M, Brinkhoff T (2011) Tropodithietic acid production in *Phaeobacter gallaeciensis* is regulated by N-Acyl homoserine lactone-mediated quorum sensing. J Bacteriol 193:6576–6585

Biggs J, Epel D (1991) Egg capsule sheath of *Loligo opalescens* berry: structure and association with bacteria. J Exp Zool 259:263–267

Boettcher KJ, Geaghan KK, Maloy AP, Barber BJ (2005) *Roseovarius crassostreae* sp nov., a member of the *Roseobacter* clade and the apparent cause of juvenile oyster disease (JOD) in cultured eastern oysters. Int J Syst Evol Microbiol 55:1531–1537

Brachmann AO, Kirchner F, Kegler C, Kinski SC, Schmitt I, Bode HB (2012) Triggering the production of the cryptic blue pigment indigoidine from *Photorhabdus luminescens*. J Biotechnol 157:96–99

Brinkhoff T, Bach G, Heidorn T, Liang LF, Schlingloff A, Simon M (2004) Antibiotic production by a *Roseobacter* clade-affiliated species from the German Wadden Sea and its antagonistic effects on indigenous isolates. Appl Environ Microbiol 70:2560–2565

Brinkmeyer R, Knittel K, Jurgens J, Weyland H, Amann R, Helmke E (2003) Diversity and structure of bacterial communities in arctic versus antarctic pack ice. Appl Environ Microbiol 69:6610–6619

Bruhn JB, Nielsen KF, Hjelm M, Hansen M, Bresciani J, Schulz S, Gram L (2005) Ecology, inhibitory activity, and morphogenesis of a marine antagonistic bacterium belonging to the *Roseobacter* clade. Appl Environ Microbiol 71:7263–7270

Bruhn JB, Gram L, Belas R (2007) Production of antibacterial compounds and biofilm formation by *Roseobacter* species are influenced by culture conditions. Appl Environ Microbiol 73:442–450

Brunvold L, Sandaa RA, Mikkelsen H, Welde E, Bleie H, Bergh O (2007) Characterisation of bacterial communities associated with early stages of intensively reared cod (*Gadus morhua*) using Denaturing Gradient Gel Electrophoresis (DGGE). Aquaculture 272:319–327

Buchan A, Gonzalez JM, Moran MA (2005) Overview of the marine *Roseobacter* lineage. Appl Environ Microbiol 71:5665–5677

Buddruhs N, Pradella S, Goker M, Pauker O, Pukall R, Sproer C, Schumann P, Petersen J, Brinkhoff T (2013) Molecular and phenotypic analyses reveal the non-identity of the *Phaeobacter gallaeciensis* type strain deposits CIP 105210(T) and DSM 17395. Int J Syst Evol Microbiol 63:4340–4349

Budihas SR, Gorshkova I, Gaidamakov S, Wamiru A, Bona MK, Parniak MA, Crouch RJ, McMahon JB, Beutler JA, Le Grice SFJ (2005) Selective inhibition of HIV-1 reverse transcriptase-associated ribonuclease H activity by hydroxylated tropolones. Nucleic Acids Res 33:1249–1256

Burkholder PR, Pfister RM, Leitz FH (1966) Production of a pyrrole antibiotic by a marine bacterium. Appl Microbiol 14:649–653

Cabello FC (2006) Heavy use of prophylactic antibiotics in aquaculture: a growing problem for human and animal health and for the environment. Environ Microbiol 8:1137–1144

Cao LX, Xie LJ, Xue XL, Tan HM, Liu YH, Zhou SN (2007) Purification and characterization of alginate lyase from *Streptomyces* species strain A5 isolated from banana rhizosphere. J Agric Food Chem 55:5113–5117

Cavicchioli R, Siddiqui KS, Andrews D, Sowers KR (2002) Low-temperature extremophiles and their applications. Curr Opin Biotechnol 13:253–261

Cavicchioli R, Charlton T, Ertan H, Omar SM, Siddiqui KS, Williams TJ (2011) Biotechnological uses of enzymes from psychrophiles. Microb Biotechnol 4:449–460

Chatterjee AK, Brown MA (1981) Chromosomal location of a gene (*Idg*) that specifies production of the blue pigment indigoidine in *Erwinia chrysanthemi*. Curr Microbiol 6:269–273

Choi DH, Cho BC (2006) *Citreimonas salinaria* gen. nov., sp nov., a member of the *Roseobacter* clade isolated from a solar saltern. Int J Syst Evol Microbiol 56:2799–2803

Collins AJ, LaBarre BA, Won BSW, Shah MV, Heng S, Choudhury MH, Haydar SA, Santiago J, Nyholm SV (2012) Diversity and partitioning of bacterial populations within the accessory nidamental gland of the squid *Euprymna scolopes*. Appl Environ Microbiol 78:4200–4208

Collins AJ, Fullmer MS, Gogarten JP, Nyholm SV (2015) Comparative genomics of Roseobacter clade bacteria isolated from the accessory nidamental gland of *Euprymna scolopes*. Front Microbiol (6):123

Cotton LA, Graham RJ, Lee RJ (2009) The role of alginate in *P. aeruginosa* PAO1 biofilm structural resistance to gentamicin and ciprofloxacin. J Exp Microbiol Immunol 13:58–62

Cottrell MT, Wood DN, Yu L, Kirchman DL (2000) Selected chitinase genes in cultured and uncultured marine bacteria in the α- and γ-subclasses of the proteobacteria. Appl Environ Microbiol 66:1195–1201

Courtois J (2009) Oligosaccharides from land plants and algae: production and applications in therapeutics and biotechnology. Curr Opin Microbiol 12:261–273

Cude WN, Buchan A (2013) Acyl-honnoserine lactone-based quorum sensing in the *Roseobacter* clade: complex cell-to-cell communication controls multiple physiologies. Front Microbiol (4):336

Cude WN, Mooney J, Tavanaei AA, Hadden MK, Frank AM, Gulvik CA, May AL, Buchan A (2012) Production of the antimicrobial secondary metabolite indigoidine contributes to competitive surface colonization by the marine *Roseobacter Phaeobacter* sp. strain Y4I. Appl Environ Microbiol 78:4771–4780

Cude WN, Prevatte CW, Hadden MK, May AL, Smith RT, Swain CL, Campagna SR, Buchan A (2015) *Phaeobacter* sp. strain Y4I utilizes two separate cell-to-cell communication systems to regulate production of the antimicrobial indigoidine. Appl Environ Microbiol 81:1417–1425

Cunliffe M (2011) Correlating carbon monoxide oxidation with cox genes in the abundant Marine *Roseobacter* Clade. ISME J 5:685–691

D'Alvise PW, Melchiorsen J, Porsby CH, Nielsen KF, Gram L (2010) Inactivation of *Vibrio anguillarum* by attached and planktonic *Roseobacter* cells. Appl Environ Microbiol 76:2366–2370

D'Alvise PW, Lillebo S, Prol-Garcia MJ, Wergeland HI, Nielsen KF, Bergh O, Gram L (2012) *Phaeobacter gallaeciensis* reduces *Vibrio anguillarum* in cultures of microalgae and rotifers, and prevents vibriosis in cod larvae. PLoS One 7:e43996

D'Alvise PW, Lillebo S, Wergeland HI, Gram L, Bergh O (2013) Protection of cod larvae from vibriosis by *Phaeobacter* spp.: a comparison of strains and introduction times. Aquaculture 384:82–86

D'Alvise PW, Magdenoska O, Melchiorsen J, Nielsen KF, Gram L (2014) Biofilm formation and antibiotic production in *Ruegeria mobilis* are influenced by intracellular concentrations of cyclic dimeric guanosinmonophosphate. Environ Microbiol 16:1252–1266

D'Alvise PW, Phippen CBW, Nielsen KF, Gram L (2015) Influence of iron on production of the antibacterial compound tropodithietic acid and its noninhibitory analog in Phaeobacter inhibens. Appl Environ Microbiol 82:502–509

D'Amico S, Claverie P, Collins T, Georlette D, Gratia E, Hoyoux A, Meuwis MA, Feller G, Gerday C (2002) Molecular basis of cold adaptation. Philos Trans Royal Soc Lond B-Biol Sci 357:917–924

Dang HY, Lovell CR (2002) Seasonal dynamics of particle-associated and free-living marine *Proteobacteria* in a salt marsh tidal creek as determined using fluorescence in situ hybridization. Environ Microbiol 4:287–295

De Ruiter GA, Rudolph B (1997) Carrageenan biotechnology. Trends Food Sci Technol 8:389–395

Douillet PA, Langdon CJ (1993) Effects of marine-bacteria on the culture of axenic oyster *Crassostrea gigas* (Thunberg) larvae. Biol Bull 184:36–51

Douillet PA, Langdon CJ (1994) Use of a probiotic for the culture of larvae of the pacific oyster (*Crassostrea gigas* Thunberg). Aquaculture 119:25–40

Elazari-Volcani B (1939) On *Pseudomonas indigofera* (Voges) *Migula* and its pigment. Arch Microbiol 10:343–358

Enticknap JJ, Kelly M, Peraud O, Hill RT (2006) Characterization of a culturable alphaproteobacterial symbiont common to many marine sponges and evidence for vertical transmission via sponge larvae. Appl Environ Microbiol 72:3724–3732

FAO (2001) Report of a joint FAO/WHO expert consultation on evaluation of health and nutritional properties of probiotics in food including powder milk with live lactic acid bacteria. Accessed July 13th, 2015: ftp://ftp.fao.org/docrep/fao/009/a0512e/a0512e00.pdf

FAO – Fishery Statistical Collections (2015) Global aquaculture production 1950-2013. Accessed July 13th, 2015: http://www.fao.org/fishery/statistics/global-aquaculture-production/query/en

Fdhila F, Vazquez V, Sanchez JL, Riguera R (2003) DD-diketopiperazines: antibiotics active against *Vibrio anguillarum* isolated from marine bacteria associated with cultures of *Pecten maximus*. J Nat Prod 66:1299–1301

Fehér D, Barlow R, McAtee J, Hemscheidt TK (2010) Highly brominated antimicrobial metabolites from a marine *Pseudoalteromonas* sp. J Nat Prod 73:1963–1966

Feller G, Gerday C (2003) Psychrophilic enzymes: hot topics in cold adaptation. Nat Rev Microbiol 1:200–208

Fernandez P (2014) Marine enzymes and food industry: insight on existing and potential interactions. Front Mar Sci 1:46

Funkhouser JD, Aronson NN (2007) Chitinase family GH18: evolutionary insights from the genomic history of a diverse protein family. BMC Evol Biol 7:96

Gallacher S, Flynn KJ, Franco JM, Brueggemann EE, Hines HB (1997) Evidence for production of paralytic shellfish toxins by bacteria associated with *Alexandrium* spp (Dinophyta) in culture. Appl Environ Microbiol 63:239–245

Garcia MJP, D'Alvise PW, Gram L (2013) Disruption of cell-to-cell signaling does not Abolish the antagonism of *Phaeobacter gallaeciensis* toward the fish pathogen *Vibrio anguillarum* in algal systems. Appl Environ Microbiol 79:5414–5417

Gardiner DM, Waring P, Howlett BJ (2005) The epipolythiodioxopiperazine (ETP) class of fungal toxins: distribution, mode of action, functions and biosynthesis. Microbiol-Sgm 151:1021–1032

Gatesoupe FJ (1991) The effect of three strains of lactic bacteria on the production rate of rotifers, *Brachionus plicatilis*, and their dietary value for larval turbot, *Scophthalmus maximus*. Aquaculture 96:335–342

Geng HF, Belas R (2010a) Expression of tropodithietic acid biosynthesis is controlled by a novel autoinducer. J Bacteriol 192:4377–4387

Geng HF, Belas R (2010b) Molecular mechanisms underlying roseobacter-phytoplankton symbioses. Curr Opin Biotechnol 21:332–338

Georlette D, Blaise V, Collins T, D'Amico S, Gratia E, Hoyoux A, Marx JC, Sonan G, Feller G, Gerday C (2004) Some like it cold: biocatalysis at low temperatures. FEMS Microbiol Rev 28:25–42

Gerday C, Aittaleb M, Bentahir M, Chessa JP, Claverie P, Collins T, D'Amico S, Dumont J, Garsoux G, Georlette D, Hoyoux A, Lonhienne T, Meuwis MA, Feller G (2000) Cold-adapted enzymes: from fundamentals to biotechnology. Trends Biotechnol 18:103–107

Giebel HA, Brinkhoff T, Zwisler W, Selje N, Simon M (2009) Distribution of *Roseobacter* RCA and SAR11 lineages and distinct bacterial communities from the subtropics to the southern ocean. Environ Microbiol 11:2164–2178

Gonzalez JM, Moran MA (1997) Numerical dominance of a group of marine bacteria in the alpha-subclass of the class *Proteobacteria* in coastal seawater. Appl Environ Microbiol 63:4237–4242

Gonzalez JM, Mayer F, Moran MA, Hodson RE, Whitman WB (1997) *Sagittula stellata* gen. nov, sp. nov, a lignin-transforming bacterium from a coastal environment. Int J Syst Bacteriol 47:773–780

Gonzalez JM, Simo R, Massana R, Covert JS, Casamayor EO, Pedros-Alio C, Moran MA (2000) Bacterial community structure associated with a dimethylsulfoniopropionate-producing North Atlantic algal bloom. Appl Environ Microbiol 66:4237–4246

Gooday GW (1990) The ecology of chitin degradation. Adv Microb Ecol 11:387–430

Gram L, Grossart HP, Schlingloff A, Kiorboe T (2002) Possible quorum sensing in marine snow bacteria: production of acylated homoserine lactones by *Roseobacter* strains isolated from marine snow. Appl Environ Microbiol 68:4111–4116

Gram L, Rasmussen BB, Wemheuer B, Bernbum N, Ng YY, Porsby CH, Breider S, and Brinkhoff T (2015) *Phaeobacter inhibens* from the *Roseobacter*-clade has an environmental niche as surface colonizer in harbours. Syst Appl Microbiol 38:483–493

Griffiths S, Melville K, Cook M, Vincent S, St Pierre M, Lanteigne C (2001) Profiling of bacterial species associated with haddock larviculture by PCR amplification of 16S rDNA and denaturing gradient gel electrophoresis. J Aquat Anim Health 13:355–363

Grigioni S, Boucher-Rodoni R, Demarta A, Tonolla M, Peduzzi R (2000) Phylogenetic characterisation of bacterial symbionts in the accessory nidamental glands of the sepioid *Sepia officinalis* (Cephalopoda: Decapoda). Mar Biol 136:217–222

Grotkjær T, Bentzon-Tilia M, D'alvise P, Dourala N, Nielsen KF, Gram L (2016) Isolation of TDA-producing Phaeobacter strains from sea bass larval rearing units and their probiotic effect against pathogenic Vibrio spp. in Artemia cultures. Syst Appl Microbiol 39:108–188

Hancock D, Besser T, Gay J, Rice D, Davis M, Gay C (2000) The global epidemiology of multiresistant *Salmonella enterica* serovar typhimurium DT104. In: Brown C, Bolin C (eds) Emerging diseases of animals. ASM Press, Washington, pp. 217–243

Harrington C, Reen FJ, Mooij MJ, Stewart FA, Chabot JB, Guerra AF, Glockner FO, Nielsen KF, Gram L, Dobson ADW, Adams C, O'Gara F (2014) Characterisation of non-autoinducing Tropodithietic Acid (TDA) production from marine sponge *Pseudovibrio* species. Marine Drugs 12:5960–5978

Heumann W, Young D, Gottlich C (1968) Leucoindigoidine formation by an *Arthrobacter* species and its oxidation to indigoidine by other micro-organisms. Biochim Biophys Acta 156:429–431

Hien NQ, Nagasawa N, Tham LX, Yoshii F, Dang VH, Mitomo H, Makuuchi K, Kume T (2000) Growth-promotion of plants with depolymerized alginates by irradiation. Radiat Phys Chem 59:97–101

Hjelm M, Bergh O, Riaza A, Nielsen J, Melchiorsen J, Jensen S, Duncan H, Ahrens P, Birkbeck H, Gram L (2004) Selection and identification of autochthonous potential probiotic bacteria from turbot larvae (*Scophthalmus maximus*) rearing units. Syst Appl Microbiol 27:360–371

Holmes AJ, Kelly DP, Baker SC, Thompson AS, DeMarco P, Kenna EM, Murrell JC (1997) *Methylosulfonomonas methylovora* gen nov, sp nov, and *Marinosulfonomonas methylotropha* gen nov, sp nov: novel methylotrophs able to grow on methanesulfonic acid. Arch Microbiol 167:46–53

Honda G, Tabata M, Tsuda M (1979) The antimicrobial specificity of tryptanthrin. Planta Med 37:172–174

Hughes CC, Prieto-Davo A, Jensen PR, Fenical W (2008) The marinopyrroles, antibiotics of an unprecedented structure class from a marine *Streptomyces* sp. Org Lett 10:629–631

Ivanova EP, Gorshkova NM, Sawabe T, Zhukova NV, Hayashi K, Kurilenko VV, Alexeeva Y, Buljan V, Nicolau DV, Mikhailov VV, Christen R (2004) *Sulfitobacter delicatus* sp nov and *Sulfitobacter dubius* sp nov., respectively from a starfish (*Stellaster equestris*) and sea grass (*Zostera marina*). Int J Syst Evol Microbiol 54:475–480

Iwamoto M, Kurachi M, Nakashima T, Kim D, Yamaguchi K, Oda T, Iwamoto Y, Muramatsu T (2005) Structure-activity relationship of alginate oligosaccharides in the induction of cytokine production from RAW264.7 cells. FEBS Lett 579:4423–4429

Iwasaki K, Matsubara Y (2000) Purification of alginate oligosaccharides with root growth-promoting activity toward lettuce. Biosci Biotechnol Biochem 64:1067–1070

Je JY, Kim SK (2006) Antimicrobial action of novel chitin derivative. Biochimica et Biophysica Acta-General Subjects 1760:104–109

Jiang Z, Boyd KG, Mearns-Spragg A, Adams DR, Wright PC, Burgess JG (2000) Two diketopiperazines and one halogenated phenol from cultures of the marine bacterium, *Pseudoalteromonas luteoviolacea*. Nat Prod Lett 14:435–440

Jonkers HM, Abed RMM (2003) Identification of aerobic heterotrophic bacteria from the photic zone of a hypersaline microbial mat. Aquat Microb Ecol 30:127–133

Jordaan A and Kliing LJ (2003) Determining the optimal temperature range of Atlantic cod (*Gadus morhua*) during early life. In: HI Browman and AB Skiftesvik (eds) The big fish bang. Proceedings of the 26th annual larval fish xonference. Institute of Marine Research, Bergen, Norway

Junge K, Imhoff F, Staley T, Deming JW (2002) Phylogenetic diversity of numerically important arctic sea-ice bacteria cultured at subzero temperature. Microb Ecol 43:315–328

Kataoka M, Hirata K, Kunikata T, Ushio S, Iwaki K, Ohashi K, Ikeda M, Kurimoto M (2001) Antibacterial action of tryptanthrin and kaempferol, isolated from the indigo plant (*Polygonum tinctorium* Lour.), against *Helicobacter pylori*-infected Mongolian gerbils. J Gastroenterol 36:5–9

Kawano Y, Nagawa H, Hakanishi H, Nakajim H, Matsuo M, Higashihara H (1997) Production of thiotropocin by a marine bacterium *Caulobacter* sp. and its antimicroalgal activities. J Mar Biotechnol 5:225–229

Kenig M, Reading C (1979) Holomycin and an antibiotic (Mm-19290) related to tunicamycin, metabolites of *Streptomyces clavuligerus*. J Antibiot 32:549–554

Keyhani NO, Roseman S (1999) Physiological aspects of chitin catabolism in marine bacteria. Biochim Biophys Acta-Gen Subj 1473:108–122

Kim HS, Lee CG, Lee EY (2011) Alginate lyase: structure, property, and application. Biotechnol Bioprocess Eng 16:843–851

Kimoto T, Hino K, Koya-Miyata S, Yamamoto Y, Takeuchi M, Nishizaki Y, Micallef MJ, Ushio S, Iwaki K, Ikeda M, Kurimoto M (2001) Cell differentiation and apoptosis of monocytic and promyelocytic leukemia cells (U-937 and HL-60) by tryptanthrin, an active ingredient of *Polygonum tinctorium* Lour. Pathol Int 51:315–325

King GA (2003) Molecular and culture-based analyses of aerobic carbon monoxide oxidizer diversity. Appl Environ Microbiol 69:7257–7265

Kintaka K, Ono H, Tsubotani S, Harada S, Okazaki H (1984) Thiotropocin, a new sulfur-containing 7-membered-ring antibiotic produced by A *Pseudomonas* Sp. J Antibiot 37:1294–1300

Krivogorsky B, Grundt P, Yolken R, Jones-Brando L (2008) Inhibition of *Toxoplasma gondii* by indirubin and tryptanthrin analogs. Antimicrob Agents Chemother 52:4466–4469

Kurachi M, Nakashima T, Miyajima C, Iwamoto Y, Muramatsu T, Yamaguchi K, Oda T (2005) Comparison of the activities of various alginates to induce TNF-alpha secretion in RAW264.7 cells. J Infect Chemother 11:199–203

Labrenz M, Collins MD, Lawson PA, Tindall BJ, Braker G, Hirsch P (1998) *Antarctobacter heliothermus* gen. nov., sp. nov., a budding bacterium from hypersaline and heliothermal Ekho lake. Int J Syst Bacteriol 48:1363–1372

Lenk S, Moraru C, Hahnke S, Arnds J, Richter M, Kube M, Reinhardt R, Brinkhoff T, Harder J, Amann R, Mussmann M (2012) *Roseobacter* clade bacteria are abundant in coastal sediments and encode a novel combination of sulfur oxidation genes. ISME J 6:2178–2187

Li L, Guenzennec J, Nichols P, Henry P, Yanagibayashi M, Kato C (1999) Microbial diversity in Nankai Trough sediments at a depth of 3,843 m. J Oceanogr 55:635–642

Long RA, Azam F (2001) Antagonistic interactions among marine pelagic bacteria. Appl Environ Microbiol 67:4975–4983

Lopez-Garcia P, Duperron S, Philippot P, Foriel J, Susini J, Moreira D (2003) Bacterial diversity in hydrothermal sediment and epsilon proteobacterial dominance in experimental microcolonizers at the Mid-Atlantic ridge. Environ Microbiol 5:961–976

Lu GF, Lomonosova E, Cheng XH, Moran EA, Meyers MJ, Le Grice SFJ, Thomas CJ, Jiang JK, Meck C, Hirsch DR, D'Erasmo MP, Suyabatmaz DM, Murelli RP, Tavis JE (2015) Hydroxylated tropolones inhibit hepatitis B virus replication by blocking viral ribonuclease H activity. Antimicrob Agents Chemother 59:1070–1079

Lund P, Lee RY, Dunsmuir P (1989) Bacterial chitinase is modified and secreted in transgenic tobacco. Plant Physiol 91:130–135

Machado H, Sonnenschein EC, Melchiorsen J, Gram L (2015) Genome mining reveals unlocked bioactive potential of marine gram-negative bacteria. BMC Genomics 16:1

Maeda M, Liao CI (1992) Effect of bacterial population on the growth of a prawn larva, *Penaeus monodon*. Bull Natl Res Inst Aquacult 21:25–29

Maloy AP, Ford SE, Karney RC, Boettcher KJ (2007) *Roseovarius crassostreae*, the etiological agent of Juvenile Oyster Disease (now to be known as Roseovarius Oyster Disease) in *Crassostrea virginica*. Aquaculture 269:71–83

Mangor-Jensen A, Adoff GR (1987) Drinking activity of the newly hatched larvae of cod *Gadus morhua* l. Fish Physiol Biochem 3:99–103

Martens T, Heidorn T, Pukall R, Simon M, Tindall BJ, Brinkhoff T (2006) Reclassification of *Roseobacter gallaeciensis* Ruiz-Ponte et al. 1998 as *Phaeobacter gallaeciensis* gen. nov., comb. nov., description of *Phaeobacter inhibens* sp nov., reclassification of *Ruegeria algicola* (Lafay et al. 1995) Uchino et al 1999 as *Marinovum algicola* gen. nov., comb. nov., and emended descriptions of the genera *Roseobacter*, *Ruegeria* and *Leisingera*. Int J Syst Evol Microbiol 56:1293–1304

Martens T, Gram L, Grossart HP, Kessler D, Muller R, Simon M, Wenzel SC, Brinkhoff T (2007) Bacteria of the *Roseobacter* clade show potential for secondary metabolite production. Microb Ecol 54:31–42

Meyer FP (1991) Aquaculture disease and health management. J Anim Sci 69:4201–4208

Micallef MJ, Iwaki K, Ishihara T, Ushio S, Aga M, Kunikata T, Koya-Miyata S, Kimoto T, Ikeda M, Kurimoto M (2002) The natural plant product tryptanthrin ameliorates dextran sodium sulfate-induced colitis in mice. Int Immunopharmacol 2:565–578

Mills HJ, Hodges C, Wilson K, MacDonald IR, Sobecky PA (2003) Microbial diversity in sediments associated with surface-breaching gas hydrate mounds in the Gulf of Mexico. FEMS Microbiol Ecol 46:39–52

Mitscher LA, Baker W (1998) Tuberculosis: a search for novel therapy starting with natural products. Med Res Rev 18:363–374

Mohamed NM, Cicirelli EM, Kan JJ, Chen F, Fuqua C, Hill RT (2008) Diversity and quorum-sensing signal production of *Proteobacteria* associated with marine sponges. Environ Microbiol 10:75–86

Mollah MZI, Khan MA, Khan RA (2009) Effect of gamma irradiated sodium alginate on red amaranth (*Amaranthus cruentus* L.) as growth promoter. Radiat Phys Chem 78:61–64

Moran MA, Buchan A, Gonzalez JM, Heidelberg JF, Whitman WB, Kiene RP, Henriksen JR, King GM, Belas R, Fuqua C, Brinkac L, Lewis M, Johri S, Weaver B, Pai G, Eisen JA, Rahe E, Sheldon WM, Ye WY, Miller TR, Carlton J, Rasko DA, Paulsen IT, Ren QH, Daugherty SC, Deboy RT, Dodson RJ, Durkin AS, Madupu R, Nelson WC, Sullivan SA, Rosovitz MJ, Haft DH, Selengut J, Ward N (2004) Genome sequence of *Silicibacter pomeroyi* reveals adaptations to the marine environment. Nature 432:910–913

Morita Y, Matsumura E, Tsujibo H, Yasuda M, Okabe T, Sakagami Y, Kumeda Y, Ishida N, Inamori Y (2002) Biological activity of 4-acetyltropolone, the minor component of Thujopsis dolabrata SIEB. et Zucc. hondai Mak. Biol Pharm Bull 25:981–985

Mullins TD, Britschgi TB, Krest RL, Giovannoni SJ (1995) Genetic comparisons reveal the same unknown bacterial lineages in Atlantic and Pacific bacterioplankton communities. Limnol Oceanogr 40:148–158

Murphy BT, Jensen PR, Fenical W (2012) The chemistry of marine bacteria. In: Fattorusso E, Gerwick WH, Taglialatela-Scafati O (eds) Handbook of marine natural products. Springer, New York, pp. 153–190

Muthusamy S, Baltar F, Gonazlez JM, Pinhassi J (2014) Dynamics of metabolic activities and gene expression in the *Roseobacter* clade bacterium *Phaeobacter* sp strain MED193 during growth with thiosulfate. Appl Environ Microbiol 80:6933–6942

Newton RJ, Griffin LE, Bowles KM, Meile C, Gifford S, Givens CE, Howard EC, King E, Oakley CA, Reisch CR, Rinta-Kanto JM, Sharma S, Sun SL, Varaljay V, Vila-Costa M, Westrich JR, Moran MA (2010) Genome characteristics of a generalist marine bacterial lineage. ISME J 4:784–798

Ngo DN, Kim MM, Kim SK (2008) Chitin oligosaccharides inhibit oxidative stress in live cells. Carbohydr Polym 74:228–234

Ngo DN, Lee SH, Kim MM, Kim SK (2009) Production of chitin oligosaccharides with different molecular weights and their antioxidant effect in RAW 264.7 cells. J Funct Foods 1:188–198

Ngo DN, Kim MM, Kim SK (2012) Protective effects of aminoethyl-chitooligosaccharides against oxidative stress in mouse macrophage RAW 264.7 cells. Int J Biol Macromol 50:624–631

Nielsen PH (2005) Life cyclce assessment supports cold-wash enzymes. Int J Appl Sci 131

Nogami K, Maeda M (1992) Bacteria as biocontrol agents for rearing larvae of the crab *Portunus trituberculatus*. Can J Fish Aquat Sci 49:2373–2376

Osama R, Koga T (1995) An investigation of aquatic bacteria capable of utilizing chitin as the sole source of nutrients. Lett Appl Microbiol 21:288–291

Park C, Shin HJ, Kim GY, Kwon TK, Nam TJ, Kim SK, Cheong J, Choi IW, Choi YH (2008) Induction of apoptosis by streptochlorin isolated from Streptomyces sp in human leukemic U937 cells. Toxicol in Vitro 22:1573–1581

Penesyan A, Tebben J, Lee M, Thomas T, Kjelleberg S, Harder T, Egan S (2011) Identification of the antibacterial compound produced by the marine epiphytic bacterium *Pseudovibrio* sp D323 and related sponge-associated bacteria. Marine Drugs 9:1391–1402

Planas M, Perez-Lorenzo M, Hjelm M, Gram L, Fiksdal IU, Bergh O, Pintado J (2006) Probiotic effect in vivo of *Roseobacter* strain 27-4 against *Vibrio* (*Listonella*) *anguillarum* infections in turbot (*Scophthalmus maximus* L.) larvae. Aquaculture 255:323–333

Porsby CH, Nielsen KF, Gram L (2008) *Phaeobacter* and *Ruegeria* species of the *Roseobacter* clade colonize separate niches in a Danish turbot (*Scophthalmus maximus*)-rearing farm and antagonize *Vibrio anguillarum* under different growth conditions. Appl Environ Microbiol 74:7356–7364

Porsby CH, Webber MA, Nielsen KF, Piddock LJV, Gram L (2011) Resistance and tolerance to tropodithietic acid, an antimicrobial in aquaculture, is hard to select. Antimicrob Agents Chemother 55:1332–1337

Powell SM, Bowman JP, Snape I, Stark JS (2003) Microbial community variation in pristine and polluted nearshore Antarctic sediments. FEMS Microbiol Ecol 45:135–145

Prado S, Montes J, Romalde JL, Barja JL (2009) Inhibitory activity of *Phaeobacter* strains against aquaculture pathogenic bacteria. Int Microbiol 12:107–114

Pukall R, Buntefuss D, Fruhling A, Rohde M, Kroppenstedt RM, Burghardt J, Lebaron P, Bernard L, Stackebrandt E (1999) *Sulfitobacter mediterraneus* sp. nov., a new sulfite-oxidizing member of the alpha-Proteobacteria. Int J Syst Bacteriol 49:513–519

Ramsey DM, Wozniak DJ (2005) Understanding the control of *Pseudomonas aeruginosa* alginate synthesis and the prospects for management of chronic infections in cystic fibrosis. Mol Microbiol 56:309–322

Rasmussen BB, Grotkjær T, D'Alvise PW, Yin G, Zhang F, Bunk B, Spröer C, Bentzon-Tilia M, Gram L (2016) Vibrio anguillarum is genetically and phenotypically unaffected by long-term continuous exposure to the antibacterial compound tropodithietic acid. Appl Environ Microbiol 82:4802–4810

Recio MC, Cerda-Nicolas M, Potterat O, Hamburger M, Rios JL (2006) Anti-inflammatory and antiallergic activity in vivo of lipophilic Isatis tinctoria extracts and tryptanthrin. Planta Med 72:539–546

Rhodes G, Huys G, Swings J, McGann P, Hiney M, Smith P, Pickup RW (2000) Distribution of oxytetracycline resistance plasmids between aeromonads in hospital and aquaculture environments: Implication of Tn1721 in dissemination of the tetracycline resistance determinant Tet A. Appl Environ Microbiol 66:3883–3890

Rüger HJ, Höfle MG (1992) Marine star-shaped-aggregate-forming bacteria – *Agrobacterium atlanticum* sp. nov.; *Agrobacterium meteori* sp. nov.; *Agrobacterium ferrugineum* sp. nov., nom. rev.; *Agrobacterium gelatinovorum* sp. nov., nom. rev.; and *Agrobacterium stellulatum* sp. nov., nom. rev. Int J Syst Bacteriol 42:133–143

Ruiz-Ponte C, Cilia V, Lambert C, Nicolas JL (1998) *Roseobacter gallaeciensis* sp. nov., a new marine bacterium isolated from rearings and collectors of the scallop *Pecten maximus*. Int J Syst Bacteriol 48:537–542

Ruiz-Ponte C, Samain JF, Sanchez JL, Nicolas JL (1999) The benefit of a *Roseobacter* species on the survival of scallop larvae. Mar Biotechnol 1:52–59

Russell NJ (1998) Molecular adaptations in psychrophilic bacteria: potential for biotechnological applications. Adv Biochem Eng Biotechnol 61:1–21

Schaefer JK, Goodwin KD, McDonald IR, Murrell JC, Oremland RS (2002) *Leisingera methylohatidivorans* gen. nov., sp nov., a marine methylotroph that grows on methyl bromide. Int J Syst Evol Microbiol 52:851–859

Schmidt TM, Delong EF, Pace NR (1991) Analysis of a marine picoplankton community by 16S ribosomal-RNA gene cloning and sequencing. J Bacteriol 173:4371–4378

Seyedsayamdost MR, Carr G, Kolter R, Clardy J (2011a) Roseobacticides: small molecule modulators of an algal-bacterial symbiosis. J Am Chem Soc 133:18343–18349

Seyedsayamdost MR, Case RJ, Kolter R, Clardy J (2011b) The Jekyll-and-Hyde chemistry of *Phaeobacter gallaciensis*. Nat Chem 3:331–335

Shiba T (1991) *Roseobacter litoralis* gen. nov., sp. nov., and *Roseobacter denitrificans* sp. nov., aerobic pink-pigmented bacteria which contain bacteriochlorophyll *a*. Syst Appl Microbiol 14:140–145

Shin HJ, Jeong HS, Lee HS, Park SK, Kim HM, Kwon HJ (2007) Isolation and structure determination of streptochlorin, an antiproliferative agent from a marine-derived *Streptomyces* sp 04DH110. J Microbiol Biotechnol 17:1403–1406

Shin DY, Shin HJ, Kim GY, Cheong J, Choi IW, Kim SK, Moon SK, Kang HS, Choi YH (2008) Streptochlorin isolated from *Streptomyces* sp induces apoptosis in human hepatocarcinoma cells through a reactive oxygen species-mediated mitochondrial pathway. J Microbiol Biotechnol 18:1862–1867

Skjermo J, Vadstein O (1999) Techniques for microbial control in the intensive rearing of marine larvae. Aquaculture 177:333–343

Skjermo J, Salvesen I, Oie G, Olsen Y, Vadstein O (1997) Microbially matured water: a technique for selection of a non-opportunistic bacterial flora in water that may improve performance of marine larvae. Aquac Int 5:13–28

Slightom RN, Buchan A (2009) Surface colonization by marine roseobacters: integrating genotype and phenotype. Appl Environ Microbiol 75:6027–6037

Smalås AO, Leiros HK, Os V, Willassen NP (2000) Cold adapted enzymes. Biotechnol Annu Rev 6:1–57

Solbak AI, Richardson TH, Mccann RT, Kline KA, Bartnek F, Tomlinson G, Tan XQ, Parra-Gessert L, Frey GJ, Podar M, Luginbuhl P, Gray KA, Mathur EJ, Robertson DE, Burk MJ, Hazlewood GP, Short JM, Kerovuo J (2005) Discovery of pectin-degrading enzymes and directed evolution of a novel pectate lyase for processing cotton fabric. J Biol Chem 280:9431–9438

Somero GN (2004) Adaptation of enzymes to temperature: searching for basic "strategies". Comp Biochem Physiol B Biochem Mol Biol 139:

Sorokin DY (1994) Influence of thiosulfate on the growth of sulfate-producing sulfur-oxidizing heterotrophic bacteria from the black-sea in continuous-culture. Microbiology 63:255–259

Soto-Gil RW, Zyskind JW (1984) Cloning of *Vibrio harveyi* chitinase and chitobiase genes in *Escherichia coli*. In: Zikakis JP (ed) Chitin, chitosan, and related enzymes. Academic Press, Inc., New York, pp. 209–223

Speitling M, Smetanina OE, Kuznetsova TA, and Laatsch H (2007) Bromoalterochromides A and A', unprecedented chromopeptides from a marine *Pseudoalteromonas maricaloris* strain KMM 636. J Antibiot 60: 36-42

Starr MP, Cosens G, Knackmuss HJ (1966) Formation of the blue pigment indigoidine by phytopathogenic *Erwinia*. Appl Microbiol 14:870–872

Suslow TV and Jones J (1988) Chitinase-producing bacteria. Patent [US 4751081 A]. 1988

Takahashi H, Kumagai T, Kitani K, Mori M, Matoba Y, Sugiyama M (2007) Cloning and characterization of a *Streptomyces* single module type non-ribosomal peptide synthetase catalyzing a blue pigment synthesis. J Biol Chem 282:9073–9081

Taubes G (2008) The bacteria fight back. Science 321:356–361

Techkarnjanaruk S, Goodman AE (1999) Multiple genes involved in chitin degradation from the marine bacterium *Pseudoalteromonas* sp. strain S91. Microbiol-Uk 145:925–934

Thiel V, Kunze B, Verma P, Wagner-Dobler I, Schulz S (2009) New structural variants of homoserine lactones in bacteria. Chembiochem 10:1861–1868

Trust TJ (1975) Antibacterial activity of tropolone. Antimicrob Agents Chemother 7:500–506

Tsubotani S, Wada Y, Kamiya K, Okazaki H, Harada S (1984) Structure of thiotropocin, a new sulfur-containing 7-membered antibiotic. Tetrahedron Lett 25:419–422

Tsujibo H, Yoshida Y, Imada C, Okami Y, Miyamoto K, Inamori Y (1991) Isolation and characterization of a chitin degrading marine bacterium belonging to the genus *Alteromonas*. Nippon Suisan Gakkaishi 57:2127–2131

Tutino ML, di Prisco G, Marino G, de Pascale D (2009) Cold-adapted esterases and lipases: from fundamentals to application. Protein Pept Lett 16:1172–1180

Van Trappen S, Mergaert J, Swings J (2004) *Loktanella salsilacus* gen. nov., sp nov., *Loktanella fryxellensis* sp. nov. and *Loktanella vestfoldensis* sp. nov., new members of the *Rhodobacter* group, isolated from microbial mats in Antarctic lakes. Int J Syst Evol Microbiol 54:1263–1269

Vo TS, Ngo DH, Ta QV, Wijesekara I, Kong CS, Kim SK (2012) Protective effect of chitin oligosaccharides against lipopolysaccharide-induced inflammatory response in BV-2 microglia. Cell Immunol 277:14–21

Wagner-Döbler I, Rheims H, Felske A, El-Ghezal A, Flade-Schorder D, Laatsch H, Lang S, Pukall R, Tindall BJ (2004) *Oceanibulbus indolifex* gen. nov., sp nov., a North Sea alphaproteobacterium that produces bioactive metabolites. Int J Syst Evol Microbiol 54:1177–1184

Wagner-Döbler I, Thiel V, Eberl L, Allgaier M, Bodor A, Meyer S, Ebner S, Hennig A, Pukall R, Schulz S (2005) Discovery of complex mixtures of novel long-chain quorum sensing signals in free-living and host-associated marine alphaproteobacteria. Chembiochem 6:2195–2206

Wang Y, Han F, Hu B, Li JB, Yu WG (2006) In vivo prebiotic properties of alginate oligosaccharides prepared through enzymatic hydrolysis of alginate. Nutr Res 26:597–603

West NJ, Obernosterer I, Zemb O, Lebaron P (2008) Major differences of bacterial diversity and activity inside and outside of a natural iron-fertilized phytoplankton bloom in the Southern Ocean. Environ Microbiol 10:738–756

Whitman WB, Coleman DC, Wiebe WJ (1998) Prokaryotes: the unseen majority. Proc Natl Acad Sci U S A 95:6578–6583

Wietz M, Gram L, Jorgensen B, Schramm A (2010) Latitudinal patterns in the abundance of major marine bacterioplankton groups. Aquat Microb Ecol 61:179–189

Wietz M, Månsson M, Vynne NG, Gram L (2013) Small-molecule antibiotics from marine bacteria and strategies to prevent rediscovery of known compounds. In: Kim SK (ed) Marine microbiology: bioactive compounds and biotechnological applications. Whiley-VCH Verlag, Veinheim, pp. 127–159

Wightman F, Lighty DL (1982) Identification of phenylacetic acid as a natural auxin in the shoots of higher-plants. Physiol Plant 55:17–24

Wilson MZ, Wang R, Gitai Z, Seyedsayamdost MR (2016) Mode of action and resistance studies unveil new roles for tropodithietic acid as an anticancer agent and the γ-glutamyl cycle as a proton sink. Proc Natl Acad Sci U S A 113:1630–1635

Wortman AT, Somerville CC, Colwell RR (1986) Chitinase determinants of *Vibrio vulnificus* – gene cloning and applications of a chitinase probe. Appl Environ Microbiol 52:142–145

Yonemoto Y, Tanaka H, Yamashita T, Kitabatake N, Ishida Y, Kimura A, Murata K (1993) Promotion of germination and shoot elongation of some plants by alginate oligomers prepared with bacterial alginate lyase. J Ferment Bioeng 75:68–70

Yoshikawa K, Adachi K, Nishida F, Mochida K (2003) Planar structure and antibacterial activity of korormicin derivatives isolated from *Pseudoalteromonas* sp F-420. J Antibiot 56:866–870

Yu ST, Chen TM, Tseng SY, Chen YH (2007) Tryptanthrin inhibits MDR1 and reverses doxorubicin resistance in breast cancer cells. Biochem Biophys Res Commun 358:79–84

Zan JD, Cicirelli EM, Mohamed NM, Sibhatu H, Kroll S, Choi O, Uhlson CL, Wysoczinski CL, Murphy RC, Churchill MEA, Hill RT, Fuqua C (2012) A complex LuxR-LuxI type quorum sensing network in a roseobacterial marine sponge symbiont activates flagellar motility and inhibits biofilm formation. Mol Microbiol 85:916–933

Zhang C and Kim SK (2010) Research and application of marine microbial enzymes: status and prospects. Mar Drugs 8: 1920-1934

Zobell CE, Rittenberg SC (1938) The occurrence and characteristics of chitinoclastic bacteria in the sea. J Bacteriol 35:275–287

Zubkov MV, Fuchs BM, Archer SD, Kiene RP, Amann R, Burkill PH (2001) Linking the composition of bacterioplankton to rapid turnover of dissolved dimethylsulphoniopropionate in an algal bloom in the North Sea. Environ Microbiol 3:304–311

Chapter 8
Iwokrama Fungal/Plant Bioprospecting Project 2000–2003 – A Model for the Future?

Ramish Pingal

Abstract In 2000 the Iwokrama International Centre for Rainforest Conservation and Development (IIC), Guyana embarked on a bioprospecting and bioinventory project of endophytic fungi and their plant hosts in partnership with other institutions. The project was called the Conservation and Sustainable Utilization of Biodiversity in the Iwokrama Forest (CSUBIF) programme and was funded by the European Commission (EC). The main objectives were (a) establishment of a bioprospecting-bioinventory laboratory in Guyana, (b) an inventory of Iwokrama's unknown endophytic fungi, (c) isolation of bioactive natural compounds from fungal endophytes and host plants, and (d) development/commercialisation of natural products. A local laboratory was established at the University of Guyana to collect, identify and analyse samples. The project was an overall success having identified (a) hitherto unknown endophytic fungi and (b) 110 fungal and 29 plant extract leads for further research and development. The major constraints of the project was its short duration, lack of Intellectual Property Rights (IPR) agreements and access to genetic resources and benefit sharing (ABS) protocols. There has never been a more important time to reassess efforts such as the Iwokrama Bioprospecting Project, given the current crisis caused by resistant bacterial and fungal diseases and the lack of effective antibiotics with which to treat them.

8.1 Background

Guyana dedicated the Iwokrama Forest (Fig. 8.1) to the international community to demonstrate how tropical forests could provide economic benefits at the 1989 Commonwealth of Nations Heads of Government meeting. This pristine forest comprises 371,000 hectares which should be utilized without compromising ecological integrity or biodiversity. The Iwokrama International Centre for Rainforest Conservation and Development (IIC) was legally established by Guyana and the

R. Pingal (✉)
Department of Chemistry, The University of the West Indies,
St. Augustine, Trinidad and Tobago
e-mail: Ramish.Pingal@sta.uwi.edu

R. Paterson, N. Lima (eds.), *Bioprospecting*, Topics in Biodiversity and Conservation 16, DOI 10.1007/978-3-319-47935-4_8

Fig. 8.1 Map of Guyana showing the location of the Iwokrama forest

Commonwealth Secretariat in 1996 as an autonomous, not-for-profit institution governed by an International Board of Trustees with the mission to:

> "Promote the conservation and sustainable and equitable use of tropical rain forest in a manner that will lead to lasting ecological, economic and social benefits to the people of Guyana and to the world in general, by undertaking research, training and the development and dissemination of technologies."

From 1993 to 2000 funding to establish IIC came from the Government of Guyana and three main donor agencies, the: (a) Global Environmental Facility (GEF) through the United Nations Development Programme (UNDP), (b) Commonwealth of Nations Secretariat and (c) International Development Research Centre (IDRC), Canada. From 2000–2007, Iwokrama's work centered on the collection of baseline information, such as biodiversity inventory surveys, to assist with the management of the reserve which included zoning of the forest, forest management and surveillance. Other research projects conducted during that time included forest utilization, surveys of the wetlands and river systems bordering the Iwokrama forest, marketing, social research and biodiversity prospecting. The IIC was supported by several donor funded programmes, the major ones being: (a) Sustainable Human Development, (Department for International Development (DFID), UK), (b) Sustainable Forest Management (International Tropical Timber Organisation (ITTO), (c) Ecotourism Development and Programme Support (Canadian International Development Agency (CIDA), and (d) Conservation and Sustainable Utilization of Biodiversity in the Iwokrama Forest (CSUBIF) (European Commission (EC). The CSUBIF project included a component concerning bioprospecting.

8.2 Conservation and Sustainable Utilisation of Biodiversity in the Iwokrama Forest Project

8.2.1 Project Conception

The CSUBIF project proposal was conceived by IIC in accordance with its Operations Plan 1998–2002 and Business Plan 1998–2007 (Baines and Warner 2000), both of which featured bioprospecting as one of the prominent potential revenue generating activities for the Centre, together with eco-forestry, certified logging, sustainable production of non-timber forest products, eco-tourism, training services and the sale of expert forest management services (Gilmour 1999). The project proposal, comprising three components - Wilderness Preserve, Bioinventory, and Bioprospecting – was presented to the EC for consideration in 1998 and was accepted on 20th January 1999. A three-year Financing Agreement was signed between IIC and the EC with funding from the Tropical Forest Budget, line B7-6201.

The Wilderness Preserve component of the project focused on the development and participative implementation of management plans for the Wilderness Preserve area of the forest and the road corridor that runs diagonally through the forest from Kurupukari (Fairview Village) in the north east to Corkwood on the south western edge of the forest (Fig. 8.2) established through extensive stakeholder consensus. Demarcation of the Iwokrama forest into two areas divided equally into the (i) Wilderness Preserve for the conservation of nature and natural processes and (ii) Sustainable Utilisation Area for the judicious use of the multiple resources of the

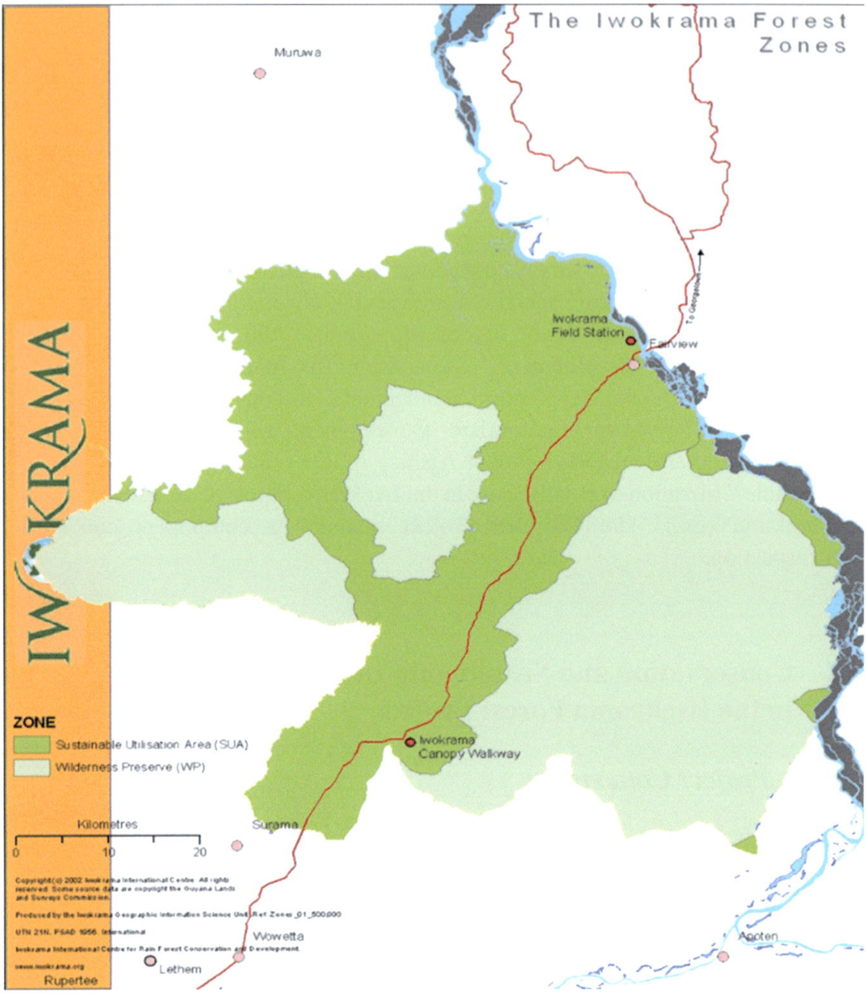

Fig. 8.2 Map of Iwokrama Forest showing the sustainable utilization area and wilderness preserve zones, road corridor, river systems bordering Iwokrama and surrounding villages

tropical rain forest to yield benefits to the peoples of Guyana and the world in general, without compromise of these resources for future generations. These were requirements of the IIC Act of 1996. Zoning of the forest was carried out with funding from DFID, EC and ITTO.

The Biodiversity Inventory component of the Iwokrama project under discussion herein was originally designed for a comprehensive inventory of the biodiversity of the Iwokrama Forest serving as a basis for sustainable forest management, further research and diversified forms of economic utilization. It was anticipated that this would create opportunities for training, capacity building, employment and the dissemination of knowledge to the wider scientific community. However, bioinventory activities were modified to complement the bioprospecting project to avoid only supplementing the existing flora inventory of over 1,075 plant species and the vertebrate inventory (90 % completed) (Smith and Kerry 1996) previously carried out by the (a) Smithsonian Institution (Clarke et al. 2001), (b) Philadelphia Academy of Sciences, (c) Royal Ontario Museum (Engstrom et al. 1999, Lim and Engstrom 2001a, b, 2005), (d) University of Kansas, (e) Florida International University, and (f) American Museum of Natural History.

A major goal of the bioprospecting component was to identify, develop and commercialize bioactive compounds. Another was to initiate pilot ventures based on clear legal foundations, making best use of national and regional laboratory facilities, and operating through equitable business partnerships with local communities, the private sector and other stakeholders. Recognizing the lack of intellectual property rights (IPR) and access to genetic resources and benefit sharing (ABS) agreements at IIC and the absence of national legislation to enact such agreements from the outset, the EC recommended that these agreements be put in place prior to natural product commercialisation or publishing. IIC proceeded with its bioprospecting project contingent on the expectation that once protocols on IPR and ABS were established by IIC and high quality leads with good potential market value were generated, national legal legislation would follow to enable these agreements to work and for commercialisation to proceed thereafter.

8.2.2 Bioprospecting-Bioinventory Project Implementation

Work on the Bioinventory and Bioprospecting components commenced on June 2000, 16 months after the contractual agreement between IIC and EC was signed. The later than anticipated commencement was due in part to the time taken to establish a unified strategy and establishing a broad enough framework for institutional collaborators to agree the project objectives, as stated in Iwokrama's CSUBIF final project report to the EC in 2003. The initial three-year funded programme was therefore given an 11 month extension by the EC to complete proposed objectives.

In 1999, IIC decided that the best way to enhance the outputs of the bioinventory work and simultaneously maximize the likelihood of commercial development of bioactive products from projects involving the forest, was to focus on lesser known smaller-bodied organisms and microorganisms well-known to be rich sources of novel bioactive compounds in some cases (Chin et al. 2006; Giddings and Newman 2013; Sneader 1996). IIC realized that it was very unlikely for a single natural product chemist (NPC) to achieve a high measure of success without some level of external collaboration with institutions having modern laboratory facilities and marketing capacities. With this strategy in mind, IIC advertised for institutional partners with a track-record of product development and commercialisation from local and international sources to work with the Centre. In August 1999 two consortiums were chosen, (a) CABI Bioscience, UK and the Royal Botanic Gardens (RBG), Kew, UK and the (b) University of the West Indies (UWI), Trinidad and Jamaica,

University of Guyana (UG), Guyana, and Institute of Applied Science and Technology (IAST), Guyana, from a field of 12 institutional and individual applicants. CABI was selected on the basis of providing training and technical support on macro and endophytic fungi. In addition, their proposal was designed to be closely linked to the parallel bioprospecting programme, in which potentially valuable biochemicals from fungi during the bioinventory activities would be screened for biological activity in collaboration with RBG Kew. The UWI-UG-IAST group, were selected on the basis of their extensive experience in plant natural products research and ability to provide local and regional laboratory facilities to the project, together with technical support for local bioprospecting staff.

Consortium members met for the first time with IIC staff in Guyana in January 2000 to decide how to utilize capacities in natural product chemistry and product development by research collaborators. The need to establish clear objectives for each collaborator was discussed. IIC agreed that the (a) CABI-RBG Kew group comprising Drs. Joan Kelley, Paul Cannon, Russell Paterson, Paul Kirk, and Tetsuo Kokubun, and Prof. Monique Simmonds, and (b) Guyana-based NPC, should focus on endophytic fungi isolated during the biodiversity inventory programme. This involved the production of extracts for screening by bioassays developed by CABI and RBG Kew. IIC established a bioprospecting lab in Guyana to facilitate the work and provide a useful output *per se*. The UWI-UG-IAST group comprising (a) Profs. Bladwin Mootoo, and Wilfred Chan, (b) Drs. Anderson Maxwell, Helen Jacobs, Marlene Cox, and David Singh and (c) Mr. John Caesar would focus on structural elucidation and biotesting of compounds isolated from host plant species to examine the interaction between host plant and endophyte chemistry.

A target of producing 2000 extracts was considered suitable to establish linkages with companies for more in depth examination and screening. Subsequent interests in commercial development of extracts produced by the project would be contingent on clear and well-constructed IPR and ABS protocols within the framework of a national IPR and ABS policy. This is stipulated under the Convention on Biological Diversity (CBD) of which Guyana is party to since 27 November 1994.

8.2.3 Recruitment and Training of Bioprospecting and Bioinventory Staff

IIC advertised internationally for applicants to undertake the bioinventory and bioprospecting components in Guyana. In June 2000, Coralie Simmons, a Guyanese national and biology graduate with experience in fungal systematics from UG, was hired as the Biodiversity Inventory Scientist (BIS). Vijay Datadin, a Guyanese national, was hired as a Geographical Information System specialist to manage the georeferenced field sample collection data from the different collection sites in the forest and to integrate this data into maps. Dr Ramish Pingal (the author of the current chapter), a Trinidadian NPC and recent PhD chemistry graduate from the

UWI, was recruited to manage the bioprospecting component in Guyana who entered the programme with over 5 years of research experience in the extraction, isolation, characterization and bioassay analysis of natural products from terrestrial plants.

Bioinventory and bioprospecting activities commenced shortly after the BIS and NPC were hired with a 4.5 month training period of the BIS on endophytic fungal isolation, identification at CABI Bioscience. The NPC also underwent training on aseptic culture and bioassay techniques, isolation of fungal metabolites and laboratory safety training split between CABI Bioscience and RBG Kew from July–November 2000. Plants collected from the Iwokrama forest prior to the departure of the BIS and NPC to the UK were used for isolating fungal endophytes during the training period in order to yield immediate results for both components. Four Guyanese nationals (a) Henry James and Linsford Lagoudou (parataxonomists, field technicians and research assistants), and (b) Melanie McTurk (BSc, Chemistry, UG), were hired in early 2001 to fill the positions of (a) two full time field/laboratory technicians and (b) laboratory technician, respectively. The main duties of the technicians were to collect plant samples from the rainforest and to prepare fungal cultures, plant extract generation and other routine activities in the lab in Guyana. Technicians were trained in-house by the BIS and NPC on plant collecting, aseptic and mycological culturing techniques, preparation of fungal and plant extracts, handling and disposal of hazardous materials, and laboratory health and safety procedures.

8.2.4 *Project Management*

The BIS and NPC were responsible for the day-to-day management of the bioinventory and bioprospecting components of the EC project, respectively. The Centre's Principal Forest Ecologist, Dr. David Hammond, served as the component head of the Wilderness Preserve aspect of the project and the team leader for the entire EC project. The BIS and NPC reported to the team leader and each component head also reported directly to the Director General of Iwokrama Mr. David Cassells July 1997–June 2001 and Dr. Kathryn Monk July 2001–May 2003.

8.2.5 *Rational for Selection of Endophytic Fungi*

Fungi were selected as the microorganisms to study because they had not been studied in Iwokrama, and some of the most valuable and widely used pharmaceuticals from natural products originate from these organisms (Kelley et al. 2003; Butler 2004, 2008; Butler et al. 2014, Newman and Cragg 2007). These include penicillin (Sneader 1996), the antifungal cyclosporins used in transplant surgeries, and cholesterol-busting drugs based on zaragozic acids (Bergstrom et al. 1995; Dreyfuss and Chapela 1994). Other fungi-derived metabolites undergoing development include the diketopiperazine, plinabulin from *Aspergillus*, which was under clinical trials for the treatment of lung cancer (Bhatnagar and Kim 2010).

Other compelling reasons that made fungi the organism of choice included (a) they are the second most diverse major organism group on Earth with species estimated at 1.5–5.1 million (Hawksworth 2012; Blackwell 2011; O'Brien et al. 2005; Hawksworth 1991; Hammond 1992; Cannon and Hawksworth 1995; Rossman 1994), (b) 10,000 to 15,000 fungi are estimated in the Iwokrama forest, of which 50–80 % were expected to be undescribed (Rossman et al. 1998; Kelley et al. 2003), (c) they have close associations with a wide range of other organisms; (d) host-fungi specificity, (e) the major functional roles fungi play in terrestrial and aquatic ecosystems; (f) their anticipated high diversity within tropical ecosystems in general and (g) the ease of maintaining ex-situ in some cases (Arx 1980; Cannon 1996; Domsch et al. 1993).

Although fungi can be challenging to isolate and culture, studying fungi in specific fungi-host associations makes the sampling process easier enabling initial collection to focus on the host rather than the fungus. This approach (a) facilitates isolation from many species where the fungal fruiting bodies are difficult to sample directly because of small size, (b) reduces the risk of sample contamination from other organisms, and (c) allows simple reproducible manipulations of culture conditions for effective isolation.

8.2.6 Field Sample Collection

Thirteen plant host species from the genera *Carapa, Catostemma, Cecropia, Chlorocardium, Eperua, Eschweilera, Euterpe, Goupia, Jacaranda, Manilkara, Mora, Sclerolobium* and *Swartzia* were selected to study their endophytic fungi. Seven bioinventory plots of one ha were established within the Sustainable Utilization Area of the Iwokrama Forest representing different (a) vegetation and edaphic types, (b) spatial distributions and (c) levels of disturbance within the forest. Selected host trees were tagged and georeferenced. The plots were located at the "3 Mile" and "8 Mile" satellite camps along the road corridor that runs through the Iwokrama Forest and Corkwood swamp, Kabokalli, White Water, Pakatau falls and Moco-Moco. A map showing the spatial distribution of the plots was generated by the Geographical Information Staff at IIC using the georeferenced data collected for each of the seven plots. A total of six field collections of healthy leaves and soft stems were carried out at six different times over the project duration: June 2000; April, October and December 2001; and February and June 2002. From the six field collections, leaf samples of 13 host plant species representing 10 families were collected. A secondary investigation into wood and bark endophytes was undertaken. Furthermore, two macro fungi collections were made from within the sample plots and other areas within the sustainable utilization area. A total of 125 macro fungi samples were collected that yielded 71 different species.

For the phytochemical work, aerial parts of all 13 host plant collections were made from two field collections in April and August 2001. A third collection was made in March 2002 of nine additional plant species from six genera (*Annona*, *Clusia*, *Dugetia*, *Piper*, *Tovomita*, and *Vismia*). This represented 22 different plant species. All plant collections were made from within the Sustainable Utilization Area and voucher specimens of each plant species collected have been lodged at the herbarium of the Centre for the Study of Biodiversity at UG.

8.2.7 Rational for Selection of Host Plants

Thirteen plant host species were selected for study from which endophytic fungi were collected based on (a) their commercial importance as timber species, (b) their abundance in the Iwokrama forest and (c) no, or very limited, phytochemical information on seven of these species. Selection of the nine additional plant species for phytochemical analysis was made on the basis of the known bioactivities of plants from the genera to which they belonged. Importantly, use was not made of ethnobotanical information or traditional knowledge in the selection of plants because of the lack of IPR and ABS protocols at Iwokrama and in Guyana more generally.

8.2.8 The Biodiversity Inventory Programme

8.2.8.1 Endophytic Fungal Isolation and Characterisation

The inventory programme focused on isolation of endophytic fungi from healthy, mature and symptomless plant leaf samples from saplings and young trees. Leaf samples of 13 host plants were collected from which a total of 912 pure fungal cultures were obtained representing over 10,000 isolations of endophytic fungi. Fungal cultures were identified using morphological techniques, including observation of cultures grown under defined conditions, and description and measurement of microscopic features. Initial fungal identification was carried out to the species aggregate level, followed by assignment to a morpho-species using a combination of cultural and micromorphological characters to make the overall process manageable.

Sixty-four fungal morpho-taxa were characterized from 2,492 cultures. Most of the endophytic fungi samples isolated belonged to four genera: *Colletotrichum, Phomopsis, Nodulisporium* and *Pestalotiopsis.* The complete strain set of *Colletotrichum* and *Pestalotiopsis*, ca. eighty strains each, was selected for further studies using inter-simple sequence repeat polymerase chain reaction (ISSR-PCR) and random amplified polymorphic deoxyribonucleic acid (RAPD) techniques. Analysis of the data from both techniques indicated that host specificity could not be detected in *Colletotrichum* nor *Pestalotiopsis* even at the strain level (Lu et al.

2004). *Colletotrichum* strains only were selected for further analysis with the aid of ribosomal deoxyribonucleic acid internal transcribed spacer (rDNA ITS) sequencing, owing to the large number of strains to be analysed within a short time, and the fact that *Colletotrichum* is an important and large genus of over 900 species comprising numerous endophytes and a few phytopathogens of tropical plants.

8.2.9 The Bioprospecting Programme

8.2.9.1 Extract Profiling and Biotesting

Over 360 endophytic fungal cultures and 17 macrofungi samples were received from the bioinventory component. These were subcultured and bulked up for extraction with a chloroform-methanol mixture followed by extraction with pure methanol to give a total of over 900 organic and aqueous extracts.

Plant Samples A total of 13 host plant species and nine additional species of plants were collected, air dried, milled and extracted with a chloroform-methanol mixture followed by extraction with methanol. A total of 60 organic and aqueous plant extracts were prepared.

Metabolic Profiling A comparison of the profiles of compounds in the fungus and plant extracts was determined by thin layer chromatography (TLC) using a range of different spray reagents and high-performance liquid chromatography (HPLC). Liquid Chromatography-Mass Spectrometry (LCMS) profile was also obtained for selected extracts. The chemical profiles of ca. 880 fungal and host plant extracts were determined at RBG Kew using HPLC equipped with a photodiode array detector.

Bioassay of Fungus and Plant Extracts Over 2800 fungal extracts obtained from 332 fungal isolates, and host plant extracts were tested by RBG. These were subjected to up to six bioassays using insects, bacteria and fungi as test organisms. Cultures of all fungal isolates extracted and assayed have been stored in Guyana and 256 of these cultures are also held in a reserve collection at CABI Bioscience Genetic Resource Collection.

A total of 60 plant extracts transferred each to UG, the UWI Trinidad and UWI Jamaica were subjected to 16 different assays which tested for antibacterial, antifungal, anti-insect, antioxidant, antiretroviral and cytotoxic activities. At the bioprospecting natural products lab in Guyana, 547 fungus extracts and 60 plant extracts were tested in cytotoxicity (using brine shrimp nauplii, antifungal and antioxidant assays).

8.2.9.2 Bioassay Results

Endophytic Fungal Extracts Of the 332 fungal isolates cultured and extracted, 185 derived from 152 fungus strains showed bioactivities in at least one of the six assays. The activities were: 13 antibacterial, 14 antifungal, 92 anti-insectal and 103 cytotoxic. Of the 185 fungus extracts that were active in at least one of the assays, 130 were of organic origin whereas 55 were aqueous. Furthermore, 33 fungal isolates had organic and aqueous extracts that were active in at least one assay. Aqueous extracts showed only cytotoxic (40 % of aqueous extracts) and anti-insect (60 % of aqueous extracts) activities whereas organic extracts showed cytotoxic, antibacterial, anti-insect and antifungal activities. From the 185 active fungus extracts identified, 110 extracts derived from 100 fungus isolates were selected as possible leads for further research to continue bioprospecting in the future. This was reported in the Iwokrama's final CSUBIF report to the EC in 2003, and the final evaluation report of the CSUBIF by Stephen Devenish, 2003 for the EC.

Plant Extracts Host plant extracts tested in the six assays at RBG Kew all showed activity in at least one assay. The bioassay results showed that a high proportion of the extracts exhibited anti-bacterial activity, whereas extracts from only two plants showed anti-insect activity. This contrasts with the high levels of anti-insect activity observed from screening the endophyte fungal extracts of organic and aqueous origin.

Of the 60 plant extracts obtained from 13 host plants and nine additional plants, tested at the lab in Guyana and by UWI-UG project collaborators, a total of 29 were selected as possible leads for further work as indicated in Iwokrama's CSUBIF final report, 2003. Only one extract showed activity in the anti-retroviral assay and in antibacterial and antifungal assays. Three extracts showed activity in cytotoxic, antioxidant and antibacterial assays, six extracts showed activity in only two assays, and two extracts each demonstrated activity against two bacterial and two fungal strains. Of the extracts tested in anti-insect assays, only four showed activity against only one insect, whereas one showed activity against two insects.

8.2.10 Analysis of Results and Summary of Achievements

8.2.10.1 Bioinventory

Although more work is required, no correlation was observed between any defined fungal communities and individual plant species, indicating that host specificity is low in the Iwokrama Forest and by extension tropical rain forests, in contrast to temperate forests. Further, molecular fingerprinting studies carried out on *Colletotrichum* and *Pestalotiopsis* did not reveal any host specificity even down to the individual strain level. Analysis of the sequences of the ITS region of the rDNA of a subset of the *Colletotrichum* strains revealed significant variation within species. Based on the analyses, several taxa appeared to be new, including at least two species of *Colletotrichum* and an apparently undescribed genus of anamorphic fungi (Lu et al. 2004).

The main achievements of the bioinventory aspect of the project were:

- First inventory of endophytic fungi and macrofungi within the Iwokrama Reserve
- Establishment of a herbarium collection at the Iwokrama field station
- Establishment of a comprehensive database on endophytic and macro fungi
- Publication of two scientific papers on endophytic fungi from the Iwokrama forest (Cannon et al. 2002; Lu et al. 2004) and presentations on Iwokrama's bioinventory programme at four international, regional and local conferences and workshops (Simmons 2002a, b, c; Simmons and Cannon 2002).

8.2.10.2 Bioprospecting

The results of the initial bioassays on the fungal and plant extracts gave very good bioactive hit rates of 3.9 % (110 bioactive fungal extracts from 2800 extracts tested) and 48 % (29 bioactive plant extracts from 60 extracts tested) respectively compared to the average 'hit' rate of 0.03 % in carrying pre-screened samples to market (Lesser and Krattiger 2007). Bioactive hit rate refers to that percentage of samples which show biological activity in a given series of chemical assays (Guérin-McManus et al. 2011). Lead rate may be defined as the proportion of expected lead compounds identified from the total number of samples screened to yield a given lead compound (Lesser and Krattiger 2007). Further research on these 139 selected bioactive extracts would involve secondary screening: bioassay directed fractionation, isolation, purification and characterization of bioactive compounds using High Performance Liquid Chromatography (HPLC), Nuclear Magnetic Resonance spectrometry (NMR) and Mass Spectrometry (MS) to rapidly identify new compounds from those that are known. Further work would also involve the (a) testing of purified bioactive compounds in several specialized assays e.g. anti-inflammatory, antioxidant, and anticancer to assess their suitability for a specific use, (b) toxicological assessment of compounds for possible drug development, and (c) structure activity relationship studies to enhance activity.

The striking difference in bioactivities exhibited by fungal and host plant extracts in antiinsect and antimicrobial assays, where a higher proportion of plant extracts showed antimicrobial activity, compared to fungal extracts which showed mainly antiinsect activity, suggests that the metabolites of fungal endophytes may be responsible for providing protection to the plants against predatory insects. The poor antiinsect activity exhibited by host plant extracts compared to endophyte extracts was corroborated in independent assays conducted by UWI-UG project collaborators and further points to a possible role of fungal endophytes in the fitness and survival strategy of their plant hosts. Early work on plant endophytes centered mainly on the endophytes of grasses and crops (Clay 1988, 1990; Márqueza et al. 2012; Saikkonen et al. 2013). Research on plant endophytes has mushroomed over the past two decades owing to the paucity of knowledge on fungal endophytes from other plant sources and the significance of endophyte-plant interactions and endophyte secondary metabolites. The emerging insecticidal (Omacini et al. 2001; Shrivastava et al. 2015; Simons et al. 2008; Zhao et al. 2011) and antimicrobial (Gutierrez et al. 2012; Guzman-Trampe et al. 2015; Lv et al. 2010; Mousa and Raizada 2013; Nisa et al. 2015) activities of endophyte metabolites and their potential for use as biopesticides and pharmaceuticals have played a major role in influencing and promoting research in this field. Despite the recent intensive work on plant endophytes, research on the endophytes of tropical rainforest plants remains limited (Arnold et al. 2001). However, data from this present study points to a low diversity of fungal endophytes and a high diversity of endophyte metabolites contrary to initial predictions. An investigation of the secondary metabolic profile of the leaves of the Iwokrama host plants with and without fungal endophytes and their resultant bioactivities is required to determine the role played by endophytes in plant-fungal associations. Wider research on endophytes of tropical forest plants is a priority to develop a more complete understanding of endophyte-plant interactions and the metabolites they produce. In summary, this project has built capacity within Guyana to carry out endophyte fungal isolations and identification, together with bioprospecting activities, and has established a collection of fungi, plants, and extracts thereof of interest to the agrochemical and pharmaceutical industries.

Major accomplishments during the implementation phase of the bioprospecting project component include:

- Establishment of a natural products bioprospecting laboratory in Guyana currently operated by UG through a MOU with Iwokrama.
- Preparation and screening of over 3400 fungal and 60 plant extracts for biological activities in 23 different assays
- Metabolic profiling of over 880 fungus and plant extracts
- Fractionation of eight bioactive extracts
- 110 fungus and 29 plant extract leads with anti-insect, anti-bacterial, anti-fungal, antioxidant and/or cytotoxic properties identified
- Establishment of a comprehensive database on fungi and plant extracts
- Antifungal, antioxidant and cytotoxicity assays established at lab in Guyana
- Laboratory technician trained in microbiological culturing, extraction, bioassay techniques and data management
- Two UG students trained in bioassay techniques and four UG students assisted with projects
- Bioprospecting project information disseminated at three international/regional conferences (Pingal 2002b; Pingal et al. 2001, 2002). Two public lectures (Pingal 2001; Pingal and McTurk 2002) were given and two national meetings with representatives of the 16 Amerindian communities that have legal rights over the Iwokrama forest were made. Two talks on biodiversity and natural product development were given at two international workshops (Pingal 2002a, c). In addition, the current chapter represents an output.

Table 8.1 Bioprospecting objectives and level of achievement (X = the level achieved)

Objectives of bioprospecting component	Level of achievement Not done	 Partially completed	 Completed
Sample selection: Select plants for endophytic fungal and phytochemical research			X
Sample preparation: Prepare endophytic fungal cultures for chemical profiling and subsequent fractionation			X
Extract preparation: Prepare at least 2000 fungal and plant extracts for metabolite profiling and analysis			X
Bioassays: Establish a set of bioassays at the bioprospecting lab in Guyana			X
Analyse & Test 2000 Extracts: Test at least 2000 fungal and plant extracts in a series of bioassays in Guyana and through regional and UK collaborators,			X
Isolate & Identify Bioactive Compounds: Isolate and characterize bioactive compounds from fungi and plant extracts		X	
Database Tracker: Establish a database to track extracts prepared			X
Establish Local Lab Facilities: Establish a bioprospecting lab in Guyana			X
Commercial Products: Develop bioactive compounds for commercialization	X		
Pilot Ventures: Initiate pilot ventures based on clear legal equitable business partnerships with local communities, private sector and other stakeholders	X		

The extent to which the original objectives and activities of the bioprospecting aspect of the project was achieved during the project's life cycle is indicated in Table 8.1.

8.3 Project Analysis

8.3.1 Aspects of the Programme That Worked Well

The project ran smoothly throughout its entire cycle despite the complexity of the relationship among individual partners comprising the consortium. Initial training of the BIS at CABI Bioscience on endophytic fungal isolation, aseptic culturing techniques and identification before the project activities started in Guyana, was essential in the provision of pure fungal isolates of the highest integrity to the bioprospecting programme. Similarly, training of the NPC at CABI Bioscience and Kew Gardens UK on aseptic culturing techniques, basic fungal identification, natural products research, lab/database management and health and safety procedures assisted tremendously in the establishment of the lab in Guyana. Preparing, analyzing and tracking extracts at all stages of the process ensured accountability. Ample support and technical assistance was provided by both groups of consortium partners over the entire course of the project facilitated by eight in country visits by eight persons and regular communication via email. CABI Bioscience assisted with ordering vital general equipment for the lab during the latter half of 2000 in order for these to arrive in Guyana in time for the establishment of the laboratory. A stock of general lab supplies and reagents sufficient for 1 year's work was ordered initially in early 2001 to avoid any delays in the progress of the project owing to the slow shipping (up to 2 months) of foreign purchased goods from Europe, the UK or EC based companies in the USA to Guyana. This core stock of materials was maintained by placing orders for spent stock at least 1 month in advance to give sufficient time for items to be delivered to Guyana on time to prevent shortages and project work delays.

There were no injuries and loss of time during the entire project cycle owing to the strict adherence to standard laboratory safety procedures as specified in IIC's bioinventory-bioprospecting laboratory safety manual, regarding the (a) safe storage, handling and disposal of biological specimens and hazardous chemical substances; (b) adequate training of all laboratory personnel on the collection of biological specimens from the forest, and the handling of plant and fungi specimens and hazardous chemicals; (c) the use of personal protective equipment e.g. respirators, gloves, safety goggles, lab coats; and (d) other safety engineering controls e.g. fume hoods, eye wash station, emergency showers and safety cabinets by all laboratory personnel. Standard laboratory health and safety rules were enforced by the BIS and NPC.

Staff at the Iwokrama Centre and Field Station provided excellent general project administration services and support in Georgetown, Guyana and in the field that enabled both components of the project to proceed without interruption. Collection of plant samples was carried out with the assistance of the field technicians, lab technician and BIS as well as the NPC upon occasion. Since collection of plant samples was restricted mainly to timber host trees ranging in height from a few metres to several metres high, collection of some leaf samples required a trained tree climber. Mr. Rodrigues Antone (now deceased) an Amerindian - a descendant of the indigenous peoples of the Americas, was hired on contract from one of the nearby communities to collect leaf samples from especially tall trees.

Linking the bioinventory activities with the bioprospecting project was a prudent decision at the time that enabled both components to work as one unit. The bioinventory component provided excellent support to the bioprospecting component by furnishing it with a continuous supply of large numbers of high quality pure strains of fungi (over 258) ready for bulking up, extraction and bioassay testing. The linking of both components eliminated the need for duplicate plant field collections for the endophyte fungal work and afforded the project significant reduction in time and resources as a major benefit.

Immediately fungal samples were received by the NPC, these were quickly sub-cultured by the lab technician and then bulked up for extraction, metabolic profiling and bioassay testing before being shipped to the UK collaborators. Similarly plant extracts were prepared quickly after collection following a period of air drying, milling, extraction, metabolic profiling and bioassay testing before delivery to UG and the UWI collaborators. So there was a steady progression at all stages of the process from plant collection, to (a) fungal isolation, (b) culturing, (c) extract preparation, (d) extract profiling, (e) extract biotesting, (f) extract tracking and (g) extract delivery to collaborators.

Semi purified extracts and isolates exported to collaborators in the UK, Trinidad and Jamaica for further analysis required the normal adherence to the regulatory procedures of the Environmental Protection Agency (EPA) of Guyana for access to genetic resources. The EPA, Guyana and the UG facilitated this process through the acquisition of relevant export permits on a timely basis. The UK and regional partners at the UWI and UG also provided valuable laboratory services in terms of testing of extracts in a series of insect and microbial assays as well as in the purification and analysis of extracts using modern analytical techniques: HPLC-Diode Array Detector, Liquid Chromatography-MS-MS and NMR spectroscopy. These analytical instruments were unavailable at the lab in Guyana. Reports on results of fungal identification and extract analysis and testing were provided by both UK and regional partners to IIC in very good time.

8.3.2 *Aspects of the Programme That Did Not Work Well*

The Iwokrama Forest lies 274 km south of Georgetown – a 7–11 h drive by road depending on road conditions. Since the majority of the road comprises unpaved laterite, road conditions especially during the May-June rainy season delayed accessibility to the forest and sample collection on a few occasions. In addition, intermittent bouts of malaria and dengue fever among field staff were a constant threat to sample collection, but did not affect the project significantly overall. There were some delays experienced in upgrading the laboratory at the UG and this coupled with power cuts/irregular power supply resulted in lost time and degradation of some fungal cultures. Contamination of sterile cultures through air borne fungi and bacteria severely hampered easy retrieval and re-culturing of fungus strains and the progress of antifungal assaying activities. The lack of basic analytical instruments e.g. Ultra Violet-Visible (UV-Vis) and Fourier Transform Infrared (FTIR) spectrometers and HPLC in the lab in Guyana prevented bioprospecting staff from preparing more extracts of higher quality.

The reference herbarium established at the Iwokrama Field Station by Bioinventory project component staff containing 140 herbarium sheets collected by the Ecology Unit at Iwokrama and over 400 sheets donated by the Smithsonian Institution through a MOU with Iwokrama, have been infested with cockroaches from time to time leading to damage to the collection. This problem needs to be rectified to preserve the remaining current collection.

Despite the fact that 50 species of macrofungi were collected and photographed for the production of a field guide to the macrofungi of the Iwokrama Forest, images obtained were not of sufficient high quality for completion of the guide. Improved photography equipment is required.

The absence of legal IPR and ABS protocols in Guyana prevented access to Amerindian and other traditional knowledge, which would have provided excellent support to sampling of plant samples for endophytic fungi and general plant samples. The exclusion of Amerindian traditional knowledge in this project prevented the targeted selection of flora and fauna which may have resulted in higher isolation of potentially useful and marketable natural products upon which bioprospecting depends, as indicated by Stephen Devenish in the final evaluation report on the CSUBIF project in 2003. One must be mindful though of the intricacies involved in the use of traditional knowledge and medicine guided selection of samples for the development and commercialisation of natural products. Although an attractive, more selective and cost-effective approach than random screening, this type of selection is fraught with challenges, relating to the long periods of time and general difficulties involved in arriving at mutually accepted legal agreements between the knowledge holders and commercial companies relating to ownership of intellectual property and equitable Benefit Sharing (BS) (Kingston 2011).

The expected pilot level commercial outputs from the bioprospecting component did not materialize. This was due to the short time frame of the project, and the speculative nature of bioprospecting. Most importantly, the lack of (a) IPR and ABS protocols at IIC and (b) national legislation to enable these agreements to work and facilitate extract leasing, development and commercialization meant commercial outputs were impossible. As such, companies could not be approached for private sector funding to continue bioprospecting activities or entering into commercial ventures based on the extracts generated from the project as highlighted by Stephen Devenish, final evaluation report of the CSUBIF project. Despite this, exploratory meetings were held between the NPC and CABI-Kew collaborators and three private sector and semi-autonomous companies in Guyana in the latter half of 2002 on possible collaborations and financial support for the bioprospecting initiative. These meetings revealed that whereas private sector and semi-autonomous companies showed a high interest in, and support for, the programme and that they would have liked to provide financial assistance, they did not have the liquidity at the time to do so.

8.4 Suggestions for Improvement: Future of Bioprospecting at Iwokrama and Lessons Learned

Funding Iwokrama should aim to secure funding of at least 10 years for future bioprospecting projects owing to the well-known (a) speculative nature of bioprospecting, (b) significant investment in time required (ten to twenty years (Conniff 2012; Guérin-McManus et al. 2011; Lesser and Krattiger 2007) and (c) money needed (US$230 million to US$1 billion (Conniff 2012; Lesser and Krattiger 2007). This would give enough time and resources to transform biota into potentially marketable products, especially in light of the on-going crisis with resistant bacterial infections. Funding should be considered from a variety of multiple financial bodies such as local and international donors and private sector. It is important that governments of large countries become involved and provide incentives for pharmaceutical companies to work in these areas. One can consider using funds obtained from multiple donors to fund one large project or individual donor funds to finance individual small projects. Although each approach would have its inherent strengths and weaknesses, the former approach is more likely to lead to the successful development and marketing of a useful product through long term and larger financial investments. One major disadvantage of multiple donor funding of a large project is the resolving of mutually acceptable legal agreements between donors and their client and any third party beneficiaries (Cordell and Colvard 2005; Kingston 2011) such as the 16 Amerindian communities that have legal rights to the Iwokrama forest in this case, before the project starts, which could take a very long time. Whereas the latter approach, where a single donor funds a smaller project, is more likely to result in partial success owing to smaller financial support and an expected shorter time period of funding. Private sector funding should be used principally to advance natural product research, and the development and commercialisation of useful bioproducts. Funding from donor agencies should be reserved for short term aspects of the project such as capacity building, outfitting the national bioprospecting lab at UG with basic analytical instruments, and maintenance of laboratory infrastructure. Not to be overlooked though is funding from academic institutions and groups engaged in bioprospecting. Although individual institutions may not have the financial leverage to fund bioprospecting initiatives they certainly have the technical expertise and modern analytical tools and facilities to identify lead compounds suitable for further development into potential marketable products if efforts and resources from a few institutions are combined (Harvey and Gericke 2011). It is important to reiterate that governments of wealthy countries especially should provide financial incentives for pharmaceutical companies to become involved in bioprospecting projects because of the current problems with resistant bacterial infections.

Collaborators In selecting collaborators to continue its bioprospecting programme, IIC should consider to advantages of working with the previous consortium or new specialist service providers. The major advantage of working with the same consortium partners is the ease with which previous work completed can be reinstated and further built upon. Furthermore, 256 of the original endophytic fungal cultures are held in reserve at CABI Bioscience Genetic Resource Collection for retrieval in the future.

Alternatively, since the bioprospecting lab is already established at UG and currently being maintained by UG, IIC can work with UG together with other organisations. These could be regional and/ or international academic institutions, industrial groups, and pharmaceutical companies, capable of conducting modern natural products research and development, coupled with business and marketing companies. For IIC to maximize its chances of finding and commercializing a bioactive compound, collaborators should be carefully selected to include those who are capable of:

(i) conducting dereplication of extracts using HPLC coupled with MS or MS/MS and IR-MS in combination with bioassays and reference libraries of natural compounds;
(ii) selectively identifying bioactive compounds early in crude extracts with the aid of modern techniques e.g. Liquid Chromatography-MS, 1D NMR and 2D NMR for faster purification and isolation of compounds;
(iii) characterizing pure isolates using automated structure elucidation software and routine techniques such as UV-Vis, FTIR, MS, 1D and 2D NMR, in addition to the latest NMR experiments e.g. Long Range Heteronuclear Sequence Quantum Multiple-Bond Correlation (LR-HSQMBC) experiment optimized to obtain very long-range ($\geq {}^4J_{\text{H-C}}$) heteronuclear correlations to assist with the elucidation of proton deficient compounds (Saurí et al. 2015). Microscale NMR techniques would be especially useful for the elucidation of metabolites of 1 mg or less (Kingston 2011), together with Comprehensive Multi-Phase NMR to characterize samples in their natural state (Monette et al. 2015);
(iv) screening extracts against a broad range of targets (e.g. antimicrobial, anticancer, insecticidal, immunological) and conducting new screening approaches even on crude extracts such as cell-based assays focusing on specific mechanisms of action and
(v) synthesizing a prospective lead natural compound to reduce the impact of collection from its natural source and to ensure that supply is unlimited (Kingston 2011).

Should a new specialist service provider be selected to take Iwokrama's bioprospecting programme to the next level, IIC should consider limiting collaboration to as few partners as is necessary per project to simplify working relationships and project implementation as recommended by Stephen Devenish in the CSUBIF final evaluation report, 2003.

National Laboratory One of the major disadvantages of the previous project was the lack of basic analytical instruments to help with the isolation and purification of natural products. The bioprospecting lab at UG should be outfitted with basic analytical instruments e.g. UV-Visible and FTIR spectrometers and HPLC to assist with preliminary extraction, isolation, purification and analysis of natural products, if this project is to continue at some stage. Acquisition of an FTIR spectrometer could be justified from its two-fold use in the analysis of natural products, and in plant research to discriminate closely related plant species through analysis of leaf samples in the Near-IR region (Durgante et al. 2013). Having these instruments would assist the lab at UG to prepare purer extracts and compounds of higher quality faster for regional and international collaborators on which to work. Essential to the isolation of high quality, value added potential leads for further development is the establishment of a set of routine and easy to maintain bioassays, such as antimicrobial, insecticidal and cytotoxic, in the field laboratory. The results of these preliminary bioassays can give collaborators an insight into the type of activities to expect and so guide further testing. In addition, facilities for local long term storage of pure fungal strains and plant samples is recommended to avoid the loss of temporarily stored samples. The avoidance of, or provisions for, power cuts is essential.

Reference Herbarium To preserve and protect the over 500 herbarium sheets lodged in the reference herbarium at the Iwokrama Field Station, the room should be sealed and air conditioned, to prevent entry by cockroaches and other destructive insects. New and adequate storage cupboards should be installed.

Traditional Knowledge Use, IPR and BS Protocols Traditional knowledge should be employed from the beginning of any new bioprospecting initiatives at the IIC to reduce the time spent on selecting flora and fauna and to improve the chances of developing a useful marketable product. The projects must be reciprocal to avoid one-way capacity building by the bioprospecting partner. According to McAfee in the conference on "Development as if Equity Mattered" Georgetown, Guyana September 24-26, 2001, any new bioprospecting projects should therefore move away from extractive research and shift towards participatory research in which the need to share benefits with indigenous communities is fundamental (Insanally 2003). This would necessitate Amerindian involvement from the beginning of the project and will require the draft IPR and ABS protocols prepared by IIC to be agreed and sanctioned according to the laws of Guyana.

In 2007 the Government of Guyana established a national policy on ABS and has enacted relevant national legislative and administrative measures to uphold the national policy through the EPA Guyana to enable agreements to work properly. The poor success by companies such as Shaman Pharmaceuticals and Phytopharm to commercialize new pharmaceuticals or botanical medicines from traditional medicines or ethnomedicine to date, have made the use of traditional knowledge and ethnomedicine in the prospecting for bioactive compounds from flora and fauna samples less appealing (Firn 2003, Harvey and Gericke 2011; Kingston 2011). In addition, the renowned Instituto Nacional de Biodiversidad (INBio) of Costa Rica had a vibrant bioprospecting programme. After over 20 years of existence with more than twenty agreements with academic institutions and industry, the most prominent being the commercial bioprospecting research collaboration agreement with Merck & Co. from 1991 to 2008, INBio's work has so far yielded over 27 patents (Lesser and Krattiger 2007), and several lead bioactive compounds, nutraceuticals and biological control microbes (Gámez 2007). However, INBio's bioprospecting efforts has failed to produce a blockbuster drug or to market a product to date (Conniff 2012). Similarly, the International Cooperative Biodiversity Group (ICBG) Projects in Suriname (1993), and Madagascar and Panama (1998), have yielded promising lead compounds. For example, (a) ipomoeassins A-F from the Suriname plant *Ipomoea squamosa* with potent antiproliferative activities, (b) a depsipeptide coibamide A from *Leptolyngbya* sp. from Panama with a unique selectivity profile in NCI 60 cancer cell lines and (c) schweinfurthins from the Madagascar plant *Macaranga alnifollia* with strong antiproliferative activity have shown potential. Nevertheless, they too are yet to be developed into commercial drugs (Kingston 2011), indicating the pitfalls of bioprospecting projects.

Despite this, IIC should consider the use of traditional knowledge and ethnomedicine in its search for new and useful bioactive compounds with market potential for two major reasons: (i) IIC, from its inception has engendered a good and long standing working and congenial relationship with the Amerindian communities within and bordering the Iwokrama forest thereby facilitating collaborations, and (ii) IIC is expected to demonstrate how tropical forests can be used in a sustainable manner to provide economic benefits to the people of Guyana and the world in general through research, training and the development and dissemination of technologies according to the Iwokrama Act. IIC is therefore presented with a good opportunity to be the first to show how traditional knowledge and ethnomedicine can be used effectively to select and commercialise new bioactive compounds into useful products or pharmaceuticals in the post CBD era. These points are even more relevant given the crisis being experienced now from resistant bacterial infections which completely changes the societal dynamics regarding bioprospecting for novel antibiotics.

Bioproduct Selection & Development By the turn of the twenty-first century most pharmaceutical companies either considerably scaled down or eliminated their investment in bioprospecting as a means of identifying new lead compounds (Conniff 2012; Harvey and Gericke 2011; Firn 2003). Developing countries with untapped pristine biodiversity resources with a high potential of new bioactive products are unable to access and utilize these resources owing to the lack of financial and technological capability to do so, thereby making bioprospecting almost nonexistent. Such biodiversity rich countries are therefore confronted with the dilemma to either 'exploit' their biological resources for short term profit or ascribe a monetary value to their biodiversity for its sustainable use and profit, an issue about which consensus is divided (Castree 2003; Lesser and Krattiger 2007; Harvey and Gericke 2011). However, the steady rise of antibiotic resistance infections caused by multi drug resistant microbial pathogens since the discovery of penicillin and the first recorded and recent emergence of a colistin-resistant strain of *E. coli* carrying the colistin-resistance gene, mcr-1 in the United States of America (McGann et al. 2016), has engaged worldwide attention for new antibiotic leads in recent times even more than before (Lewis 1995, ECDC/EMEA Joint Technical Report 2009; Overbye and Barrett 2005; Braine 2011; Garner and Brown 2015). Natural products, an underexploited source of potentially novel leads with new modes of action, may be discovered through bioprospecting initiatives (Koehn 2008a, 2008b; Li and Vederas 2009; Nobili et al. 2009; Xiong et al. 2015). There has never been a better time to reconsider the work undertaken in this Iwokrama Bioprospecting project, and to develop it further with the lessons learned given the urgency and the highly increased awareness of the problems of multi drug resistant bacteria.

Should IIC decide to further develop the 110 lead fungal extracts in future from the present project, the repatriation of the 100 viable fungal strains held in long term storage at CABI from which the extracts were obtained would be required, since these fungal strains held at the lab in Guyana may no longer be viable. However, the Guyana strains require checking for viability as soon as possible as indeed do the CABI strains. A diverse collection of products should be carefully selected to study to enable IIC to improve the probability of developing a useful product to market. The selection of products can usefully be divided into two categories:

Category 1; Essential oils, resins and waxes, nutraceuticals, teas, health and so called "well-being" products, cosmetics, herbal or food supplements, and repellants requiring a short time for product development; and

Category 2; Bio-pesticides, food ingredients, botanical medicines and bioactive lead components for pharmaceuticals and which require a longer time to develop.

In making this distinction, the category 1 products that do not need the same strict regulatory testing as category 2 products and could be commercialized quickly and provide immediate financial support to IIC. Whereas the category 2 products could be commercialized over a longer period funded by resources obtained from the category 1 products.

8.5 Iwokrama Post 2003

By the time Iwokrama's initial bioprospecting project finished in March 2003, the project provided IIC with a solid start. Experience, technical knowledge, laboratory facilities, local, regional and international collaborators, skilled local staff and sufficient lead extracts were also obtained. These deserve further development and possible commercialisation by a collaborator with adequate funds to advance its programme to the next stage. Unfortunately, bioprospecting activities at IIC also ended in March 2003, since new donors or private sector funding was not procured. Iwokrama was unable to capitalize on the momentum generated by the initial project owing to financial constraints.

All staff hired under the CSUBIF project left Iwokrama in search of alternative employment. The bioinventory aseptic lab and the bioprospecting lab established at the University of Guyana was formally handed over to the University shortly after the project terminated through the establishment of a MOU between IIC and UG. The UG was given custodial responsibility of the facility with the understanding that IIC will resume bioprospecting activities, should funding become available. Completion dates for the two remaining and active major donor funded programmes (ITTO, December 2003 and CIDA, March 2004) was imminent. IIC was at the crossroads of being an organization that was primarily funded by external donors through its various programmes, to one that was expected to be self-financing. Unable to realize the goal of becoming financially self-reliant by the end of its donor funded programmes, activities at the IIC was reduced considerably over the next 5 years and IIC made a radical shift in focus to short term commercial projects such as reduced impact and sustainable logging.

An acting Director General was reserved to oversee a short transition period before a new Director General was appointed in January 2005 to manage the IIC for a 3 year term where major emphasis was placed on sustainable income generating activities and business development through the sustainable use of forest resources. Following the resignation of the Director General in October 2007, IIC appointed the CEO, Mr. Dane Gobin shortly thereafter to manage its operations. Between 2005 and 2014, IIC was actively engaged in nine large projects that complemented its research goals and among these were sustainable timber operations. Phase I logging activities in the Iwokrama forest commenced in 2008 and continued until 2012, to give sufficient time to prepare for phase 2 logging activities to commence from the third quarter of 2015 following finalization of a contractual agreement in 2014.

Three main research themes exemplify the work currently being carried out by IIC: (i) Environmental Resilience, (ii) Human, Social and Cultural Capital and (iii) Ecosystem Service Values. These three themes were carefully selected to establish a harmonious partnership among (a) sixteen local communities within and bordering the Iwokrama Forest, (b) scientists, and (c) sustainably managed business operations to reinforce Iwokrama as a global leader in forestry management. Oversight for onsite research themes is the responsibility of the Iwokrama Science Committee established by IIC in 2009 in accordance with Article 14 of the Iwokrama Act 1996. The committee includes representatives from (a) Newcastle University, UK, (b) UG and (c) UWI, amongst other institutions.

IIC is currently managed by its CEO and staff in Georgetown and at the Iwokrama River Lodge and Research Centre with support and strategic policy guidance of IIC's International Board of Trustees. Financial support is provided by the Government of Guyana, the Commonwealth Secretariat, the Commonwealth Foundation and the Commonwealth Forestry Association. The three broad research themes are currently achieved through seven work programmes: (i) community development, (ii) conservation and monitoring, (iii) science and research, (iv) sustainable timber operations, (v) eco-tourism, (vi) learning services and (vii) information and communication.

The Learning Services programme has only recently become a promising area of revenue generation and the sustainable timber operations programme is only expected to become a significant contributor to revenue earnings when phase 2 operations commences in the latter half of 2015, although ecotourism continues to be a profitable activity for the Centre. IIC is likely to continue to experience major funding challenges in the short term to meet its core annual costs owing to current international financial volatility and reduced donor subsidies. In light of this reality, IIC issued a Request for Proposals representing a wide range of scientific research from private and public companies, institutions and universities, donors as well as other organisations engaged in tropical forest conservation and research in 2014, in order to initiate and catalyse the revenue generation potential of IIC's Research and Science Programme. IIC is seeking scientific research proposals on sustainable forest management, climate change, hydrology/geochemistry interactions, ecosystems services, community impacts, biodiversity and *bioprospecting*. Proposals are also being actively sought from pharmaceutical companies interested in developing products from medicinal plants with the anticipation of long term income generation for the Centre. At present IIC is collaborating with UG and other institutions to develop another bioprospecting project building on the work that was done in the earlier project. This is crucial, amongst other things, in light of the current crisis caused by the lack of novel antibiotic leads and shortage of compounds to control multi drug resistant pathogenic bacteria.

8.6 Conclusions

In the current author's opinion, IIC's bioprospecting-bioinventory project was over optimistic and ambitious in its design considering the (a) capricious and risky nature of bioprospecting, (b) large investment in time and development costs required to successfully commercialize a product and (c) legal and ABS issues involved in commercializing a product developed from the Iwokrama Forest. The over optimism in the initial design of the project can be attributable to several reasons: (a) it was one of the few low impact and sustainable projects with the potential of earning significant financial rewards for the Centre at a time when IIC was expected to become financially self-sufficient and less dependent on donor funding, (b) IIC regarded this as a rare opportunity to initiate its bioprospecting programme and to achieve a great deal within the limited timeframe and resources allocated, given the shortage of donors and private sector organizations willing to fund bioprospecting, (c) the lack of interest worldwide by large pharmaceutical companies in bioprospecting as a viable means of identifying new bioactive lead compounds, and (d) an inadvertent response to conform to donor funding requirements and expectations in an effort to secure limited but vital funding for the project. Iwokrama's bioprospecting project almost met 50 % of the original project objectives, and resulted in the identification of more than 25 host plant extract leads and 110 fungal extract leads worthy of further development. The objectives on commercialisation of bioactive compounds and the establishment of pilot ventures to generate revenue for IIC were near impossible from the beginning in the absence of any IPR and ABS protocols prior to 2003.

The combined performance of Iwokrama's bioinventory and bioprospecting projects is a much better picture though, having achieved five out of the seven objectives of the project with an overall success rate of 70 %. Linking the bioinventory activities with the bioprospecting project was therefore a prudent decision that enabled both components to work in unison and achieve better success overall. Nevertheless, the bioprospecting project may have achieved more, with greater efforts placed on identification, purification and bioactivity of compounds of interest, if the time and financial support for the project was focused less on bioinventory aspects. This was a matter of judgment: time and resources would still have to be committed to the selection, collection and identification of plants and fungi. In any case, the project still would not have been able to lease extracts, commercialize bioactive compounds and establish pilot ventures as explained before and the bioprospecting project would have only been able to achieve a 50 % success rate at best. Further, if the project was simply centered on bioprospecting, the EC and other donors may have been more reluctant to fund such a high risk project and it may not have materialized in the first place. It is a matter of obtaining the correct balance.

The bioprospecting project, after only 33 months of research, performed admirably in comparison to other bioprospecting initiatives. However, compared to the International Cooperative Biodiversity Group (ICBG) projects in Suriname (1993), and Madagascar and Panama (1998), and the much publicized INBio of Costa Rica project, more could have been done. Technical backstopping, support and extract analyses provided by the consortium of collaborators was essential to the success of the Iwokrama programme overall.

In its future bioprospecting initiatives, Iwokrama should aim to achieve the following:

- secure funding of at least 10 years for future bioprospecting projects to give enough time to transform genetic resources into potentially marketable products
- procure funding from a variety of multiple financial bodies such as national, regional and international donors, private sector and academic institutions and other groups engaged in bioprospecting. Iwokrama could use private sector funding principally to advance natural product research, and the development and commercialisation of useful bioproducts. Funding from donor agencies should be reserved for short term aspects of the project such as capacity building, outfitting the national bioprospecting lab at UG with basic analytical instruments, and maintenance of laboratory infrastructure
- work with either the previous consortium of partners or UG together with new specialist service providers such as regional and/or international academic institutions, industrial groups, and biotechnology and small pharmaceutical companies, capable of conducting modern natural products research and development, as well as business and marketing companies to market products
- re-fit the bioprospecting lab at UG with basic analytical instruments to assist with preliminary extraction, isolation, purification and analysis of bioactive compounds
- establish a set of routine bioassays and long term storage facilities for fungal strains, flora and fauna specimens at the bioprospecting lab at UG; and undertake the repatriation of the 100 viable fungal strains held in long term storage at CABI

- upgrade the reference herbarium at the Iwokrama Field station that will serve as a reference collection of all biological specimens collected from the Iwokrama forest
- ensure that from the establishment of any new bioprospecting project, there is continued strong participatory involvement of stakeholders of those Amerindian communities with legal rights to the Iwokrama forest as has been practicing from inception. Ensure that traditional knowledge is employed in the selection of flora and fauna, in accordance with the Amerindian Act 2005, Iwokrama Act 1996, Iwokrama's Policy on IPR, Access to Genetic Resources and BS. Guyana's National Policy on Access to Genetic Resources and the Fair and Equitable Sharing of Benefits Arising from their Utilization, 2007 should be adhered to.
- identify and conduct research on a diverse array of biological resources ranging from raw and semi purified natural products such as exudates, essential oils, and herbal food supplements to biopesticides.
- in particular, emphasis needs placing on novel pharmaceuticals with emphasis on antibiotic lead compounds to tackle multi drug resistant bacteria in light of the increase in antibiotic resistant infections worldwide.

Finally, IIC should actively seek out and engage the attention of national, regional and international donors, natural products researchers and private sector companies to re-establish its unique programme on bioprospecting that involves an integrated and strong partnership among Amerindian stakeholders, scientists and businesses to realize lasting ecological, social and economic benefits to the peoples of Guyana and the world. Iwokrama is therefore poised to be the first to demonstrate to the world how traditional knowledge and ethnomedicine can be used effectively to select and successfully transform new natural products from the Iwokrama forest into marketable products or pharmaceuticals.

Acknowledgement The EC is gratefully acknowledged for its funding of the Conservation and Sustainable Utilisation of Biodiversity in the Iwokrama Forest EC Project B7-6021/98-02/VIII/FOR Guyana of which bioprospecting was a component.

Disclaimer The views expressed in this article do not necessarily reflect those held by the European Commission nor the International Board of Trustees, management and staff of the Iwokrama Centre for Rain Forest Conservation and Development.

References

Arnold AE, Maynard Z, Gilbert GS (2001) Fungal endophytes in dicotyledonous neotropical trees: patterns of abundance and diversity. Mycol Res 105(12):1502–1507

Baines G, Warner R (2000) Guyana 1999 – Iwokrama Rain Forest Programme – Project evaluation report GEF ASSISTANCE TO THE IWOKRAMA RAIN FOREST PROGRAMME Project Number GUY/92/G31/A/1G/99 Draft Report of the Final Project Evaluation Mission Final Project Evaluation, GUY/92/G31/A/1G/99. https://www.academia.edu/2085038/Guyana_1999_-_Iwokrama_Rain_Forest_Programme_-Project_evaluation_report. Accessed 26 June 2015

Bergstrom JD, Dufresne C, Bills GF, Nallin-Omstead M, Byrne K (1995) Discovery, biosynthesis and mechanism of action of the zaragozic acids: potent inhibitors of squalene synthase. Annu Rev Microbiol 49:607–639

Bhatnagar I, Kim SK (2010) Immense essence of excellence: marine microbial bioactive compounds. Mar Drugs 8(10):2673–2701

Blackwell M (2011) The fungi: 1, 2, 3... 5.1 million species? Am J Bot 98(3):426–438

Braine T, (2011) Race against time to develop new antibiotics. In: Bulletin of the World Health Organization 89:88-89 doi:10.2471/BLT11.030211.http://www.who.int/bulletin/volumes/89/2/11-030211/en/. Accessed 11 July 2015

Butler MS (2004) The role of natural product chemistry in drug discovery. J Nat Prod 67(12):2141–2153

Butler MS (2008) Natural products to drugs: natural product-derived compounds in clinical trials. Nat Prod Rep 25(3):475–516. doi:10.1039/b514294f

Butler MS, Robertson AAB, Cooper MA (2014) Natural product and natural product derived drugs in clinical trials. Nat Prod Rep 31(11):1612–1661

Cannon PF (1996) Filamentous fungi. In: Hall GS (ed) Methods for the examination of organismal diversity in soils and sediments. CABI, Wallingford, pp. 125–143

Cannon PF, Hawksworth DL (1995) The diversity of fungi associated with vascular plants: the known, the unknown and the need to bridge the knowledge gap. In: Andrews JH, Tommerup IC (eds) Advances in plant pathology, vol 11. Academic Press, London, pp. 277–302

Cannon PF, Reid A, Simmons CM (2002) Diversity and host preference of leaf endophytic fungi in the Iwokrama forest reserve, Guyana. Mycologia 94(2):210–220

Castree N (2003) Bioprospecting: from theory to practice (and back again). Trans Inst Br Geogr 28(1):35–55. doi:10.1111/1475-5661.00076

Chin Y-W, Balunas MJ, Chai HB, Kinghorn AD (2006) Drug discovery from natural sources. AAPS J 8(2):E239–E253

Clarke HD, Funk VA, Hollowell T (2001) Using checklists and collections data to investigate plant diversity I: a comparative checklist of the plant diversity of the Iwokrama forest, Guyana. In: BL Lipscomb (ed) SIDA, Botanical Miscellany, no 21 Botanical Research Institute of Texas, Forth Worth, pp 1–86

Clay K (1988) Clavicipitaceous fungal endophytes of grasses: coevolution and the change from parasitism to mutualism. In: Pirozynski KA, Hawksworth DL (eds) Coevolution of fungi with plants and animals. Academic Press, London, pp. 79–105

Clay K (1990) Fungal endophytes of grasses. Annu Rev Ecol Syst 21:275–297

Conniff R (2012) A bitter pill. Conservation 13(1):18–23

Cordell GA, Colvard MD (2005) Some thoughts on the future of ethnopharmacology. J Ethnopharmacol 100(1-2):5–14

Domsch KH, Gams W, Anderson T-H (1993) Compendium of soil fungi, 2nd edn, 2 vols. Academic Press, London

Dreyfuss MM, Chapela IH (1994) Potential of fungi in the discovery of novel, low-molecular weight pharmaceuticals. In: Gullo VP (ed) Discovery of natural products with therapeutic potential. Butterworth-Heinemann, Boston, pp. 49–80

Durgante FM, Higuchi N, Almeida A, Vicentini A (2013) Species spectral signature: discriminating closely related plant species in the Amazon with near-infrared leaf-spectroscopy. For Ecol Manag 291:240–248

Engstrom MD, Lim BK, Reid FA (1999) Guide to the mammals of Iwokrama. http://www.iwokrama.org/mammals/. Accessed 26 June 2015.

European Centre for Disease Prevention and Control (2009) ECDC/EMEA joint technical report the bacterial challenge: time to react. A call to narrow the gap between multidrug-resistant bacteria in the EU and the development of new antibacterial agents. http://ecdc.europa.eu/en/publications/Publications/0909_TER_The_Bacterial_Challenge_Time_to_React.pdf. Accessed 14 Dec 2015

Firn RD (2003) Bioprospecting – why is it so unrewarding? Biodivers Conserv 12(2):207–216

Gámez R (2007) The link between biodiversity and sustainable development: lessons from INBio's bioprospecting program in Costa Rica. In: McManis CR (ed) Biodiversity and the law: intellectual property, biotechnology and traditional knowledge. Earthscan, London, pp. 77–90

Garner C, Brown D (2015) Spotlight on antibiotic research UK. Biochemist 37(3):36–37

Giddings L-A, Newman DJ (2013) Microbial natural products: molecular blueprints for antitumor drugs. J Ind Microbiol Biotechnol 40(11):1181–1210

Gilmour D (1999) Rainforest pioneer. Millennium trailblazers 3: David Cassell. People & the Planet/IPPF, UNFPA, IUCN 8(4):20

Guérin-McManus M, Famolare L, Bowles IA, Malone SAJ, Mittermeier R, and Rosenfeld AB (2011) Bioprospecting in practice: a case study of the suriname ICBG project and benefits sharing under the convention on biological diversity. Montreal: Report for Convention on Biological Diversity, 11 2001

Gutierrez RMP, Gonzalez AMN, Ramirez AM (2012) Compounds derived from endophytes: a review of phytochemistry and pharmacology. Curr Med Chem 19(18):2992–3030

Guzman-Trampe S, Rodriguez-Peña K, Espinosa-Gómez A, Sánchez-Fernández RE, Macias-Rubalcava ML, Flores-Cotera LB, Sánchez S (2015) Endophytes as a potential source of new antibiotics. In: Sanchez S, Demain AL (eds) Antibiotics: current innovations and future trends. Caister Academic Press, Norfolk, pp. 175–204

Hammond P (1992) Species inventory. In: Groombridge B (ed) Global biodiversity: status of the earth's living resources. Chapman and Hall, London, pp. 17–39

Harvey AL, Gericke N (2011) Bioprospecting: creating a value for biodiversity. In: Pavlinov IY (ed) Research in biodiversity – models and applications. Available via INTECH. http://www.intechopen.com/books/research-in-biodiversity-models-and-applications/bioprospecting-creating-a-value-for-biodiversity. Accessed 20 Dec 2015

Hawksworth DL (1991) The fungal component of biodiversity: magnitude, significance and conservation. Mycol Res 95(6):641–655

Hawksworth DL (2012) Global species numbers of fungi: are tropical studies and molecular approaches contributing to a more robust estimate? Biodivers Conserv 21(9):2425–2433

Insanally R (ed) (2003) The cropper foundation report on the conference "Development as if Equity Mattered". Georgetown Guyana September 24-26, 2001, Port of Spain, Trinidad and Tobago

Iwokrama International Centre for Rain Forest Conservation and Development (2015) News Archives. http://iwokrama.org/2014/07/iwokrama-issues-request-for-proposals-for-its-science-and-research-programme/. Accessed 16 June 2015

Kelley J, Cannon P, Simmonds M, Pingal R and Simmons C (2003) The Iwokrama programme: an approach to the sustainable exploitation of genetic resources. In: Biological resource centres and the use of microbes. Lima N, Smith D (eds) XXII ECCO Meeting Proceedings Book 185–195

Kingston DGI (2011) Modern natural products drug discovery and its relevance to biodiversity conservation. J Nat Prod 74(3):496–511

Koehn FE (2008a) High impact technologies for natural products screening. In: Petersen F, Amstutz R (eds) Natural compounds as drugs volume I, Progress in drug research, vol 65. Birkhäuser, Basel, pp. 175–210

Koehn FE (2008b) New strategies and methods in the discovery of natural product anti-infective agents: the mannopeptimycins. J Med Chem 51(9):2613–2617

Lesser WH, A Krattiger (2007) Valuation of bioprospecting samples: approaches, calculations, and implications for policy-makers. In: (Krattiger A, Mahoney RT, Nelsen L, et al.) Intellectual Property Management in Health and Agricultural Innovation: A Handbook of Best Practices, vols 1 and 2 Chap 9.4, pp. 861-876 MIHR: Oxford, U.K., and PIPRA: Davis, U.S.A. (accessed December 5 2015) Available online at www.ipHandbook.org

Lewis R (1995) The rise of antibiotic-resistant infections. FDA Consum 29(7):11–15

Li JW-H, Vederas JC (2009) Drug discovery and natural products: end of an era or an endless frontier? Science 325(5937):161–165

Lim BK, Engstrom MD (2001a) Species diversity of bats (Mammalia: Chiroptera) in Iwokrama forest, Guyana, and the Guianan subregion: implications for conservation. Biodivers Conserv 10(4):613–657
Lim BK, Engstrom MD (2001b) Bat community structure at Iwokrama forest, Guyana. J Trop Ecol 17(5):647–665. doi:10.1017/S0266467401001481
Lim BK, Engstrom MD (2005) Mammals of Iwokrama forest. Proceedings of the Academy of Natural Sciences of Philadelphia 154(1):71-108doi: http://dx.doi.org/10.1635/0097-3157(2004)154[0071:MOIF]2.0.CO;2
Lu G, Cannon PF, Reid A, Simmons CM (2004) Diversity and molecular relationships of endophytic Colletotrichum isolates from the Iwokrama forest reserve, Guyana. Mycol Res 108(1):53–63
Lv Y-L, Zhang F-S, Chen J, Cui J-L, Xing Y-M, Li X-D, Guo S-X (2010) Diversity and antimicrobial activity of endophytic fungi associated with the alpine plant Saussurea involucrata. Biol Pharm Bull 33(8):1300–1306
Márqueza SS, Bills GF, Herrero N, Zabalgogeazcoaa Í (2012) Non-systemic fungal endophytes of grasses. Fungal Ecol 5(3):289–297
McGann P, Snesrud E, Maybank R, Corey B, Ong AC, Clifford R, Hinkle M, Whitman T, Lesho E, Schaecher KE (2016) Escherichia coli Harboring mcr-1 and blaCTX-M on a novel IncF plasmid: fFirst report of mcr-1 in the USA. Antimicrob Agents Chemother. doi:10.1128/AAC.01103-16
Monette M, Simpson A, Masoom H, Liaghati Y, Mobarhan Courtier-Murias D, Fortier-McGill B, Soong R, KL Colson (2015) Characterization of environmental samples in their natural state using comprehensive multiphase NMR. In: 2015 Annual Meeting of the ASP, July 25th–29th, 2015, Copper Mountain, CO, USA
Mousa WK, Raizada MN (2013) The diversity of anti-microbial secondary metabolites produced by fungal endophytes: an interdisciplinary perspective. Front Microbiol 4:65
Newman DJ, Cragg GM (2007) Natural products as sources of new drugs over the last 25 years. J Nat Prod 70(3):461–477
Nobili S, Lippi D, Witort E, Donnini M, Bausi L, Mini E, Capaccioli S (2009) Natural compounds for cancer treatment and prevention. Pharmacol Res 59(6):365–378
Nisa H, Kamili AN, Nawchoo IA, Shafi S, Shameem N, Bandh SA (2015) Fungal endophytes as prolific source of phytochemicals and other bioactive natural products: a review. Microb Pathog 82:50–59
O'Brien HE, Parrent JL, Jackson JA, Moncalvo J-M, Vilgalys R (2005) Fungal community analysis by large-scale sequencing of environmental samples. Appl Environ Microbiol 71(9):5544–5550
Omacini M, Chaneton EJ, Ghersa CM, Muller CB (2001) Symbiotic fungal endophytes control insect host-parasite interaction webs. Nature 409(6816):78–81
Overbye KM, Barrett JF (2005) Antibiotics: where did we go wrong? Drug Discov Today 10(1):45–52
Pingal R (2001) From tropical rain forest to pharmaceuticals and other bioactive compounds: the process of discovery. Paper presented at Iwokrama Brown Bag Lunch Seminar, Iwokrama Centre, Bel Air, Georgetown, Guyana 14 February 2001
Pingal R (2002a) Challenges in using biodiversity. Paper presented at the international workshop on national biodiversity systems: building inter-generational capabilities for conservation and sustainable use, Georgetown, Guyana, July 2002
Pingal R (2002b) Iwokrama's bodiversity prospecting programme. Paper presented at the 5th International inter-Guianas conference on the development of the Guianas: building capacity to meet the emerging challenges in the Guianas, University of Guyana, Turkeyen, Guyana, 22-25 October 2002
Pingal R (2002c) Physical and chemical properties of crabwood oil: requirements for export. Paper presented at the international workshop on critical issues in the market and sustainable and equitable use of crabwood oil in Guyana, Lake Mainstay Resort, Essequibo, Guyana, 23-24 November 2002

Pingal R, McTurk M (2002) Biodiversity prospecting in the Iwokrama forest. Paper presented at the Friends of Iwokrama Public Seminar, Iwokrama Centre, Bel Air, Georgetown, Guyana, 7 June 2002
Pingal R, Simmons C, Cannon P, Paterson R, Simmonds M, Kokubun T, Mootoo B, Maxwell A and Jacobs H (2001) Iwokrama's bioprospecting programme: Paper presented at the Caribbean Academy of Sciences 12th General Meeting and Conference, Georgetown, Guyana 4-9 June 2001
Pingal R, Simmons C, Cannon P, Paterson R, Simmonds M, Kokubun T, Mootoo B, Maxwell A and Jacobs H (2002) Bioprospecting for bioactive metabolites from endophytic fungi and plants of the Iwokrama forest. Paper presented at the 19th Mona Symposium on Natural Products and Medicinal Chemistry, the University of the West Indies, Mona, Jamaica, 7-10 January 2002
Rossman AY (1994) A strategy for an all-taxa inventory of fungal biodiversity. In: Peng C-I, Chou CH (eds) Biodiversity and terrestrial ecosystems, *a*cademia sinica monograph series no 14, Institute of Botany, Tapiei, p 1691-1694
Rossman AY, Tulloss RE, O'Dell TE, Thorn RG (1998) Protocols for an all taxa biodiversity inventory of fungi in a Costa Rican conservation area. Parkway, Boone
Saikkonen K, Gundel PE, Helander M (2013) Chemical ecology mediated by fungal endophytes in grasses. J Chem Ecol 39(7):962–968
Saurí J, Chan STS, Buevich V, Gustafson KR, Williamson T, Martin GE (2015) Structure elucidation of a proton-deficient natural product using LR-HSQMBC supported by DFT calculations. Paper presented at the ASP 2015 annual meeting, Copper Mountain, Colorado 25-29 July 2015
Shrivastava G, Ownley BH, Auge RM, Toler H, Dee M, Vu AK, Tobias G, Chen F (2015) Colonization by arbuscular mycorrhizal and endophytic fungi enhanced terpene production in tomato plants and their defense against a herbivorous insect. Symbiosis 65(2):65–74
Simmons C (2002a) Iwokrama's bioinventory/bioprospecting programme. Paper presented at the 5th international inter-Guianas conference on the development of the Guianas: building capacity to meet the emerging challenges in the Guianas, University of Guyana, Turkeyen, Guyana, 22-25 October 2002
Simmons C (2002b) Iwokrama – efforts towards the conservation and sustainable utilization of the Iwokrama forest. Paper presented at the Caribbean Herbs Business Forum, Montego Bay, Jamaica, 2 – 5 December 2002
Simmons C (2002c) Medicinal plants of the Iwokrama reserve. Paper presented at the Natural Products Institute University of the West Indies, Mona Campus 6 – 7 December 2002
Simmons C, Cannon PF (2002) Diversity and host preference of leaf endophytic fungi in the Iwokrama forest reserve, Guyana. Paper presented at 7th International Mycological Congress, University of Oslo, Norway, 11 – 17 August 2002
Simons L, Bultman TL, Sullivan TJ (2008) Effects of methyl jasmonate and an endophytic fungus on plant resistance to insect herbivores. J Chem Ecol 34(12):1511–1517
Smith PG, Kerry SM (1996) The Iwokrama rain forest programme for sustainable development: how much of Guyana's bat (Chiroptera) diversity does it encompass? Biodivers Conserv 5(7):921–942
Sneader W (1996) Drug prototypes and their exploitation. Wiley, Chichester 788 pp
von Arx JA (1980) The genera of fungi sporulating in pure culture, 3rd edn. Lubrecht & Cramer, Vaduz
Xiong M, Guo Z, Han B, Chen M (2015) Combating multidrug resistance in bacterial infection by targeting functional proteome with natural products. Nat Prod Res 29(17):1624–1629
Zhao J, Shan T, Mou Y, Zhou L (2011) Plant-derived bioactive compounds produced by endophytic fungi. Mini-Rev Med Chem 11(2):159–168

Chapter 9
Bioprospecting with Brazilian Fungi

João Vicente Braga de Souza, Diego Rayan Teixeira de Sousa, Jessyca dos Reis Celestino, Walter Oliva Pinto Filho Segundo, and Érica Simplício de Souza

Abstract Fungi produce important substances for industrial utilization. Among these substances, colorants, biosurfactants, antibacterial compounds and enzymes are of particular relevance. Bioprospecting studies are important in order to identify fungal producers of these substances. Understanding that good producers of these substances can be found in places with high diversity and microbial competition is recognized widely and Brazil is perhaps the most biodiverse country for this type of work. The aim of this chapter is to present relevant research involving bioprospecting with Brazilian fungi.

9.1 Introduction

Brazilian researchers have initiated bioprospecting efforts to identify fungi that produce substances of industrial interest. Among these compounds, colorants, biosurfactants, antibacterial compounds and enzymes are of particular relevance. This chapter describes how the bioprospecting of these substances is being carried out in Brazil.

Brazil is the 5th largest country in the world and occupies 47 % of South America. This large territory contains different ecosystems such as the (a) Amazon rainforest (recognized as having the greatest biological diversity in the world), (b) Atlantic forest, (c) Cerrado savanna, (d) Caatinga (a desert in northeast Brazil) and (e) Araucaria forest, a temperate forest in the south. These conditions make Brazil a mega-diverse country (Brasil 2015).

The study of this diversity and its technological potential is mainly being performed by government-run universities and research institutes. Most of these institutions are located in the southeast region of Brazil, which is the most economically

J.V.B. de Souza (✉) • D.R.T. de Sousa • J.d.R. Celestino
W.O.P.F. Segundo • É.S. de Souza
Laboratório de Micologia, Instituto Nacional de Pesquisas da Amazônia - INPA, Av. André Araújo 2936, 69080-97 Manaus, Amazonas, Brazil
e-mail: joao.souza@inpa.gov.br

R. Paterson, N. Lima (eds.), *Bioprospecting*, Topics in Biodiversity and Conservation 16, DOI 10.1007/978-3-319-47935-4_9

developed region of the country. The results generated by these researchers typically have academic value, but unfortunately, very few of these findings have been applied to the industrial sector. Brazil significantly contributes to the scientific community via manuscripts, books, etc.; however, the country generates few patents (Glänzel et al. 2006; Leta et al. 2006).

Many of the chemicals that exhibit technological potential and could result in new patents, products and services are fungal and fungi play a very important role in environmental dynamics. The following discussion aims to examine the importance of these substances and presents some of the major related studies that have been conducted in Brazil.

9.2 Colorants

9.2.1 General

Colorant additives are used to provide color to otherwise dull substances. They can be classified as insoluble pigments or soluble dyes; however, these terms are typically used interchangeably (Saron and Felisberti 2006; Mapari et al. 2010). Dyes have been used worldwide as food additives to enhance the marketability of products by making their color more attractive (Uenojo et al. 2007; Volp et al. 2009). This technique is utilized because the loss or reduction of a food product's natural color during processing or storage, which it is assumed, lessens their appeal to the consumer (Serdar and Knežević 2009).

Currently there is growing interest in the discovery of dyes of natural origin, because synthetic dyes have been reported as carcinogenic and mutagenic, as well as causing allergies (Gunasekaran and Poorniammal 2008; Polônio and Peres 2009). Examples are the amaranth and erythrosine dyes, which have been shown to be genotoxic although they are presumably safe (and permitted) at low concentrations (Düsman et al. 2012). Therefore, synthetic dyes face more severe legislation, which has reduced the number of substances that can be used in food due to their adverse health effects in the short and/or long term, further fueling the search for biocolorants.

Bacteria, yeasts, filamentous fungi and algae, can synthesize pigments, but fungi stand out for their high productivity and extracellular release of such metabolites (Mapari et al. 2010; Hailei et al. 2011). *Blakeslea trispora* is a non-toxigenic filamentous fungus isolated from tropical plants that can synthesize high concentrations of the yellow-orange carotene pigment (Dufossé 2006). There are other producing species, such as *Phycomyces blakesleeanus* and *Mucor circinelloides* and the yeast, *Rhodotorula* (Dufossé 2006; Oh et al. 2009; Takahashi and Carvalho 2010). Riboflavin, or vitamin B2, can be synthesized by the fungus *Ashbya gossypii* and is also used as a food colorant (Braga et al. 2011).

Bioprospecting of fungal-derived dyes has gained increasing prominence due to their reduced toxic properties and added medicinal values (Chengaiah et al. 2010). In the textile industry, replacing synthetic dyes with other non-synthetic sources is feasible and would also decrease environmental toxicity (Mirjalili et al. 2011).

Synthetic dye residues can contaminate soil and natural water sources and are responsible for causing serious problems of environmental pollution (Ali 2010; Mirjalili et al. 2011). Fungal pigments, in turn, are more readily biodegradable and are potential dyes for industrial application.

Among the producing fungi the literature has highlighted, *Aspergillus*, *Penicillium*, *Paecilomyces* and *Monascus* are prominent (Gunasekaran and Poorniammal 2008; Méndez et al. 2011). *Monascus* is a cosmopolitan genus, and species such as *M. ruber* and *M. purpureous* are known for production of orange and red pigments. Some *Penicillium* species are capable of producing chemicals of various colors, including azaphilones. The sclerotiorin pigment produced by *P. sclerotiorum* and chromophores of the anthraquinone-type red pigment, such as Red Arpink produced by *P. oxalicum* are used in the food industry (Gunasekaran and Poorniammal 2008; Petit et al. 2009; Celestino et al. 2014; Kumar et al. 2015).

9.2.2 Brazilian Situation

In Brazil, many fungi have been studied for their synthesis of pigments during fermentation as follows:

A *P. sclerotiorum* strain isolated from soil samples of the Brazilian Cerrado produced the yellow-orange pigment sclerotiorin (Hamano and Kilikian 2006; Kanokmedhakul et al. 2006; Lucas et al. 2010). Sclerotiorin was also obtained from *P. sclerotiorum* isolated from Amazonian soil that showed increased production of the metabolite when modified sources of carbon and nitrogen were added to the growth medium (Celestino et al. 2014).

The endophyte fungus *Phoma sorghina* found in association with *Tithonia diversifolia* (Asteraceae), produced anthraquinones with orange and yellow colors (Borges and Pupo 2006). Polyketides of red, yellow and lilac shades have been found in Ascomycetes isolated from Amazonian soil, flowers and sawdust (Durán et al. 2002). Another report listed three Amazonian strains of *Penicillium simplicissimum*, *Penicillium melinii* and *Aspergillus sclerotiorum* that produced xantoepocinapigments(yellow),atrovenetina(yellow)andneoaspergilicoacid(yellow–green), respectively. These have high economic value and low toxicity (Teixeira et al. 2012).

9.3 Biosurfactants

9.3.1 General

Surfactants have industrial applications including detergency, emulsification, lubrication, foaming capacity, "wettability", solubilization and dispersion. The use of these substances is increasing particularly in cleaning products (soaps and detergents), oils, cosmetics and toiletries (Nitschke and Pastore 2002). Commercially available surfactants are synthetic and are obtained from petroleum products, although they present toxicity and are non-biodegradable (Soberón-Chávez and Maier 2011). The growing environmental concerns among consumers, combined with new environmental control laws, have led to the search for alternative biosurfactants (Nitschke and Pastore 2002).

Biosurfactants are a structurally diverse group of surface-active substances produced by living organisms. These substances are amphiphilic and composed of a hydrophilic and a hydrophobic group. The hydrophilic group consists of mono, oligo or polysaccharides, peptides or proteins and typically contains hydrophobic mid-chain saturated hydrocarbons or unsaturated fatty acids (Marchant and Banat 2012).

The composition and variations of biosurfactants are classified according to their chemical composition and microbial origin, because they have different chemical structures, especially those produced in the presence of hydrocarbons. These can belong to seven groups: glycopeptides, lipopeptides, phospholipids, fatty acids, neutral lipids, surfactants and polymeric surfactants particulates (Shekhar et al. 2014).The production of biosurfactants by microorganisms is well studied and has been published in studies using bacteria and filamentous fungi (Bhardwaj 2013).

9.3.2 Brazilian Situation

In Brazil, the most investigated topics are the use of agro-industrial waste as substrates for biosurfactant production and the bioprospecting of fungi for biosurfactants. It is expected that new and innovative products will be generated as a result. Although not bioprospecting in terms of using novel fungi, Luna et al. (2012) evaluated the use of industrial wastes from processing corn and peanuts as substrates for the production of biosurfactants from *Candida sphaerica* and obtained yields of up to 9 g/l. Katerine et al. (2013) investigated the potential use of waste from the bioethanol and fuel industry in the production of biosurfactants by mixed cultures and also obtained good yields. Two recent publications (Santos et al. 2013, 2014) reviewed the use of animal fat and corn steep liquor as substrates for biosurfactant production from *Candida lipolytica*. Silva et al. (2014) investigated the use of residues from corn and soybean processing industries with *Cunninghamella echinulata*, a fungus isolated from the Brazilian Cerrado. They showed that the biosurfactant could reduce and increase the vicosity of hydrophobic substrates and their molecules, suggesting a candidate for oil recovery. The residual glycerol from biodiesel production has been studied as a substrate by *Yarrowia lipolytica* for biosurfactant production (Ribeiro et al. 2013). Luna et al. (2012) investigated the environmental applications of biosurfactants produced by *C. sphaerica* and concluded that these biosurfactants have potential for use as adjuvants in the remediation/treatment of oily industrial effluents. Solid-state fermentation was also investigated for the production of biosurfactants by *Aspergillus fumigatus* (Castiglioni et al. 2013) on paddy rice bran and verified the importance of inducing substrates such as diesel oil.

With optimal nutritional conditions, *A. fumigatus* showed a good emulsifying activity and in experimental conditions, was able to provide a novel alternative for process optimizations in biosurfactants production. However, this fungus may be a human pathogen and should be avoided. In general, the data from these studies indicate that agro-industrial waste can be used for production of biosurfactants that have the potential to generate economically-promising bioprocesses.

Bioprospecting studies of biosurfactant-producing filamentous fungi have been conducted (Teixeira et al. 2012). Suzana et al. (2014) and Da Silva et al. (2014) isolated (a) *Pichia* strains from industrial effluents and (b) endophytic fungi using the biopanning technique (peptides affinity selection) in the plant *Myrcia guianensis* respectively. Both studies revealed high emulsification indexes, and that the strains were able to produce biosurfactants, demonstrating the potential of these organisms for bioremediation under a wide range of environmental conditions.

9.4 Antibacterials

The misuse of antibiotics for the treatment and prevention of infectious diseases has led to an increase in antimicrobial resistance (Michael et al. 2014). Millions of years ago, pathogenic bacteria modified their virulence to adapt to the host defense system (Beceiro et al. 2013). Evidence suggests that the development of antibiotic resistance by bacteria, over time, is a natural process, occurring in the absence of humans and animals (Arias and Murray 2009). While there was a marked decrease in the discovery of new antimicrobial agents in the last 30 years due to lack of research and development by large drug companies, the rate of bacteria resistant to multiple drugs (MDR) has alarmingly increased, resulting in a serious worldwide problem with consequences for the treatment of infectious diseases (Wright et al. 2014).

Bacterial resistance is a consequence of the evolution of bacteria and worsened with the ease of mobility of easy international travel. The (a) increasing world population; (b) misuse of antibiotics in human medicine, veterinary medicine and agriculture; (c) constant loss of antimicrobial efficacy and (d) decrease of new antimicrobial agents (Wright et al. 2014; Shaikh et al. 2015) contribute to the situation. In the 1980s and 1990s, many pharmaceutical companies refocused their research programs for new antimicrobial agents in more profitable areas, primarily focusing on gram-positive bacteria, due to the rapid rise of *Staphylococcus aureus* resistant to methicillin (MRSA). The increase in MDR gram-negative bacteria intensified the search for new antibiotics, as these also promised a good financial return for pharmaceutical companies (Theuretzbacher 2009).

Antimicrobials are generally classified by their molecular structure and mechanism of action in the bacterial cell. The β-lactams target Penicillin Binding Proteins PBPs, inhibiting the synthesis of peptidoglycans and the formation of the cell wall in susceptible bacteria. These glycopeptides act on the D-ala-D-wing of lipid II, inhibiting peptidoglycan synthesis. Macrolides, lincosamides, chloramphenicol and oxazolidones act on the 50S subunit of the ribosome and inhibits protein synthesis. Tetracycline and aminoglycosides affect the 30S ribosomal subunit, thus inhibiting protein synthesis. The fluoroquinolones inhibit topoisomerases (DNA gyrase and topoisomerase IV), thus inhibiting DNA replication (Silver 2011).

Fungi are used for the discovery of new bioactive natural products because they are a source of compounds with different biological activities and can produce antiviral, antimicrobial and insecticidal substances with relevance in the industrial, agricultural and pharmaceutical sectors (Vieira et al. 2011). Most of the classes of antimicrobial agents used today were discovered from actinomyces in the soil (Aminov 2010). However, there are many antimicrobials produced by fungi currently used in therapy, including (a) cephalosporins produced by *Cephalosporium acremonium*; (b) penicillins produced by *Penicillium chrysogenum*, *Aspergillus nidulans* and *Cephalosporium acremonium*; (c) pleuromutilin produced by *Pleurotus mutilus* and *P. passeckerianos*; and (d) fusidic acid produced by *Fusidium coccineum* and *Acremonium fusidioide*s (Wright et al. 2014).

9.4.1 Brazilian Situation

Brazil is carrying out bioprospecting of antibiotics produced by endophytes fungi, the production of nanoparticles with antimicrobial activity and optimizing antimicrobial activity. Using a bioprospecting approach, Orlandelli et al. (2012) investigated the production of antimicrobials, including terpenes by the endophytic fungus *Piper hispidum*. They observed that some of the isolates produced antimicrobials and three produced terpenes. Vaz et al. (2012) investigated the endophytic fungi on plants belonging to Brazilian flora (i.e. *Myrciaria floribunda*, *Alchornea castaneifolia* and *Eugenia* aff. *bimarginata*) and *Emericellopsis donezkii* and *Colletotrichum gloesporioides* produced an antimicrobial with an MIC similar to that of conventionally used antimicrobials. Santos et al. (2015) investigated the fungi from the leaves of *Indigofera suffruticose* Miller. (Fabaceae) where *Nigrospora sphaerica* and *Pestalotiopsis maculans* showed antimicrobial activity against gram positive (*Staphylococcus aureus*, *Bacillus subtilis*) and gram negative (*Escherichia coli*, *Klebsiella pneumonia*, *Pseudomonas aeruginosa*) bacteria. Flores et al. (2013) investigated the production of 3-nitropropionic acid by endophytic *Phomopsis longico* from *Trichilia elegans* A. JUSS spp. and found that it had activity against *Mycobacterium tuberculosis*.

Important investigations have also been carried out in the biogenesis of nanoparticles. Rodrigues et al. (2013) demonstrated the ability of *Aspergillus tubingensis* and *Bionectria ochroleuca* to produce silver nanoparticles with antimicrobial activity. Ishida et al. (2014) used a similar approach and obtained similar results by using *Fusarium oxysporum*.

The influence of the bioprocess factors on the production of metabolites with antimicrobial activity has also been studied. Bracarense and Takahashi (2014) using *A. parasiticus,* investigated modulation in the production of antibiotics, including kojic acid, showing growth inhibition against *A. flavus, C. albicans*, *E. coli* and *S. aureus*. However, *A. parasiticus* is a well-known aflatoxin producing fungus and should be avoided for bioprospecting procedures. Pigments of Brazilian fungal origin were also assessed for biological activity. Teixeira et al. (2012) investigated the biological activity of the dyes produced by *Aspergillus* and *Penicillium* isolated from the Amazon forest and found that many of these showed biological activity, with the pigments produced by *A. sclerotiorum* and *P. simplicissimum* being particularly important.

The cited studies present a panorama of bioprospecting and screening of endophytic fungi; however, clinical evaluation of these substances is required to determine the therapeutic potential for these preliminary findings.

9.5 Enzymes

The enzyme industry is part of biotechnology that has developed rapidly, especially during the previous four decades. Since ancient times, enzymes found in nature have been used in the production of foodstuffs such as cheese, beer, wine and vinegar (Saxena 2015). The use of fungi for the production of enzymes has led to a highly diverse industry with significant economic importance.

Fungi produce enzymes that are critical to their survival. These can act extracellularly or intracellularly to contribute to the digestion of food or in defense (Lange et al. 2012). These enzymes have not escaped the eyes of bio-engineering researchers. They are still being studied and occupy important positions in various industries, including the food, pharmaceutical and chemical industries (Kirk et al. 2002; Choi et al. 2015).

Fungal enzymes have been produced by the biotechnology industry in large quantities and low cost, and these enzymes can be modified according to desired characteristics. Enzymes of animal and plant origin have more complicated procurement mechanisms and modification procedures (Freedonia 2015). Multidisciplinary teams of chemists, microbiologist, biochemical engineers, biochemists and experts in other areas have come together to complement the knowledge that each area has on enzymes to improve their practices and develop technological innovations (Monteiro and Silva 2009).

The consumer markets are based on enzymes intended for industrial fabrics and cleaning products, foods and drinks and animal feed. The main industrial enzymes are proteases, amylases, lipases, cellulases, xylanases and phytases, and the largest producers are often European, e.g. International (Finland), Gist-Brocades (the Netherlands), and Novo Nordisk (Denmark), with Genencor, USA also a major player (Mussatto et al. 2007). Novo Nordisk controls about half of the global market where costs for production are decreasing, while the demand continually increases (Sanderson 2011; Jemli et al. 2014).

Brazil has an enormous diversity of microorganisms that can be exploited for the production of different enzymes of industrial interest in various areas (Table 9.1).

Table 9.1 Studies on enzyme production by fungi in Brazil

Industrial enzymes	Microorganisms	Authors	Methodology
Amylase	Filamentous fungi	De Castro et al. (2010)	Solid State Fermentation
Amylase	*Lichtheimia ramosa*	Silva et al. (2013)	Solid State Fermentation
Amylase	Filamentous fungi	Pasin et al. 2014)	Submerged Fermentation
Amylase	*Candida parapsilosis, Rhodotorula mucilaginosa, Candida glabrata*	De Oliveira et al. (2015)	Solid State Fermentation
Amylase	*Pycnoporus sanguineus*	Onofre et al. (2015)	Semi Solid Fermentation
Cellulase	*Aspergillus niger*	Cunha et al. (2012)	Submerged Fermentation, Semi Solid Fermentation
Cellulase	*Aspergillus fumigatus*	Moretti et al. (2012)	Submerged Fermentation
Cellulase	*Acremonium strictum*	Goldbeck et al. (2013)	Submerged Fermentation
Cellulase	*Penicillium funiculosum*	Maeda et al. (2013)	Submerged Fermentation
Cellulase	*Lasiodiplodia theobromae, Trichoderma* sp. *Fusarium* sp.	Faheina Junior et al. (2015)	Submerged Fermentation
Lipase	*Penicillium sp.*	Griebeler et al. (2009)	Solid State Fermentation
Lipase	*Penicillium sp.*	Rigo et al. (2010)	Solid State Fermentation
Lipase	Yeast	Bussamara et al. (2010)	Submerged Fermentation
Lipase	*Aspergillus sp.*	Colla et al. (2010)	Submerged Fermentation
Lipase	*Aspergillus candidus*	Cyndy et al. (2015)	Solid State Fermentation
Phytase	Filamentous fungi	Guimarães et al. (2006)	Submerged Fermentation
Phytase	*Aspergillus niger*	Spier et al. (2011)	Solid State Fermentation
Phytase	*Paecilomyces variotii*	Madeira et al. (2011)	Solid State Fermentation
Phytase	*Lichtheimia blakesleeana*	Neves et al. (2011)	Solid State Fermentation
Phytase	*Penicillium chrysogenum*	Ribeiro Corrêa et al. (2015)	Recombinant Expression
Protease	*Myceliophthora sp.*	Zanphorlin et al. (2010)	Solid State Fermentation, Submerged Fermentation
Protease	*Duddingtonia flagrans*	Braga et al. (2011)	Submerged Fermentation
Protease	Filamentous fungi Yeast	Rodarte et al. (2011)	Solid State Fermentation
Protease	*Mucor hiemalis*	Ribeiro et al. (2015)	Submerged Fermentation
Protease	Filamentous fungi	Mendes et al. (2015)	Solid State Fermentation
Xylanase	Filamentous fungi	Simões et al. (2009)	Submerged Fermentation, Solid State Fermentation
Xylanase	*Aspergillus sp.*	Peixoto-Nogueira et al. (2009)	Semi Solid Fermentation
Xylanase	*Neosartorya spinosa*	Alves-Prado et al. (2010)	Solid State Fermentation
Xylanase	*Lichtheimia blakesleeana*	Neves et al. (2011)	Solid State Fermentation
Xylanase	*Myceliophtora thermophile*	Moretti et al. (2012)	Solid State Fermentation
Xylanase	*Rhizomucor* sp. *Myceliophthora* sp.	Cassia Pereira et al. (2015)	Solid State Fermentation

However, enzyme technology is clearly overdue in the country, which is paradoxical. According to the Bio-Economy Agenda of Brazil, the enzyme industry is of great importance to the Brazilian economy, being directly linked to the "Third Industrial Revolution". Brazil is one of the countries that can benefit from the development of a national enzyme technology because it has a huge amount of renewable raw materials that can be transformed enzymatically into products with high added value and would be useful for strategic sectors of the economy (Harvard Business Review 2013).

A study by the US Research Industry Freedonia group estimated that world demand for enzymes will grow 6.3 % annually to $7 billion by 2017. The increase in per capita income in countries such as China and India will support consumer demand for higher value products, which can be achieved with enzymes such as detergents and foodstuffs. Advances in biotechnology will also boost demand for enzymes (Freedonia 2015).

With the advent of biofuels, studies related to the production of these compounds involving enzymes has become increasingly common (Cadete et al. 2014; Damaso et al. 2014; Aguieiras et al. 2015; Carvalho et al. 2015; Duarte et al. 2015). In addition, it is possible to obtain enzymes of industrial interest using certain waste (or byproducts) as substrates. The need for these enzymes by the world market has spurred studies in several parts of Brazil that go beyond the basic techniques of fermentation and genetic engineering to meet the future demand for renewable energy (Delabona et al. 2012; Valencia and Chambergo 2013; Ióca et al. 2014; Katoch et al. 2014; Souza et al. 2014). Brazil is underexplored for the production of enzymes of industrial interest. The country imports most of the enzymes it uses. Imports were $119 million, while exports reached $52 million. The Brazilian market for enzymes was estimated in 2011 at approximately $200 million (Ministério Do Desenvolvimento, Indústria e Comércio Exterior 2012). From 2007 to 2011, imports have tripled, while exports grew only moderately. To address this, Decree 6041/2007 established the Biotechnology Development Policy and includes the production and industrial use of enzymes (Bon et al. 2008). The current focus on enzymes research in Brazil has been applied in the food industry, antibiotic production, products for cleaning industries, effluent treatment and biofuel production e.g. biodiesel, bioethanol and biogas.

9.6 Conclusions

Brazil is experiencing tremendous growth in the biotech sector, as the main challenge of the current policy in science and technology is to ensure domestic firms participate more intensively in conducting and funding research activities to engender technological autonomy for the country. Innovative companies are being created that seek international competitiveness (Rezaie et al. 2008; Resende 2012). The enormous potential of the Brazilian biodiversity means that new substances and products can be discovered, resulting in the improvement of society's quality of life (Ribeiro and Raiher 2013; Corrêa et al. 2014; Ferreira et al. 2015). In this context, the scientific community, society and governing bodies should strengthen their relationships in a shared vision to invest in the development of technologies to expand domestic production and export of enzymes.

References

Aguieiras ECG, Cavalcanti-Oliveira ED, Freire DMG (2015) Current status and new developments of biodiesel production using fungal lipases. Fuel 159:52–67. doi:10.1016/j.fuel.2015.06.064

Ali H (2010) Biodegradation of synthetic dyes—a review. Water Air Soil Pollut 213:251–273. doi:10.1007/s11270-010-0382-4

Alves-Prado HF, Pavezzi FC, Leite RSR et al (2010) Screening and production study of microbial xylanase producers from Brazilian Cerrado. Appl Biochem Biotechnol 161:333–346. doi:10.1007/s12010-009-8823-5

Aminov RI (2010) A brief history of the antibiotic era: lessons learned and challenges for the future. Front Microbiol 1:134. doi:10.3389/fmicb.2010.00134

Ana PADO, Maria AS, Heloiza FA-P et al (2015) Bioprospecting of yeasts for amylase production in solid state fermentation and evaluation of the catalytic properties of enzymatic extracts. Afr J Biotechnol 14:1215–1223. doi:10.5897/AJB2014.14062

Arias CA, Murray BE (2009) Antibiotic-resistant bugs in the 21st century — a clinical super-challenge. N Engl J Med 360:439–443. doi:10.1056/NEJMp0804651

Beceiro A, Tomás M, Bou G et al (2013) Antimicrobial resistance and virulence: a successful or deleterious association in the bacterial world? Clin Microbiol Rev 26:185–230. doi:10.1128/CMR.00059-12

Bhardwaj G (2013) Biosurfactants from fungi: a review. J Pet Environ Biotechnol 04:1–6. doi:10.4172/2157-7463.1000160

Bon EPS, Jr NP, Gottschalk LMF et al (2008) Bioprocessos para produção de enzimas. In: Enzimas em Biotecnologia – Produção, Aplicações e Mercado, 1rd edn. Interciência, Rio de Janeiro, pp. 95–122

Borges WDS, Pupo MT (2006) Novel anthraquinone derivatives produced by *Phoma sorghina*, an endophyte found in association with the medicinal plant *Tithonia diversifolia* (Asteraceae). J Braz Chem Soc 17:929–934. doi:10.1590/S0103-50532006000500017

Bracarense AAP, Takahashi JA (2014) Modulation of antimicrobial metabolites production by the fungus *Aspergillus parasiticus*. Braz J Microbiol 321:313–321. doi:10.1590/s1517-83822014000100045

Braga FR, Araújo JV, Soares FEF et al (2011) Optimizing protease production from an isolate of the nematophagous fungus *Duddingtonia flagrans* using response surface methodology and its larvicidal activity on horse cyathostomins. J Helminthol 85:164–170. doi:10.1017/S0022149X10000416

Brasil (2015) Instituto Brasileiro de Geografia e Estatística. In: Ministério do Planejamento, Orçamento e Gestão. http://www.sidra.ibge.gov.br. Accessed 7 Jan 2015

Bussamara R, Fuentefria AM, De Oliveira ES et al (2010) Isolation of a lipase-secreting yeast for enzyme production in a pilot-plant scale batch fermentation. Bioresour Technol 101:268–275. doi:10.1016/j.biortech.2008.10.063

Cadete RM, Fonseca C, Rosa CA (2014) Novel yeast strains from Brazilian biodiversity: biotechnological applications in lignocellulose conversion into biofuels. In: Biofuels in Brazil: fundamental aspects, recent developments, and future perspectives 1rd edn. Springer International Publishing, New York, pp 255–279

Carvalho AKF, Rivaldi JD, Barbosa JC, de Castro HF (2015) Biosynthesis, characterization and enzymatic transesterification of single cell oil of *Mucor circinelloides* – a sustainable pathway for biofuel production. Bioresour Technol 181:47–53. doi:10.1016/j.biortech.2014.12.110

Cassia Pereira J, Paganini Marques N, Rodrigues A et al (2015) Thermophilic fungi as new sources for production of cellulases and xylanases with potential use in sugarcane bagasse saccharification. J Appl Microbiol 118:928–939. doi:10.1111/jam.12757

Castiglioni G, Stanescu G, Rocha LAO, Costa JAV (2013) Analytical modeling and numerical optimization of the biosurfactants production in solid-state fermentation by *Aspergillus fumigatus*. Acta. Sci Technol 36:61–67. doi:10.4025/actascitechnol.v36i1.17818

Celestino JDR, De Carvalho LE, Lima MDP et al (2014) Bioprospecting of Amazon soil fungi with the potential for pigment production. Process Biochem 49:569–575. doi:10.1016/j.procbio.2014.01.018

Chengaiah B, Rao KM, Kumar KM et al (2010) Medicinal importance of natural dyes-a review. Int J Pharm Tech Res 2:144–154

Choi J-M, Han S-S, Kim H-S (2015) Industrial applications of enzyme biocatalysis: current status and future aspects. Biotechnol Adv 33:1443–1454. doi:10.1016/j.biotechadv.2015.02.014

Colla LM, Rizzardi J, Pinto MH et al (2010) Simultaneous production of lipases and biosurfactants by submerged and solid-state bioprocesses. Bioresour Technol 101:8308–8314. doi: 10.1016/j.biortech.2010.05.086

Corrêa RCG, Rhoden SA, Mota TR et al (2014) Endophytic fungi: expanding the arsenal of industrial enzyme producers. J Ind Microbiol Biotechnol 41:1467–1478. doi:10.1007/s10295-014-1496-2

Cunha FM, Esperança MN, Zangirolami TC et al (2012) Sequential solid-state and submerged cultivation of *Aspergillus niger* on sugarcane bagasse for the production of cellulase. Bioresour Technol 112:270–274. doi:10.1016/j.biortech.2012.02.082

Cyndy MF, Odacy CDS, Minelli AS et al (2015) High-level lipase production by *Aspergillus candidus* URM 5611 under solid state fermentation (SSF) using waste from *Siagrus coronata* (Martius) Becari. African. J Biotechnol 14:820–828. doi:10.5897/AJB2014.14339

Da Silva M, Nascimento C, Junior S, Albuquerque P (2014) Biosurfactant production by *Myrcia guianensis* endophytic fungi. BMC Proc 8:213. doi:10.1186/1753-6561-8-S4-P213

Damaso MCT, Machado CMM, Rodrigues DDS et al (2014) Bioprocesses for biofuels: an overview of the Brazilian case. Chem Biol Technol Agric 1:1–8. doi:10.1186/s40538-014-0006-0

De Castro AM, De Andréa TV, Dos Reis CL, Freire DMG (2010) Use of mesophilic fungal amylases produced by solid-state fermentation in the cold hydrolysis of raw babassu cake starch. Appl Biochem Biotechnol 162:1612–1625. doi:10.1007/s12010-010-8942-z

De Oliveira APA, Silvestre MA, Alves-Prado HF et al (2015) Bioprospecting of yeasts for amylase production in solid state fermentation and evaluation of the catalytic properties of enzymatic extracts. Afr J Biotechnol 14:1215–1223. doi:10.5897/ajb2014.14062

Delabona PDS, Pirota RDPB, Codima CA (2012) Using Amazon forest fungi and agricultural residues as a strategy to produce cellulolytic enzymes. Biomass Bioenergy 37:243–250. doi:10.1016/j.biombioe.2011.12.006

Dos Santos IP, Da Silva LCN, Da Silva MV et al (2015) Antibacterial activity of endophytic fungi from leaves of *Indigofera suffruticosa* Miller (Fabaceae). Front Microbiol. doi:10.3389/fmicb.2015.00350

Duarte SH, del Peso Hernández GL, Canet A et al (2015) Enzymatic biodiesel synthesis from yeast oil using immobilized recombinant *Rhizopus oryzae* lipase. Bioresour Technol 183:175–180. doi:10.1016/j.biortech.2015.01.133

Dufossé L (2006) Microbial production of food grade pigments. Food Technol Biotechnol 44:313–323

Durán N, Teixeira MFS, De Conti R, Esposito E et al (2002) Ecological-friendly pigments from fungi. Crit Rev Food Sci Nutr 42:53–66. doi:10.1080/10408690290825457

Düsman E, Berti AP, Soares LC, Vicentini VEP et al (2012) Principais agentes mutagênicos e carcinogênicos de exposição humana. SaBios-Revs Saúde e Biol 7:66–81

Faheina Junior GDS, Amorim MVDFS, Souza CG De et al (2015) Strategies to increase cellulase production with submerged fermentation using fungi isolated from the Brazilian biome. Acta Sci Biol Sci 37:15-22. doi: 10.4025/actascibiolsci.v37i1.23483

Ferreira MC, Vieira MDLA, Zani CL et al (2015) Molecular phylogeny, diversity, symbiosis and discover of bioactive compounds of endophytic fungi associated with the medicinal Amazonian plant *Carapa guianensis* Aublet (Meliaceae). Biochem Syst Ecol 59:36–44. doi:10.1016/j.bse.2014.12.017

Flores AC, Pamphile JA, Sarragiotto MH, Clemente E (2013) Production of 3-nitropropionic acid by endophytic fungus *Phomopsis longicolla* isolated from *Trichilia elegans* A. JUSS ssp. elegans and evaluation of biological activity. World J Microbiol Biotechnol 29:923–932. doi:10.1007/s11274-013-1251-2

Freedonia (2015) World enzymes: market environment. The freedonia Group. http://www.freedoniagroup.com/brochure/28xx/2824smwe.pdf. Acessed 03 Dec 2015

Glänzel W, Leta J, Thijs B (2006) Science in Brazil part 1: A macro-level comparative study. Scientometrics 67:67–86. doi:10.1007/s11192-006-0055-7

Goldbeck R, Ramos MM, GAG P, Maugeri-Filho F (2013) Cellulase production from a new strain *Acremonium strictum* isolated from the Brazilian biome using different substrates. Bioresour Technol 128:797–803. doi:10.1016/j.biortech.2012.10.034

Griebeler N, Polloni AE, Remonatto D et al (2009) Isolation and screening of lipase-producing fungi with hydrolytic activity. Food Bioprocess Technol 4:578–586. doi:10.1007/s11947-008-0176-5

Guimarães LHS, Peixoto-Nogueira SC, Michelin M et al (2006) Screening of filamentous fungi for production of enzymes of biotechnological interest. Braz J Microbiol 37:474–480. doi:10.1590/S1517-83822006000400014

Gunasekaran S, Poorniammal R (2008) Optimization of fermentation conditions for red pigment production from *Penicillium sp.* under submerged cultivation. Afr J Biotechnol 7:1894–1898. doi:10.4314/ajb.v7i12.58846

Hailei W, Zhifang R, Ping L et al (2011) Improvement of the production of a red pigment in *Penicillium sp.* HSD07B synthesized during co-culture with *Candida tropicalis*. Bioresour Technol 102:6082–6087

Hamano PS, Kilikian BV (2006) Production of red pigments by *Monascus ruber* in culture media containing corn steep liquor. Braz J Chem Eng 23:443–449

Harvard Business Review (2013) Bioeconomia: Uma Agenda Para o Brasil. Brasilia: CNI 40p

Ióca LP, Allard PM, Berlinck RGS (2014) Thinking big about small beings – the (yet) underdeveloped microbial natural products chemistry in Brazil. Nat Prod Rep 31:646–675. doi:10.1039/c3np70112c

Ishida K, Cipriano TF, Rocha GM et al (2014) Silver nanoparticle production by the fungus *Fusarium oxysporum*: nanoparticle characterisation and analysis of antifungal activity against pathogenic yeasts. Mem Inst Oswaldo Cruz 109:220–228. doi:10.1590/0074-0276130269

Jemli S, Ayadi-Zouari D, Hlima HB, Bejar S (2014) Biocatalysts: application and engineering for industrial purposes. Crit Rev Biotechnol 36:246–258. doi:10.3109/07388551.2014.950550

Kanokmedhakul S, Kanokmedhakul K, Nasomjai P et al (2006) Antifungal azaphilones from the *Fungus Chaetomium* cupreum CC3003. J Nat Prod 69:891–895. doi:10.1021/np060051v

Katerine A, Lobato DCL, Almeida AF et al (2013) Biosurfactant production from industrial residues using microorganisms isolated from oil wells. Int Rev Chem Eng 5:310–316. doi:10.3109/07388551.2014.950550

Katoch M, Salgotra A, Singh G (2014) Endophytic fungi found in association with *Bacopa monnieri* as potential producers of industrial enzymes and antimicrobial bioactive compounds. Braz Arch Biol Biotechnol 57:714–722. doi:10.1590/s1516-8913201402502

Kirk O, Borchert TV, Fuglsang CC (2002) Industrial enzyme applications. Curr Opin Biotechnol 13:345–351. doi:10.1016/S0958-1669(02)00328-2

Kumar A, Vishwakarma HS, Singh J, Kumar M (2015) Microbial pigments: production and their applications in various industries. Int J Pharm Chem Biol Sci 5:203–212

Lange L, Bech L, Busk PK et al (2012) The importance of fungi and of mycology for a global development of the bioeconomy. IMA Fungus 3:87–92. doi:10.5598/imafungus.2012.03.01.09

Leta J, Glänzel W, Thijs B (2006) Science in Brazil. Part 2: sectoral and institutional research profiles. Scientometrics 67:87–105. doi:10.1007/s11192-006-0051-y

Lucas E, Machado Y, Ferreira A et al (2010) Improved production of pharmacologically-active sclerotiorin by *Penicillium sclerotiorum*. Trop J Pharm Res 9:365–371. doi:10.4314/tjpr.v9i4.58930

Luna JM, Rufino RD, Campos-Takaki GM, Sarubbo LA (2012) Properties of the biosurfactant produced by *Candida sphaerica* cultivated in low-cost substrates. Chem Eng Trans 27:67–72. doi:10.3303/CET1227012

Madeira JV, Macedo JA, Macedo GA (2011) Detoxification of castor bean residues and the simultaneous production of tannase and phytase by solid-state fermentation using *Paecilomyces variotii*. Bioresour Technol 102:7343–7348. doi:10.1016/j.biortech.2011.04.099

Maeda RN, Barcelos CA, Anna LMMS, Pereira N (2013) Cellulase production by *Penicillium funiculosum* and its application in the hydrolysis of sugar cane bagasse for second generation ethanol production by fed batch operation. J Biotechnol 163:38–44. doi:10.1016/j.jbiotec.2012.10.014

Mapari SAS, Thrane U, Meyer AS (2010) Fungal polyketide azaphilone pigments as future natural food colorants? Trends Biotechnol 28:300–307. doi: http://dx.doi.org/10.1016/j.tibtech.2010.03.004

Marchant R, Banat IM (2012) Microbial biosurfactants: challenges and opportunities for future exploitation. Trends Biotechnol 30:558–565. doi: http://dx.doi.org/10.1016/j.tibtech.2012.07.003

Mendes MMGS, Pereira SA, Oliveira RL et al (2015) Screening of Amazon fungi for the production of hydrolytic enzymes Afr J Microbiol Res 9:741–748. doi: 10.5897/A

Méndez A, Pérez C, Montañéz JC et al (2011) Red pigment production by *Penicillium purpurogenum* GH2 is influenced by pH and temperature. J Zhejiang Univ Sci B 12:961–968. doi:10.1631/jzus.B1100039

Michael CA, Dominey-Howes D, Labbate M (2014) The antimicrobial resistance crisis: causes, consequences, and management. Front Public Heal 2:145–145. doi:10.3389/fpubh.2014.00145

Ministério Do Desenvolvimento Indústria E Comércio Exterior (2012) Diagnóstico de referência sobre serviços de escalonamento de biotecnologias no Brasil. Fundação Bio-Rio, Rio de Janeiro

Mirjalili M, Nazarpoor K, Karimi L (2011) Eco-friendly dyeing of wool using natural dye from weld as co-partner with synthetic dye. J Clean Prod 19:1045–1051. doi:10.1016/j.jclepro.2011.02.001

Monteiro VN, Silva RN (2009) Aplicações Industriais da Biotecnologia Enzimática. Revista processos químicos 3:9–23

Moretti MMS, Bocchini-Martins DA, Silva DR et al (2012) Selection of thermophilic and thermotolerant fungi for the production of cellulases and xylanases under solid-state fermentation. Braz J Microbiol:1062–1071. doi:10.1590/s1517-83822012000300032

Mussatto SI, Fernandes M, Milagres AMF (2007) Enzimas: Poderosas ferramentas na indústria. Ciência Hoje 41:28–33

Neves MLC, Da Silva MF, Souza-Motta CM et al (2011) *Lichtheimia blakesleeana* as a new potencial producer of phytase and xylanase. Molecules 16:4807–4817. doi:10.3390/molecules16064807

Nitschke M, Pastore M (2002) Biossurfactantes: Propriedades e aplicações. Quím Nova 25:772–776. doi:10.1590/s0100-40422002000500013

Oh J, Jeong H, Oh S (2009) Characterization of optimal growth conditions and carotenoid production of strain *Rhodotorula mucilaginosa* HP isolated from larvae of *Pieris rapae*. Entomol Res 39:380–387. doi:10.1111/j.1748-5967.2009.00250.x

Onofre SB, Santos ZMQ, Kagimura FY, Mattiello SP et al (2015) Cellulases produced by the endophytic fungus *Pycnoporus sanguineus* (L .) Murrill. Afr J Agric Res 10:1557–1564. doi:10.5897/AJAR2015.9487

Orlandelli RC, Alberto RN, Almeida TT et al (2012) In vitro antibacterial activity of crude extracts produced by endophytic fungi isolated from *Piper hispidum* Sw. J Appl Pharm Sci 2:137–141. doi:10.7324/JAPS.2012.21027

Pasin TM, Benassi VM, Moreira EA et al (2014) Prospecting filamentous fungi for amylase production: standardization of *Aspergillus japonicus* culture conditions. British. Biotechnol J 4:482–498. doi:10.9734/bbj/2014/7659

Peixoto-Nogueira SC, Michelin M, Betini JHA et al (2009) Production of xylanase by Aspergilli using alternative carbon sources: application of the crude extract on cellulose pulp biobleaching. J Ind Microbiol Biotechnol 36:149–155. doi:10.1007/s10295-008-0482-y

Petit P, Lucas EMF, Abreu LM et al (2009) Novel antimicrobial secondary metabolites from a *Penicillium sp.* isolated from Brazilian cerrado soil. Electron J Biotechnol. doi:10.2225/vol12-issue4-fulltext-9

Polônio MLT, Peres F (2009) Consumo de aditivos alimentares e efeitos à saúde: desafios para a saúde pública brasileira. Cad Saúde Pública 25:1653–1666. doi:10.1590/S0102-311X2009000800002

Resende V (2012) The biotechnology market in Brazil. US Dept Commer 9. https://www.nccommerce.com/Portals/5/Documents/ITD/Biotech%20Market%20in%20Brazil.pdf. Acessed 03 Dec 2015

Rezaie R, Frew SE, Sammut SM et al (2008) Brazilian health biotech fostering crosstalk between public and private sectors. Nat Biotechnol 26:627–644. doi:10.1038/nbt0608-627

Ribeiro MDFDS, Raiher AP (2013) Potentialities of energy generation from waste and feedstock produced by the agricultural sector in Brazil: the case of the state of Paraná. Energ Policy 60:208–216. doi:10.1016/j.enpol.2013.05.004

Ribeiro Corrêa TL, De Queiroz MV, De Araújo EF (2015) Cloning, recombinant expression and characterization of a new phytase from *Penicillium chrysogenum*. Microbiol Res 170:205–212. doi:10.1016/j.micres.2014.06.005

Ribeiro J, Da Lima JRC, Maria L, Lopes DA (2013) Utilization of crude glycerol by *Yarrowia lipolytica* IMUFRJ 50678 in bioproduct production. J Chem Chem Eng 7:1087–1093

Ribeiro RCDS, Ribeiro TRDS, De Souza-Motta CM et al (2015) Production and partial characterization of proteases from *Mucor hiemalis* URM3773. Acta Sci Biol Sci 37:71. doi:10.4025/actascibiolsci.v37i1.21016

Rigo E, Ninow JL, Di Luccio M et al (2010) Lipase production by solid fermentation of soybean meal with different supplements. LWT Food Sci Technol 43:1132–1137. doi:10.1016/j.lwt.2010.03.002

Rodarte MP, DR D, DM V, RF S (2011) Proteolytic activities of bacteria, yeasts and filamentous fungi isolated from coffee fruit (*Coffea arabica* L.). Acta Sci Agron 33:457–464. doi:10.4025/actasciagron.v33i3.6734

Rodrigues AG, Ping LY, Marcato PD et al (2013) Biogenic antimicrobial silver nanoparticles produced by fungi. Appl Microbiol Biotechnol 97:775–782. doi:10.1007/s00253-012-4209-7

Sanderson K (2011) Chemistry: enzyme expertise. Nature 471:397–398. doi:10.1038/nj7338-397a

Santos DKF, Rufino RD, Luna JM et al (2013) Synthesis and evaluation of biosurfactant produced by *Candida lipolytica* using animal fat and corn steep liquor. J Pet Sci Eng 105:43–50. doi:10.1016/j.petrol.2013.03.028

Santos DKF, Brandão YB, Rufino RD et al (2014) Optimization of cultural conditions for biosurfactant production from *Candida lipolytica*. Biocatal Agric Biotechnol 3:48–57. doi:10.1016/j.bcab.2014.02.004

Saron C, Felisberti I (2006) Ação de colorantes na degradação e estabilização de polímeros. Quím Nova 29:124–128. doi:10.1590/s0100-40422006000100022

Saxena S (2015) Microbial enzymes and their industrial applications. Appl Microbiol:121–154. doi:10.1007/978-81-322-2259-0_9

Serdar M, Knežević Z (2009) Simultaneous LC analysis of food dyes in soft drinks. Chromatographia 70:1519–1521. doi:10.1365/s10337-009-1340-4

Shaikh S, Fatima J, Shakil S et al (2015) Antibiotic resistance and extended spectrum beta-lactamases: types, epidemiology and treatment. Saudi J Biol Sci 22:90–101. doi: http://dx.doi.org/10.1016/j.sjbs.2014.08.002

Shekhar S, Sundaramanickam A, Balasubramanian T (2014) Biosurfactant producing microbes and their potential applications: a review. Crit Rev Environ Sci Technol 45:1522–1554. doi:10.1080/10643389.2014.955631

Silva CADA, Lacerda MPF, Leite RSR, Fonseca GG (2013) Production of enzymes from *Lichtheimia ramosa* using Brazilian savannah fruit wastes as substrate on solid state bioprocessess. Electron J Biotechnol doi: 10.2225/vol16

Silva NAR, Luna M, Santiago A et al (2014) Biosurfactant-and-bioemulsifier produced by a promising *Cunninghamella echinulata* Isolated from caatinga Soil in the Northeast of Brazil. Int J Mol Sci 15:15377–15395. doi:10.3390/ijms150915377

Silver LL (2011) Challenges of antibacterial discovery. Clin Microbiol Rev 24:71–109. doi:10.1128/CMR.00030-10

Simões MLG, Tauk-tornisielo SM, Tapia DM (2009) Screening of culture condition for xylanase production by filamentous fungi. Afr J Biotechnol 8:6317–6326

Soberón-Chávez G, Maier R (2011) Biosurfactants: a general overview. In: Soberón-Chávez G (ed) Biosurfactants SE. Springer, Berlin/Heidelberg, pp. 1–11

Souza FHM, Meleiro LP, Machado CB et al (2014) Gene cloning, expression and biochemical characterization of a glucose- and xylose-stimulated β-glucosidase from Humicola insolens RP86. J Mol Catal B Enzym 106:1–10. doi:10.1016/j.molcatb.2014.04.007

Spier MR, Fendrich RC, Almeida PC et al (2011) Phytase produced on citric byproducts: purification and characterization. World J Microbiol Biotechnol 27:267–274. doi:10.1007/s11274-010-0455-y

Suzana CSM, Viviane OA, Claudia MM (2014) Pichia spp. yeasts from brazilian industrial wastewaters: physiological characterization and potential for petroleum hydrocarbon utilization and biosurfactant production. Afr J Microbiol Res 8:664–672. doi:10.5897/AJMR2013.6037

Takahashi JA, Carvalho SA (2010) Nutritional potential of biomass and metabolites from filamentous fungi. Curr Res Technol Educ Top Appl Microbiol Microb Biotechnol 2:1126–1135

Teixeira MFS, Martins MS, da Silva JC et al (2012) Amazonian biodiversity: pigments from *Aspergillus* and *Penicillium*-characterizations, antibacterial activities and their toxicities. Curr Trends Biotechnol Pharm 6:300–311

Theuretzbacher U (2009) Future antibiotics scenarios: is the tide starting to turn? Int J Antimicrob Agents 34:15–20. doi: http://dx.doi.org/10.1016/j.ijantimicag.2009.02.005

Uenojo M, Roberto M, Junior M, Pastore M (2007) Carotenóides: Propriedades, aplicações e biotransformação para formação de compostos de aroma. Quím Nova 30:616–622. doi:10.1590/s0100-40422007000300022

Valencia EY, Chambergo FS (2013) Mini-review: Brazilian fungi diversity for biomass degradation. Fungal Genet Biol 60:9–18. doi:10.1016/j.fgb.2013.07.005

Vaz ABM, Brandão LR, Vieira ML et al (2012) Diversity and antimicrobial activity of fungal endophyte communities associated with plants of Brazilian savanna ecosystems. African J. Microbiol Res 6:3173–3185. doi:10.5897/AJMR11.1359

Vieira MLA, Hughes AFS, Gil VB et al (2011) Diversity and antimicrobial activities of the fungal endophyte community associated with the traditional Brazilian medicinal plant *Solanum cernuum* Vell. (Solanaceae). Can J Microbiol 58:54–66. doi: 10.1139/w11-105

Volp ACP, Renhe IRT, Stringueta PC (2009) Pigmentos naturais bioativos. Aliment e Nutr 20:157–166

Wright PM, Seiple IB, Myers AG (2014) The evolving role of chemical synthesis in antibacterial drug discovery. Angew Chem Int Ed 53:8840–8869. doi:10.1002/anie.201310843

Zanphorlin LM, Facchini FDA, Vasconcelos F et al (2010) Production, partial characterization, and immobilization in alginate beads of an alkaline protease from a new thermophilic fungus *Myceliophthora sp*. J Microbiol 48:331–336. doi:10.1007/s12275-010-9269-8

Chapter 10
Secondary Metabolites of Mine Waste Acidophilic Fungi

Andrea A. Stierle and Donald B. Stierle

Abstract Microorganisms isolated from an acid mine waste lake have proven an unexpected source of novel, bioactive metabolites. Berkeley Pit Lake is part of the largest Superfund site in North America. Despite its low pH, high E_h, and high metal concentration, it harbors extremophilic microbes that have been grown in the laboratory in liquid culture using an array of physicochemical conditions. Bioassay-guided fractionation has directed the purification of small molecule inhibitors of enzymes associated with inflammation (caspase-1) and epithelial-mesenchymal transition (matrix-metalloproteinase-3, MMP-3). Caspase-1 plays an important role in chronic inflammation. It is activated upon binding to the inflammasome, a multiprotein complex that plays a key role in innate immunity. Activated caspase-1 in turn activates pro-inflammatory cytokines. Up-regulation of caspase-1 and concomitant chronic inflammation have been associated with leukemia, melanoma, glioblastoma, pancreatic cancers and breast cancer. MMP-3 promotes tumor cell invasion through the loss of cellular adhesion and promotion of epithelial-mesenchymal transition, which is associated with the metastatic spread of cancer. Inflammation and metastasis are interconnected and provide important targets for chemotherapy. Novel compounds have been tested by the National Cancer Institute-Developmental Therapeutics Program, Memorial Sloan Kettering Cancer Center and Eisai, Inc. for anti-proliferative activity against specific and established human cancer cell lines. In these studies, attention has been paid to the isolation and characterization of enzyme inhibitors as well as inactive analogues to facilitate assessment of structure/activity relationships. Compounds isolated using this methodology have demonstrated activity against specific cancer cell lines, and some of these compounds will be described.

A.A. Stierle (✉) • D.B. Stierle
Department of Biomedical and Pharmaceutical Sciences, University of Montana, Missoula, MT, USA, 59812
e-mail: andrea.stierle@mso.umt.edu; donald.stierle@mso.umt.edu

R. Paterson, N. Lima (eds.), *Bioprospecting*, Topics in Biodiversity and Conservation 16, DOI 10.1007/978-3-319-47935-4_10

10.1 Background

10.1.1 Bioprospecting in Berkeley Pit Lake

Our preliminary investigation of Berkeley Pit Lake (BPL) microbes began in 1995 when we were asked to identify a green slime coating a piece of wood submerged in the waters of this abandoned open-pit copper mine. BPL was presumed to be sterile due to its low pH (2.5) and high metal concentrations. However, when phycologist Grant Mitman identified the green slime as the alga *Euglena chlorella*, we hypothesized that this hostile environment could potentially support other types of microbial life, so we initiated a pilot project to investigate that possibility (Stierle and Stierle 2005). We obtained a sample of surface water from BPL and streaked it onto nutrient agar plates incubated at 25° C, and within a few days yeast colonies and fungal hyphae began to proliferate. The microbes were established in pure culture and identified as the filamentous fungi *Pithomyces* sp. and *Penicillium chrysogenum*, and the yeast *Hansenula anomala* (Stierle and Stierle 2005).

As natural products chemists our primary interest in these three fungi was their secondary metabolism under defined culture conditions. Secondary metabolites are organic compounds that are not required for growth, development or reproduction of an organism and are often produced by a particular genus or species (Fraenkel 1959). These compounds often exhibit biological activities that can translate into pharmaceutical potential (Fraenkel 1959). Many antibiotics and anticancer agents including penicillin, streptomycin and paclitaxel are produced by fungi, bacteria, or higher plants. Such organisms produce hundreds of secondary metabolites, so the discovery of compounds with drug-like potential is often dependent on bioassay systems that focus and guide isolation efforts towards a particular biological activity. In 1995, our primary interest was the discovery of new antibiotic agents, so our initial efforts were guided by standard disk antimicrobial assays. Following our usual laboratory protocols, we grew each of the three fungi in liquid media and extracted the resulting cultures with organic solvents at specific time points (Stierle and Stierle 2005). Unfortunately, none of the extracts exhibited antimicrobial activity, so we turned our attention from BPL microbes to other target organisms from other environments.

Despite the lack of antimicrobial activity in the pilot study, however, the BPL microbes continued to intrigue us. Microbes from unexplored ecosystems hold the promise of novel secondary metabolites with potential medicinal, industrial or agrochemical applications. But before we turned our attention back to BPL microbes, it would be necessary to find an appropriate bioassay to direct compound isolation.

In 2002 a small pilot grant funded our efforts to find small molecule inhibitors of specific signaling enzymes. Our research focus had shifted from new antibiotic agents to compounds that could mitigate pathways associated with carcinogenesis. We therefore directed our search towards inhibitors of matrix metalloproteinase-3 (MMP-3) and caspase-1 for reasons that will be described later in the chapter. The organic extracts of several of the BPL microbes inhibited one or both of these enzymes *in vitro* and our formal investigation of BPL microbes commenced (Stierle and Stierle 2005).

In this chapter we will describe the isolation and characterization of small molecules produced by BPL extremophiles that inhibit discrete enzyme pathways associated with inflammation and epithelial-mesenchymal transition. We will demonstrate the correlation of this enzyme inhibitory activity with two orthogonal cell line assays: cytotoxicity against specific cancer cell lines, and mitigation of the production of proinflammatory cytokines in an induced inflammasome assay. Fundamental to this work, however, are the microbes themselves and the environment from which they were isolated. As the extreme environment itself selected for these microbes, we will provide a profile of BPL and a brief history of its evolution from an abandoned open-pit copper mine into a dynamic ecosystem.

10.1.2 Life in Extreme Environments

Before we focus on the microbes of BPL, it is important to note that they are not the only organisms that have been isolated from an inhospitable ecosystem. Indeed, life exists and often flourishes in a wide range of environments. These include natural systems such as deep-sea vents, salt brines, thermal pools, volcanic lakes, and frigid ice fields. Other extreme environments, like BPL, are anthropogenic and can be the result of the extractive resources industry, industrial accidents, or munitions testing. Nobel laureate Paul Crutzen suggested that this current era should be called the "anthropocene", with humans and human activities acting as a major geological force on the planet (Crutzen and Stoermer 2000). Whether man-made or natural, extreme environments can harbor life forms called *extremophiles*. Bacteriologist Thomas Brock and his colleague, Hudson Freeze demonstrated the importance of extremophiles in the 1960s when they isolated bacteria from a 70 °C thermal pool in Yellowstone National Park (Brock and Freeze 1969). One of those bacteria, *Thermus aquaticus*, produces a heat tolerant polymerase (Taq polymerase) that revolutionized the polymerase chain reaction (Saiki et al. 1988; Mullis 1990).

Although the search for new extremophiles conjures the image of daring adventurers exploring dramatic terrestrial, lunar, or martian landscapes, BPL was created in a mineral rich formation in the scenic Rocky Mountains in the heart of Butte, Montana. Its evolution from an open-pit copper mine to an acid mine waste lake took less than 20 years. Rain and ground water continually flow into BPL causing the water level to rise, and it is estimated that in 2023 the water level will reach the "critical point" where the bedrock meets the alluvial interface and 160 billion L of contaminated water will begin to flow into Silver Bow Creek, a major tributary of the Columbia River system (Duaime 2006). Because of its size, toxicity, dynamics and location, BPL is part of the largest U.S. Environmental Protection Agency Superfund site in the North America and a potential ecological time bomb. To the scientist, however, it is also an evolving, dynamic ecosystem and a classic by-product of the industrial age when mining regulations – or lack thereof – allowed such environmental devastation to occur. Although conditions within BPL are toxic for "normal" aquatic biota, the extremophilic microbes have adapted well to the environment.

MacElroy coined the term *extremophile* in 1974 to describe microorganisms that thrive under conditions that would be considered extreme from a human perspective (MacElroy 1974). Of course, extreme is a relative term. Obligate anaerobic microbes have long been known and are not viewed as extremophiles, yet life without oxygen would certainly be a challenge to most of us. In essence, the term extremophile is used to describe microbes that thrive in environments where most microbes cannot grow or thrive because of extremes in temperature, salinity, pH, or pressure. Extremophiles can be classified according to the environments in which they thrive (Barnes 2013):

Acidophiles thrive in an acidic environment, usually at an optimum pH of 2–3
Alkaliphiles thrive in an alkaline environment, usually at a minimum pH of 10
Halophiles require a salty environment, with a minimum salt concentration of 0.2 M
Barophiles thrive at high pressures
Radioresistant microbes survive doses of radiation 500 times greater than the lethal dose for humans
Psychrophiles thrive in a cold environment
Thermophiles thrive in a hot environment, with an optimum growth temperature of ≥45° C
Hyperthermophiles flourish at even higher temperatures, between 80–100° C
Endoliths survive within solid rock or deep within the Earth's crust
Xerophiles grow and reproduce in conditions with very little water available

The BPL microbes studied to date are acidophilic, thriving at a pH of 2.5, as well as metal-tolerant, withstanding high levels of copper, iron, zinc, cadmium and other metal sulfates (Duaime 2006; Nordstrom and Southam 1997). Microbes isolated from the lower levels of BPL are also barophilic, as they are exposed to pressures as great as 25 atm. at the bottom of the lake. These conditions have selected for a population of microbes that have evolved a number of survival strategies. It was our hypothesis that these survival mechanisms include the production of small molecules that regulate pathways associated with low pH and high metal concentrations. It was our goal to demonstrate that these small molecules might also be deployed to regulate pathways in mammalian systems that are associated with an acidic microenvironment, including inflammatory diseases and certain cancers.

10.1.3 The Evolution and Characterization of Berkeley Pit Lake

Although there is nothing natural about BPL, its inception and evolution are part of the tradition of copper, gold and silver mining that shaped the western United States. Too often the legacy of mining is typified by the ravaged landscape that dominates northeastern Butte, Montana, USA. A half century of deep shaft mining, followed by 30 years of open-pit mining, left a 4000 km network of tunnels as well as the Berkeley Pit (BP), a mile wide crater, that was dewatered by massive mechanical pumps to facilitate the mining operation. When copper prices dropped and the mine was closed, the pumps were decommissioned and BP began to slowly fill with water. Rain and ground water percolated through the mineral rich overburden and followed the natural down-gradient into the BP. Thirty years later, 160 billion L of an acidic, heavy metal contaminated, aqueous solution fills the pit to a depth of 275 m (Duaime 2006; Nordstrom and Southam 1997).

A single mineral species, iron pyrite, plays a dominant role in the geology of the area, and defines the geochemistry of BPL. The metal pyrite walls of the pit continually react with air and water to generate sulfuric acid, which further dissolves the minerals in the surrounding rocky overburden. Because of this dynamic interplay, the water is acidic (pH 2.5–2.7) and contaminated with high concentrations of metal sulfates including iron (1200 ppm), copper (190 ppm), aluminum (290 ppm), cadmium (2 ppm), manganese (230 ppm) and zinc (650 ppm) (Duaime 2006). It also has a positive E_h (oxidation potential) that ranges from +800 mV at the surface to +600 mV at a depth of 270 m. This is an unusually high E_h for such a deep lake. This combination of low pH and high E_h provided some interesting correlations with biological activity that will be explored later in the chapter (Duaime 2006; Nordstrom and Southam 1997).

As oxygen concentration typically decreases with depth, pyrite oxidation and resulting acid generation would also be expected to decrease. However, oxidation of pyrite by dissolved ferric iron can take place at a rapid rate in acidic waters, even in the complete absence of oxygen. The rate of ferrous iron oxidation by O_2 increases many orders of magnitude in the presence of certain acidophilic bacteria, chiefly *Acidithiobacillus ferrooxidans* (Nordstrom and Southam 1997). These coupled processes continually generate sulfuric acid, which further dissolves the mineral rich ore body. (Duaime 2006; Nordstrom and Southam 1997). As the pit lake slowly evolved from an empty crater to an acid mine waste lake, little attention was paid to the biological aspects of the water because it was considered too toxic to support life – until our discovery in 1995.

10.2 Microbial Diversity

10.2.1 Profiling Microorganisms in an Extreme Environment

The discovery of Taq polymerase inspired other scientists to search not only for other novel extremozymes (enzymes that are stable under extreme environmental conditions) but also for new bioactive secondary metabolites produced by extremophiles (Stierle and Stierle 2005). For the Stierle laboratory, the extremophilic microbes of BPL promised new chemistry that could be accessed through fermentation and iterative chromatographic separations. Although the rapidly developing tools of molecular biology have focused attention on population genomics and proteomics, the tools of classical microbiology still play an important role in the investigation of a microbial population. The two methodologies can give a very different view of microbial diversity in a given environment and the choice of approaches is often determined by the ultimate goal of the research endeavor. If a comprehensive profile of the population of a site is the goal, then the tools of molecular biology are preferable to those of classical microbiology. If the goal is the purification and characterization of bioactive secondary metabolites from these microbes, then microbial isolation and fermentation is still an important means to the discovery of new chemotherapeutic agents.

10.2.2 Extremophiles Associated with Berkeley Pit Lake

Unfortunately, many microbes are either difficult to culture or grow very slowly under any physicochemical conditions employed. Despite these problems, the subpopulation of BPL microbes that has been established in pure culture and that grows under standard conditions has provided excellent opportunities for study. Even if the more fastidious microbes resist standard cultivation techniques, the culturable microbes of BPL have proven themselves capable of producing interesting new chemistry. These include several species that were present at multiple depths and several species that were isolated only at a specific depth, in sediment samples, or associated with an alga. For example, the fungus identified as *Penicillium* sp. was associated with the alga *Chlorella mutabilis* (Stierle et al. 2006). *P. rubrum* was isolated from surface water, and from depths of 3 m, 50 m and 275 m, although only the deep water isolate of *P. rubrum* has been extensively studied to date (Stierle et al. 2004, 2007, 2008, 2011). Likewise, *Phialomonium curvatum* was also isolated from various depths down to 225 m as well as from sediments, while *Oidiodendron tenuissimum*, a *Chaetomium* sp. and a *Pithomyces* sp. were only isolated from surface water. BPL bacteria exhibited a similar distribution pattern. *Stenotrophomonas maltophilia* was isolated from sediment samples and from water samples taken at the surface, 10 m and 225 m, while *Methylobacterium fujisawaense* was only isolated from surface water and *Streptomyces griseoplanus* was only isolated from sediment samples.

Certain microbes, including all of the *Penicillium* sp., grew equally well in both a low pH, high metal concentration broth and in a more generalist medium like potato dextrose broth, although the production of secondary metabolites often varied in different media. *Phialemonium curvatum* actually demonstrated more robust growth under more extreme conditions. These microbes were designated as *thrivers* – microbes uniquely adapted to the environment in which they were found. Others, particularly the bacteria, grew very slowly at low pH and it was clear that these microbes were "extremotolerant" rather than truly extremophilic. Most of our subsequent work focused on the thrivers, the true extremophiles. Although we propose to launch a complete genomic analysis of the waters and sediments of BPL in the future, our current studies require the actual collection and maintenance of a microbial library. We have isolated several new, bioactive compounds from these robust, acid mine waste fungi, several of which will be described later in this chapter (Stierle and Stierle 2005; Stierle et al. 2006, 2004, 2007, 2008, 2011, 2012a, b; Stierle and Stierle 2013; Stierle et al. 2014; Stierle and Stierle 2014).

10.3 Selected Physiological Effects of External and Internal Environmental Stress

10.3.1 Correlation of Low pH and an Oxidative Environment with the Tumor Microenvironment and Inflammation

The bioassay-guided isolation of secondary metabolites is biased by the choice of assay systems. Our previous studies of plant endophytic fungi demonstrated that these organisms often produce antibacterial or antifungal agents to enhance their abilities to compete for vital resources. Therefore, at the onset of this project, antimicrobial activity was used to guide isolation and purification of small molecules from liquid cultures of BPL extremophiles. However, the crude organic and aqueous extracts of these organisms did not exhibit strong antibacterial or antifungal activity. Evidently, the harshness of the environment itself, not interspecies competition for resources, kept population densities at very low levels in BPL.

As we evaluated the appropriateness of different assay systems to guide compound isolation, we considered the uniqueness of BPL microbes and their environment and how that could apply to human pathologies. Two factors seemed particularly relevant: iron concentration and pH. While these factors clearly dominate the environment of BPL microbes, they have also been implicated in pathological conditions that affect human metabolism (Toyokuni 2009). Iron is the most abundant transition metal in the human body. In humans, redox cycling of iron is closely associated with the generation of reactive oxygen or nitrogen species (ROS and RNS) and the induction of oxidative stress, which occurs when the generation of excess free radicals and active intermediates exceeds the system's ability to eliminate them. Oxidative stress can lead to DNA damage and promote numerous pathologies including hepatocarcinoma (Toyokuni 2009).

Elevated rates of ROS have been detected in almost all cancers and they promote tumor progression and development through a variety of mechanisms (Liou and Storz 2010). ROS also play an important role in the progression of inflammatory disorders (Mittal et al. 2014). One of the mechanisms by which ROS promotes inflammation is through the induction of the NLR protein 3 (NLRP3) inflammasome, an important multiprotein complex that binds and subsequently activates caspase-1 (Davis and Ting 2010). Upon activation, caspase-1 stimulates production of several proinflammatory cytokines including IL-1β and IL-18.

Low pH is also associated with both inflammation and carcinogenesis. An acidic environment is a hallmark of the tumor microenvironment due to the production of acidic metabolites caused by anaerobic glycolysis under the hypoxic conditions associated with tumors (Warburg et al. 1924; Kato et al. 2013). According to a theory known as the "Warburg effect" tumors tend to produce lactate because they use the anaerobic glycolytic pathway rather than oxidative phosphorylation for energy production, even in the presence of sufficient oxygen (Warburg et al. 1924; Kato et al. 2013). Low pH in inflammatory tissue is also due to production of acids including retinoic acid and quinolinic acid by macrophages (Sanders et al. 2014; Smith et al. 2001).

10.3.2 Fungi as a Source of Bioactive Compounds in Response to ROS-Induced Stress

The connection between low pH and ROS-induced oxidative stress within both the tumor microenvironment and inflammation – and BPL – intrigued us. Although mammalian systems respond to stress inducers through numerous pathways, fungi often rely on their secondary metabolism to ameliorate the effects of adverse conditions. In a similar manner, we proposed that BPL microbes, particularly the "thrivers", would synthesize small molecules to help them deal with the effects of low pH and ROS. We further proposed that these molecules could potentially target enzyme pathways in mammalian systems that were up-regulated by these same stressors. Indeed, fungi produce many different types of molecules that have demonstrated antioxidant activity and many of these compounds are produced in response to ROS or RNS-induced oxidative stress. These compounds include dimerumic acid (DMA), an antioxidant produced by a *Monascus* sp. that inhibits ROS-induced oxidative stress *in vitro* and *in vivo* (Lee and Pan 2013). DMA has been shown to prevent the invasion of human adenocarcinoma cell line SW620 into surrounding tissues by suppressing the hydrogen peroxide-mediated activation of two *mitogen-activated protein kinases* (MAPK): *Jun N-terminal kinase* (JNK) and *extracellular-regulated kinase* (ERK). Activation of JNK and ERK ultimately lead to the production of

MMP-7, one of the metalloproteinases that (a) promotes epithelial-mesenchymal transition and (b) is particularly up-regulated in colon cancers (Ho 2011).

Fungi also produce numerous compounds that are redox active because they contain a bridged polysulfide piperazine ring that that can cycle between the oxidized and reduced forms depending on the redox state of the surrounding tissues (Lee et al. 2001). One of the most well-studied of these molecules is gliotoxin, an epipolythiodioxopiperazine produced by several fungi including *Trichoderma hamatum*, *T. virens*, *Aspergillus fumigatus* and *Penicillium* sp (Lee et al. 2001). Gliotoxin inhibits angiogenesis (Mason and Kidd 1951) and mitigates the proinflammatory response and has been shown to be cytotoxic and immunoinhibitory (Mason and Kidd 1951; Vigushin et al. 2004). Gliotoxin demonstrated *in vivo* activity against human breast cancer MCF-7 cells in culture and on several solid human cancer xenograft tissues in SCID mice (Pan and Harday 2007). There are over 120 fungal metabolites that incorporate a bridged disulfide moiety including the sporodesmins, the sirodesmins, the verticillins, the chaetocins and chaetomin (Boyer et al. 2013). All of these compounds are redox active and bioactive, and most exhibit varying degrees of cytotoxicity against human cancer cell lines (Jiang and Guo 2011).

10.4 Mechanistic Pathways Associated with Cancer

10.4.1 Correlations between Cancer and Inflammation

While carcinogenesis and inflammation can be influenced by ROS and low pH, they can also have a synergistic effect on each other. Inflammation has long been considered an important component of tumor progression, and many researchers are now finding that inflammatory cells orchestrate much of the tumor microenvironment (Lu et al. 2006). Tumor cells have also co-opted some of the signaling molecules of the innate immune system, such as selectins, chemokines and their receptors for invasion, migration and metastasis (Yabu et al. 2011). In addition, lactate, which is one of the most important by-products of anaerobic glycolysis in tumor cells, also functions as an intrinsic inflammatory mediator. It increases production of interleukin-17A (IL-17A) by T-cells and macrophages, resulting in the promotion of chronic inflammation in the tumor microenvironment (Warburg et al. 1924; Kato et al. 2013; Yabu et al. 2011).

Inflammation and cancer are complex phenomena and are controlled by numerous mechanisms (Lu et al. 2006). When inflammation is initiated by infection, wounding, or irritation a wide array of immune cells are recruited to the site. This leads to the release of various pro-inflammatory cytokines and other agents, often through the activation of the NLRP3 inflammasome (Davis and Ting 2010). These molecules orchestrate the initiation of an inflammatory cascade which is precisely timed and often resolves itself. Unfortunately, if unresolved, chronic inflammation may develop which results in tissue damage, cell proliferation and the generation of

ROS and RNS, the hallmarks of the cancer-prone microenvironment. Inflammatory cells (macrophages, neutrophils, eosinophils and basophils) and molecules (cytokines and chemokines) persist after tumor initiation and may infiltrate into the tumor site. They may protect tumor cells from the host immune response or directly facilitate angiogenesis, tumor growth, invasion, and metastasis by themselves or by inducing other effector molecules, such as matrix metalloproteinases (MMPs) (Davis and Ting 2010).

Interestingly, certain compounds with anti-inflammatory activity have also been found to have potent anticancer activity (Orlikova et al. 2014). These include several nonsteroidal anti-inflammatory drugs (NSAIDS) that mitigate inflammation through cyclooxygenase-2 dependent and independent pathways. Among these NSAIDs, sulindac sulfide and tolfenamic acid have been well investigated for their anti-tumorigenic activity in many different types of cancer (Liggetta et al. 2014). Although the NSAID aspirin has also been shown to have anticancer activity, its most significant effect might be as a chemopreventative agent against colorectal cancer (CRC). Five cardiovascular-prevention randomized controlled trials (RCTs) evaluating the use of aspirin were also analyzed for effects on cancer outcomes. Daily aspirin use at any dose reduced the risk of CRC by 24 % and of CRC-associated mortality by 35 % (Garcia-Albeniz and Chan 2011).

10.4.2 Inflammation, the Inflammasome and Caspase-1

Despite its connection to carcinogenesis, inflammation is an important natural defense reaction instigated by tissue damage or the presence of foreign proteins or pathogens, and is characterized by redness, heat, swelling, and pain. The primary objective of inflammation is to localize and eradicate the irritant and repair the surrounding tissue. For the survival of the host, inflammation is a necessary and beneficial process, but it can be difficult to control (Understanding Autoimmune Diseases 1998). When the body's own immune responses are directed against its own tissues, autoimmune disorders, characterized by prolonged inflammation and subsequent tissue destruction can result (Understanding Autoimmune Diseases 1998). Autoimmune disorders can cause immune-responsive cells to attack the linings of the joints, resulting in rheumatoid arthritis, or trigger immune cells to attack the insulin-producing islet cells of the pancreas leading to insulin-dependent diabetes (Understanding Autoimmune Diseases 1998).

Caspase-1 plays an important role in inflammation, and in chronic inflammation, through the production of specific cytokines. It was the first of a novel type of cysteine proteases responsible for converting IL-1β to its mature form in monocytes. Caspase-1 is activated upon binding to the NLRP3 inflammasome, a multiprotein complex that plays a key role in innate immunity. Once caspase-1 is activated, it cleaves (activates) the precursors of IL-1β, IL-18, and IL-33 (Davis and Ting 2010; Franchi et al. 2009). The up-regulation of caspase-1 and concomitant chronic inflammation has been associated with a number of different pathologies. Although

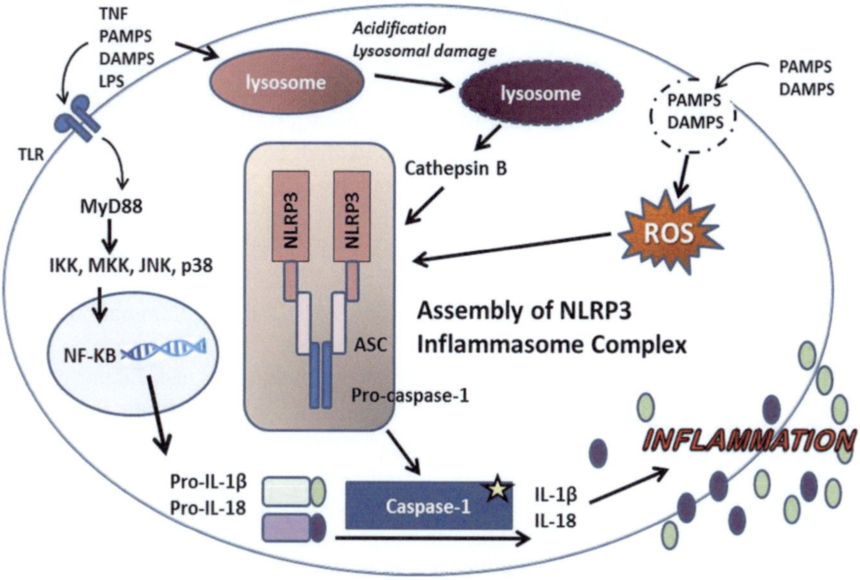

Fig. 10.1 IL-1β and IL-18 are produced by the NLRP3 inflammasome which is activated upon binding to and cleaving procaspase-1. Inflammasome assembly is stimulated by several factors including ROS. The production of proIL-1β and -18 are regulated by nuclear factor-kappa B (NF-κB) which is activated by exposure of the cell to tumor necrosis factor (TNF), and pathogen- and danger-associated molecular patterns (PAMPS and DAMPS) including bacterial LPS

elevated levels of caspases are important in inducing cell death in many types of cancer cells, the overexpression of caspase-1 and/or IL-1β have been found in certain cancers (Sternlicht et al. 1999) including acute myelogenous leukemia (Granot et al. 2006), melanoma (Okamoto et al. 2010a; Voronov et al. 2003a), certain glioblastomas (Paugh et al. 2009), pancreatic cancers (Schlosser et al. 2001; Muerkoster et al. 2006) and certain breast cancers (Jin et al. 1997), all of which are exacerbated by chronic inflammation associated with activation of the inflammasome.

As shown in Fig. 10.1, the production of pro-IL-1β and IL-18 are regulated by *nuclear factor-kappa B* (NF-κB). NF- κB can be activated by exposure of the cell to *tumor necrosis factor* (TNF), and *pathogen-associated* or *danger-associated molecular patterns* (PAMPS and DAMPS), including bacterial lipopolysaccharides (LPS). These same factors also stimulate the assemblage of the NLRP3 inflammasome (Franchi et al. 2009). Proinflammatory cytokines can lead to chronic inflammation and the production of ROS that can induce oxidative damage to DNA, and consequently lead to the initiation and progression of carcinogenesis (Franchi et al. 2009; Radisky et al. 2005). As caspase-1 plays such a pivotal role in the production of proinflammatory cytokines, we decided to direct part of our isolation efforts towards caspase-1 inhibitors.

10.4.3 Correlating Inflammation and Metastasis

Although inflammation is associated with many aspects of tumorigenesis, its role in metastasis is of particular interest not only to the Stierle laboratory but to many other researchers (Coffelt and de Visser 2014). A high concentration of inflammatory cells, particularly tumor-associated macrophages (TAMs), is commonly found at the invasive fronts of advanced carcinoma. Macrophages are key cells in chronic inflammation and TAMs produce a wide variety of growth factors and cytokines (such as TNFα, IL-6, IL-1, and interferons) that stimulate the growth, motility, and invasiveness of tumor cells (Wu and Zhou 2009). Mitigating inflammation could therefore play an important role in preventing metastasis.

The treatment of metastatic cancer is the most challenging task facing cancer researchers and oncologists. According to the National Cancer Institute, the majority of people who die of cancer, die of metastatic cancer (American Cancer Society. Cancer Facts & Figures 2014). In spite of progress in the understanding of the onset and progression of cancers and in the treatment of localized malignancies, metastatic disease is often incurable. The American Cancer Society reported that the five-year survival rate for localized breast cancer (invasive cancer confined entirely to the organ of origin) is 99 %, but the survival rate drops to 24 % when the cancer spreads to parts of the body remote from the primary tumor either by direct extension or by discontinuous metastasis to distant organs or tissues. Metastasis takes a similar toll in many other cancers. Lung and bronchus cancers account for more deaths than any other cancer in men and women. The 5 year survival rate for localized disease is 54 %, but for metastatic disease the survival rate drops to 4 %. Current treatment strategies for metastatic cancer largely rely on the use of systemic cytotoxic agents, which often have severe side effects and, in many cases, have limited long-term success (American Cancer Society. Cancer Facts & Figures,2014).

Preventing metastatic dissemination would greatly enhance cancer survival rates. Small molecule inhibitors of the enzyme pathways associated with metastasis would be useful tools to study the phenomenon itself. Moreover, they have the potential to serve as chemotherapies that could effectively prevent the spread of cancers to remote sites.

10.4.4 Correlating Epithelial-Mesenchymal Transition, Metastasis and MMP-3

Metastasis requires mobilization of cancer cells and invasion of healthy tissue, both of which are facilitated by epithelial-mesenchymal transition (EMT). EMT typifies cells undergoing proliferation and differentiation and is characterized by loss of cell adhesion, repression of E-cadherin (E-cad) expression, and increased cell mobility (Radisky et al. 2005). Although these processes are an important component of embryogenesis, neural tube development, wound healing, and angiogenesis, they are also associated with tumor cell invasion and metastatic spread of many types of cancer (Radisky et al. 2005). If this process could be circumvented, then the risk of metastatic spread would be mitigated. In spite of progress in the understanding of the onset and progression of cancers and in the treatment of localized malignancies, metastatic disease continues to be intractable and often incurable.

Although several enzymes are involved in EMT, matrix metalloproteinase-3 (MMP-3) has been shown to play a pivotal role in multiple aspects of the phenomenon. Furthermore, MMP-3 is up-regulated in many cancers, especially metastatic cancers (Radisky et al. 2005). MMP-3 is part of the family of zinc endopeptidases that are required for the degradation of extracellular matrix components during normal embryo development, morphogenesis and tissue remodeling (Nagase 1996). Tumor cell invasion is considered to be a disregulated physiologic invasion. Investigators have likened the molecular events involved in the process to events such as angiogenesis and wound healing. MMP activity is a common denominator in pathologic conditions and in normal responses. Within the body, endogenous *tissue inhibitors of metalloproteinases* (TIMPs) precisely regulate the levels and metabolic activities of MMPs (Nagase 1996; Coussens and Werb 1996). Disruption of this balance results in diseases such as arthritis, atherosclerosis, tumor growth and metastasis (Coussens and Werb 1996). Several studies suggest that the inhibition of MMP activity may prevent tumor cell dissemination (Stetler-Steveson et al. 1996).

10.4.4.1 Previous Attempts to Use MMP Inhibitors as Chemotherapeutics

The search for specific inhibitors of MMPs is not new. Several clinical trials of MMP inhibitors in the 1990s focused on compounds that were specifically designed to target MMP catalytic sites, with disappointing results. In retrospect, careful evaluation of these trials indicated two serious problems with many of the trials:

- The majority of Phase II and III clinical trials involved late-stage (III and IV) cancers that had already metastasized throughout the patients' bodies (Coussens et al. 2002; Zucker et al. 2000; Mannello et al. 2005). Many MMPs promote carcinogenesis by increasing EMT and subsequent metastasis. Chemotherapies that target metastasis should be prescribed in the early stages of cancer to prevent the progression to metastatic disease.
- It was assumed that key MMPs promoted EMT only through direct catalytic processes; therefore all early MMP inhibitors targeted the deep cleft catalytic sites of various MMPs either selectively or as broad spectrum therapies (Coussens et al. 2002; Zucker et al. 2000; Mannello et al. 2005). Recent studies have shown that MMP-3 induces EMT and metastatic cancer through multiple pathways and its catalytic and non-catalytic HPEX domains (Correia et al. 2013).

New insights into the complex interactions of MMPs with other molecular pathways and the role of non-catalytic domains in EMT suggest that MMP inhibitors deserve further consideration.

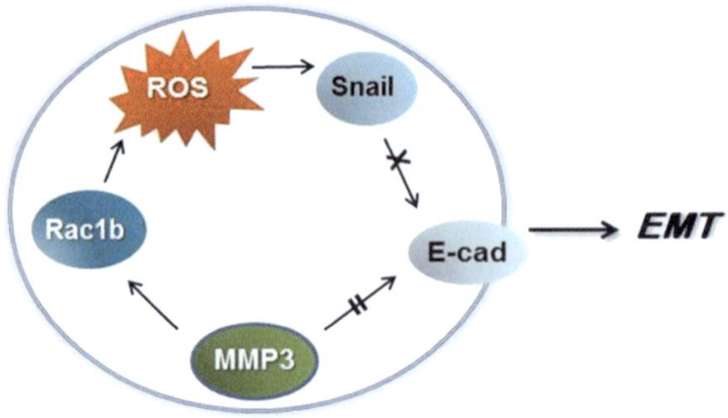

Fig. 10.2 MMP-3 up-regulates Rac1b which leads to enhanced production of ROS. The result is genomic instability and increased expression of the Snail transcription factor. While MMP-3 cleaves E-cadherin, Snail inhibits E-cadherin synthesis, thereby disrupting cell-cell contacts and fostering acquisition of a less differentiated, more aggressive mesenchymal phenotype and EMT

10.4.4.2 Multiple Mechanisms of EMT Promotion by MMP-3

MMP-3 promotes EMT through many different mechanisms including direct cleavage of the cell adhesion protein E-cad, repression of E-cad synthesis, and activation of MMP-9, a Type IV collagenase. MMP-9 is one of the Rosetta 70 genes and serves as a poor prognosis signature for patients with breast cancer (Vandooren et al. 2013). Once it is activated by MMP-3, the MMP-9 catalytic domain also directly promotes EMT through the disruption of extracellular basal lamina. These two proteins work synergistically to dissolve endothelial matrices (Vandooren et al. 2013). These are characteristic features of cells undergoing proliferation. The tumor microenvironment itself is actually a potent carcinogen, not only by facilitating cancer progression and activating dormant cancer cells, but also by stimulating tumor formation through induction of even higher levels of enzymes like MMP-3 Radisky et al. 2005.

Higher levels of MMP-3 in tumor tissue subsequently induce the expression of an alternatively spliced form of Rac-1 (called Rac-1b). Rac-1 is a member of the Rho family of GTPases and is a regulator of many cellular processes, including the cell cycle, epithelial differentiation, and cell-cell adhesion. Rac-1b causes an increase in ROS production. ROS stimulate the expression of the transcription factor **Snail**, which directly represses E-cad synthesis. ROS also cause oxidative damage to DNA and genomic instability. In essence, the tumor itself induces changes that alter cellular structure in culture and tissue structure *in vivo*, leading to EMT and malignant transformation (Fig. 10.2) (Radisky et al. 2005).

In 2013 however, mammary cell carcinomas studies in Bissell's laboratory demonstrated that the non-catalytic HPEX domain of MMP-3 is also involved in the promotion of EMT. MMP-3-HPEX binds to *extracellular* heat shock protein-90b (HSP90b), which is often over-expressed in cancer and is associated with more aggressive, larger tumors with more lymph node involvement and decreased patient survival (Correia et al. 2013). Preventing the formation of this complex could

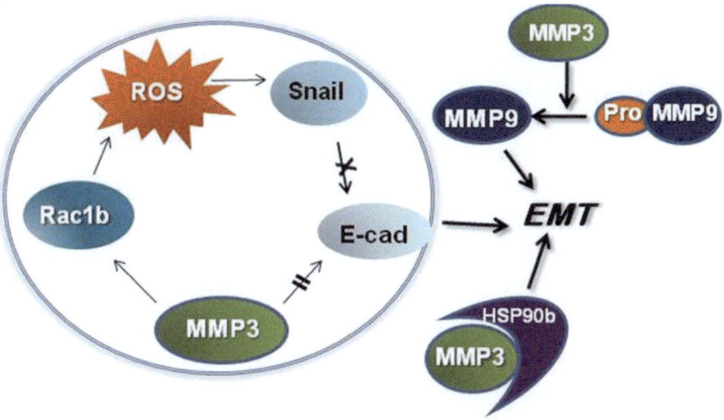

Fig. 10.3 MMP-3 promotes EMT through multiple mechanisms. In addition to its direct effects, it promotes EMT by the activation of MMP-9 which subsequently cleaves basal lamina. It binds extracellular HSP90b which leads to EMT through an unknown mechanism. Our efforts to mitigate EMT will involve inhibition of MMP-3 and/ or MMP-9 at their catalytic sites or HPEX domains, or of HSP90b at its catalytic site

mitigate EMT and thereby metastasis. Small molecule inhibitors of this complex could bind to the MMP-3-HPEX domain or the catalytic site of HSP90b.

Other studies have demonstrated the importance of the HPEX domains of other MMPs and consequently, have enhanced their importance as targets for drug discovery. MMP-9-HPEX has also been shown to mediate EMT cell migration (Dufour et al. 2011). In the only reported study of small molecule inhibitors of MMP-9-HPEX, specific HPEX inhibitors prevented the homodimerization of MMP-9, which effectively blocked MMP-9-mediated cancer cell proliferation, migration, and invasion to reduce tumor metastasis. These new insights are facilitating our search for compounds that can bind to MMP-3 at either the catalytic site or the HPEX domain, or to HSP90b, and thereby mitigate EMT and subsequent processes that lead to metastatic cancer (Fig. 10.3) (Thiery et al. 2009). Small molecule inhibitors of MMP-3 would block the action of both MMP-3 and MMP-9 and help maintain epithelial integrity.

10.4.5 The Inflammation - EMT Molecular Pathways Connection

10.4.5.1 The Induced Inflammasome Assay

Although specific MMP-3 and caspase-1 inhibition assays guide compound isolation in our lab, these *in vitro* assays must be validated by orthogonal *in vitro* assays. In 2009, immunologist/toxicologist Girtsman introduced the *induced inflammasome assay* (IIA) into our research protocol. The IIA is used to assess the ability of crude extracts, column fractions or pure compounds to block production of proinflammatory cytokines in *human monocytic leukemia cells* (THP-1 cells), a cellular system that is analogous to tumor-associated macrophages (TAMs) (Stierle et al. 2011, 2012a, b). TAMs represent the predominant population of inflammatory cells in solid tumors. When exposed to titanium nanowires and bacterial lipopolysaccharides (LPS), THP-1 cells produce large numbers of inflammasomes and proinflammatory cytokines including pro-IL-1β and IL-18. The inflammasome binds to and activates caspase-1, which in turn cleaves the cytokines to produce their active forms, IL-1β and IL-18. In carefully controlled experiments, induced THP-1 cells were exposed to test compounds at concentrations of 100, 10 and 1 μM, and the concentrations of IL-1β, IL-18 and TNFα post-exposure were determined (Stierle et al. 2011, 2012a, b). Test compounds included MMP-3 inhibitors, caspase-1 inhibitors and MMP-3/caspase-1 inhibitors.

As expected, all caspase-1 inhibitors isolated from BPL microbes were active in the inflammasome assay. As caspase-1 is required to activate proinflammatory cytokines, caspase-1 inhibition should prevent cytokine production in a cellular system and an enzyme system – if the compound can cross cell membranes. Compounds that were dual enzyme inhibitors were also active in the assay. What was surprising, however, was that several compounds that were not effective caspase-1 inhibitors (in the enzyme assay), mitigated cytokine production in the IIA. Unlike an enzyme assay, in a cellular system signaling crosstalk is possible, so MMP-3 inhibitors could affect proinflammatory cytokine production through a *non-caspase mediated* pathway.

10.4.5.2 Synergy Among Caspase-1 Activation, the Inflammasome and MMP-3

Several studies have demonstrated that cytokines that are activated upon cleavage by caspase-1 induce production of MMP-3. IL-1β and TNFα stimulate production of MMP-3 through the activation of cellular signaling pathways involving *mitogen-activated protein kinases* (MAPKs), NF-κB and *activating protein-1* (AP-1) (Kelley et al. 2007; Tsuzaki et al. 2003). The resulting up-regulation of MMP-3 leads to EMT as a direct result of the inflammatory pathway. It is noteworthy that the natural product cordycepin, a potent inhibitor of IL-1β production, effectively inhibits MMP-3 expression (Kim et al. 2006; Noh et al. 2009).

In a reciprocal fashion, MMP-3 also stimulates production of pro-IL-1β and pro-IL-18 which are activated ultimately by caspase-1. Macrophage exposure to MMP-3 or MMP-1 triggers a rapid release of TNFα, which induces the production of pro-IL-1β and pro-IL-18 through nuclear factor kappa-B (NF-κβ), an important transcription regulator. MMP-3 inhibitors blocked this activity, effectively reducing the concentration of pro-IL-1β and -18, and subsequently preventing synthesis of IL-1β and other cytokines (Steenport et al. 2009). These data could account for the efficacy of MMP-3 inhibitors in reducing the production of IL-1β in the *induced inflammasome assay*. MMP-3-induced production of ROS induces the assemblage of

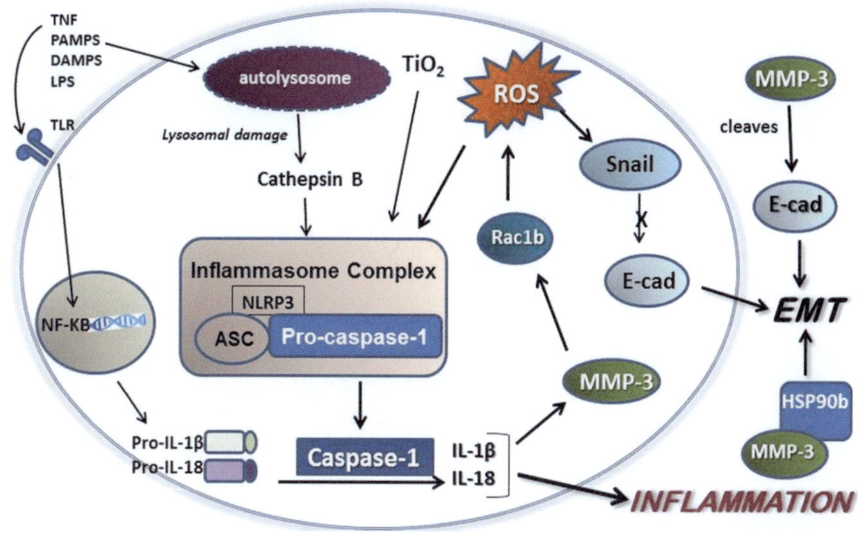

Fig. 10.4 Inflammation and EMT are two key cellular processes implicated in various pathologies. These processes are synergistic. Up-regulated MMP-3 leads to production of ROS which induces the assemblage of the NLRP3 inflammasome and the subsequent activation (cleavage) of procaspase-1. Caspase-1 activates cytokines 1L-1β and IL-18 which leads to the up-regulation of MMP-3. MMP-3 promotes EMT through multiple pathways

inflammasomes in macrophages (Harijith and Ebenezer 2014; Martinon 2010; Tschopp and Schroeder 2010). MMP-3 not only orchestrates events that promote EMT, it also induces further inflammation in a cell-cyclic series of events (Fig. 10.4).

The importance of extracellular HSP90b as a promoter of EMT was mentioned earlier in this chapter. Recent studies have shown that intracellular HSP90 is crucial for the stabilization of the NLRP3 protein that forms the core of the assembling inflammasome, and inhibition of HSP90 prevents the assemblage of the inflammasome. In the presence of the HSP90 (and HSP90b) inhibitor geldanamycin, the inflammasome cannot form properly (Mayor et al. 2007).

The inflammasome is becoming an important therapeutic target as studies continue to link inflammation and various types of cancer including colorectal, melanoma, and certain types of breast cancer (Elinav et al. 2013; Kolb et al. 2014; Okamoto et al. 2010b). Furthermore, in mice, potential therapeutic agents that target the inflammasome and mitigate the production of IL-1β have been shown to reduce the inflammatory response following thromboembolic stroke (Voronov et al. 2003b) and to improve histopathology following traumatic brain injury (Voronov et al. 2003b).

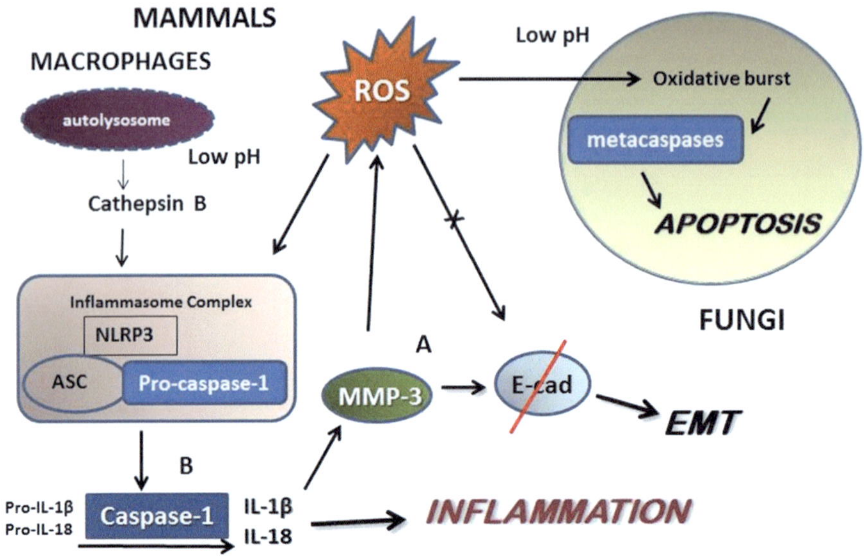

Fig. 10.5 Low pH and ROS induce caspase-1 mediated responses in mammalian macrophages through the NLRP3 inflammasome and metacaspase- mediated responses in fungi. Both caspase and metacaspases can be inhibited by the same small molecules

10.4.6 Mammalian Caspases and Fungal Metacaspases

An interesting biochemical connection exists between fungi and mammals exposed to oxidative stress. As previously noted, in mammalian cells, low pH and ROS are intimately associated with inflammation. Within macrophages, ROS stimulates the activation of the inflammasome, which leads to the activation of caspase-1 and the subsequent activation of proinflammatory cytokines.

In an analogous manner, fungi produce metacaspases in acidic environments when stimulated by ROS-mediated oxidative bursts (Fig. 10.5) (Tsiatsiani et al. 2011; Carmona-Gutierrez et al. 2010; Mousavi and Robson 2003). Like the mammalian caspases -3,-6,-7,-8, and -9, metacaspases promote the progression to apoptosis (McIlwain et al. 2013). Although they are not structurally related to caspases, they cleave some of the same substrates including *Tudor Staphylococcus nuclease* (Carmona-Gutierrez et al. 2010) and are inhibited by the pan-caspase inhibitors zVAD-fmk and zVEID-fmk (Tsiatsiani et al. 2011). We speculated that a fungus consistently exposed to an oxidative, acidic environment would be a logical source of compounds that would regulate metacaspase activities and that these same inhibitors would also function as caspase inhibitors.

Fungi produce secondary metabolites for a number of reasons and their importance to the producing organism is not always clear. Although more attention is often paid to potential applications of fungal secondary metabolites than to their importance in fungal metabolism, the roles of certain compounds have been determined. Some of these small molecules are involved in fungal morphological differentiation. For example, zealarenone, an estrogenic mycotoxin produced by *Fusarium graminearum* is involved in the sexual development of the fungus and enhances perithecial production (Wolf and Mirocha 1973). Butyrolactone I, produced by *Aspergillus terreus*, is a selective inhibitor of fungal cyclin-dependent kinases (cdks). It increases hyphal branching, sporulation, and production of lovastatin, another secondary metabolite produced by this fungus (Schimmel et al. 1998). Bridged polysulfide piperazines like gliotoxin are believed to participate in redox regulation in the fungus (Lee et al. 2001) and to play a key role in fungal virulence (Sugui et al. 2007). These compounds illustrate the regulatory role of certain secondary metabolites in fungi. We proposed that this previously unstudied population of extremophiles would produce novel bioactive secondary metabolites. Moreover, we hypothesized that the analogous roles of mammalian caspases and fungal metacaspases, as well as the up-regulation of metacaspases in acidic, oxidative environments, heightened the probability of discovering fungal metabolites that could function as caspase inhibitors. It is beyond the scope of our work to determine how BPL fungal caspase inhibitors affect host metabolism. However, their effects on caspase-mediated inflammation and human cancer cell lines are the focus of ongoing research. Several of these molecules will be described later in the chapter.

10.5 Results of Studies in the Stierle Laboratory

Our studies have shown that mechanisms associated with inflammation and EMT are interconnected. At the cellular level, small molecule inhibitors of inflammation-associated enzymes (caspase-1) block pathways associated with EMT, and small molecule inhibitors of enzymes that are associated with EMT (MMP-3) inhibit inflammation. In the past, EMT and inflammation were studied as discreet phenomena, but recent studies have highlighted their synergistic relationship. In this chapter we have described some of the key points of this synergy which involves the following processes:

- MMP-3 induces formation of ROS through the Rac1b pathway and subsequent up-regulation of Snail (Radisky et al. 2005)
- Snail suppresses E-cad production which leads to EMT by disrupting cell–cell adhesion (Radisky et al. 2005)
- ROS also stimulates the assemblage of the NLRP3 inflammasome which activates caspase-1 (Mittal et al. 2014; Davis and Ting 2010)
- Caspase-1 activates the proinflammatory cytokines IL-1β and IL-18 which leads to inflammation (Mittal et al. 2014; Davis and Ting 2010; Franchi et al. 2009)
- IL-1β promotes MMP-3 production which directly cleaves E-cad and leads to EMT (Kelley et al. 2007; Tsuzaki et al. 2003; Kim et al. 2006; Noh et al. 2009)
- MMP-3 also activates MMP-9 which exacerbates EMT through dissolution of the basal lamina (Vandooren et al. 2013)

In 2002, we began a targeted search for MMP-3 and caspase-1 inhibitors produced by BPL microorganisms in broth culture (Stierle and Stierle 2005). Compounds that were isolated based on their inhibition of MMP-3, caspase-1 or both enzymes have demonstrated selective activity against specific human cancer cell lines when tested by the National Cancer Institute-Developmental Therapeutics Program (NCI-DTP) and by Memorial Sloan Kettering Cancer Center (MSKCC). Caspase-1 inhibitors including berkazaphilone B and C have shown selective cytotoxicity towards leukemia or melanoma cell lines and have demonstrated the ability to mitigate the production of pro-inflammatory cytokines in induced inflammasome assays. Enzyme inhibition assay-guided fractionation has resulted in the isolation of several active, novel secondary metabolites from the complex organic extracts of acid mine waste microbes grown in liquid cultures. Each of the fungi in this study were grown under various physicochemical conditions and generally produced very different chemistry when grown in still culture versus shake culture or potato dextrose broth acidified to pH 2.5 or pH 6.0. Promising fungi have been grown repeatedly to provide adequate material for compound isolation and eventual structure elucidation and testing. These compounds will be discussed later in this section.

10.5.1 Selected Secondary Metabolites of BP Lake Fungal Extremophiles

We have published three review articles that highlight many of the active compounds isolated to date, the BPL fungus from which they were derived, and their biological activities (Stierle and Stierle 2005, 2013, 2014). Only a fraction of the BPL microbes have been thoroughly studied to date: a *Pithomyces* sp., a *Penicillium* sp., a deep water isolate of *P. rubrum*, *P. solitum*, *P. clavigerum*, *Pleurostomophora* sp., *Oidiodendrum tenuissimum*, *Trichoderma virens*, and the actinomycete *Streptomyces griseoplanus*. Many of the compounds reported to date were isolated from iterative fermentations of *Penicillium* sp., *P. solitum*, *P. clavigerum*, and *P. rubrum*. Figure 10.6 provides an overview of the culturable fungi and bacteria isolated from different depths in BPL. Organisms highlighted in yellow have been studied extensively and have yielded bioactive metabolites that have been reported; organisms highlighted in green are currently being studied.

We have long observed that a single organism can produce a wide array of secondary metabolites if the culture conditions are varied, which has indeed been the case with BPL fungi (Stierle and Stierle 2005). For example, when grown in acidified potato dextrose broth ($PDBH^+$, pH 2.5–2.7) as a shake culture for 5 days followed by 21 days as a still culture, a deep water isolate of *P. rubrum* yielded berkeleydione and berkeleytrione (Stierle et al. 2004), berkeleyones A-C (Stierle et al. 2011), preaustinoid A and A1 (Stierle et al. 2011). However, the same fungus produced the berkazaphilones (Stierle et al. 2012b), berkeleyamides A-D (Stierle et al. 2008), and berkeleyacetals A-C (Stierle et al. 2007) when grown in $PDBH^+$ for 24 h shaken/21 days still (Table 10.1).

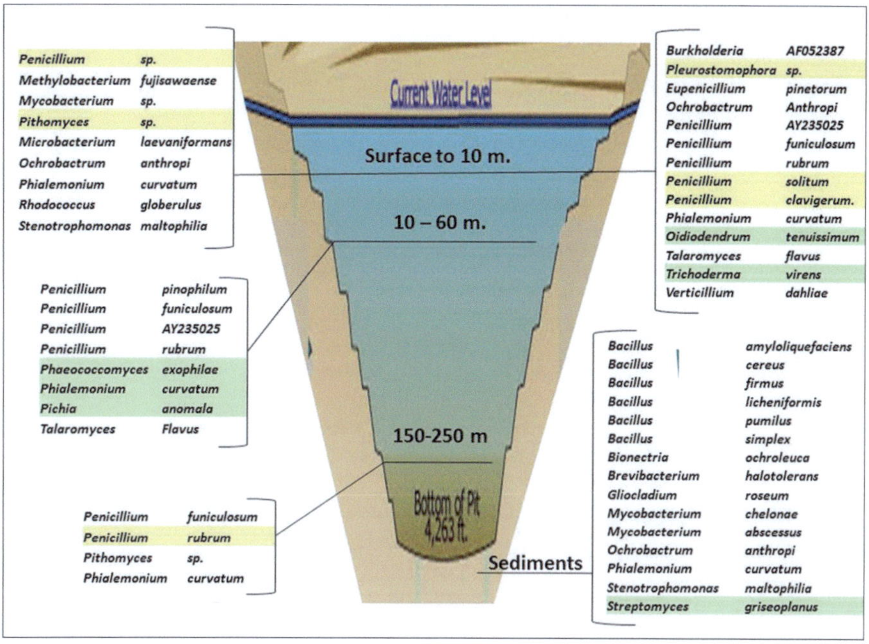

Fig. 10.6 Overview of fungi and bacteria isolated from different depths of the Berkeley Pit. Organisms highlighted in yellow have been studied extensively and have yielded bioactive metabolites that have been reported. Organisms highlighted in green are currently being studied

The 2D and 3D structures of the existing collection of natural product inhibitors of caspase-1 and MMP-3 have been carefully determined through spectroscopic and x-ray crystallographic methods (Stierle and Stierle 2005; Stierle et al. 2006, 2004, 2007, 2008, 2011, 2012a, b; Stierle and Stierle 2013; Stierle et al. 2014; Stierle and Stierle 2014). Each of these molecules was isolated from complex organic extracts of microbial fermentations using MMP-3 or caspase-1 inhibition assays to direct each step of the chromatographic separations. Each pure compound was tested to establish IC_{50} values for enzyme inhibition.

10.5.1.1 MMP-3 Inhibitors

We have begun the process of determining the mode of inhibition of each of the enzyme inhibitors isolated from BPL fungi. Repeated enzyme assays have consistently demonstrated the MMP-3 inhibitory activity of certain compounds (Fig. 10.7), and collaborators at NCI-DTP and MSKCC have determined that each of these compounds is active against specific cancer cell lines. The mechanism of action and binding interactions of these compounds as MMP-3 ligands is critical to the development of targeted, "intelligent" synthetic analogues with enhanced activity.

Table 10.1 Overview of BPL fungi and metabolite production under defined culture conditions

Penicillium rubrum	
Culture Conditions	PDBH$^+$, 5 days shaken/21 days still
Metabolites	Berkeleydione (5); Berkeleytrione (6); Berkeleyones A (13), B (15), C (16); Preaustinoid A (11); Preaustinoid A1 (12)
Culture Conditions	PDBH$^+$, 24 h shaken/21 days still
Metabolites	Berkazaphilones A, B (27), C (28); Berkeleyamides A (7), B (8), C (9), D (10); Berkeleyacetals A, B, C (14), D
Penicillium sp.	
Culture Conditions	PDBH$^+$, 5 days shaken/21 days still
Metabolites	Berkelic acid (1); Berkebisabolanes A (2), B (3), C (4)
Penicillium clavigerum	
Culture Conditions	PDB, 7 days shaken/21 days still
Metabolites	Phomopsolides A (17), B (18), C (19), E (20)
Penicillium solitum	
Culture Conditions	Mycological broth, 22 days still
Metabolites	Berkedrimanes A (25), B (26)
Pleurostomophora sp.	
Culture Conditions	PDB, 10 days shaken
Metabolites	Berkchaetoazaphilone A (21), B (22), Berkchaetorubramine (24)

10.5.1.2 Exploring Structure Activity Relationships of MMP-3 Inhibitors

These compounds have facilitated some preliminary observations about the relationship between structure and biological activities. For example, berkchaetoazaphilone A (**21**), which was isolated from *Pleurostomophora* sp., is a moderately active MMP-3 inhibitor, while B (**22**) is a potent inhibitor of MMP-3 and has demonstrated μM activity against human retinoblastoma, leukemia and melanoma cell lines (Stierle and Stierle 2014, 2015). The only difference between **21** and **22** is a single epoxide moiety. Other compounds with built-in structure-activity relationships (SAR) include the berkeleyacetals. Moderately potent inhibitor berkeleyacetal C (**14**) lacks an acetate moiety that is present in inactive berkeleyacetal B (Stierle et al. 2007). The phomopsolides have also facilitated SAR evaluation. Phomopsolides A (**17**) and C (**19**) have exhibited the most potent activity against MMP-3 and are also active against several cancer cell lines. Phomopsolides A and C are E-Z stereoisomers of each other but demonstrate comparable activities (Stierle et al. 2014).

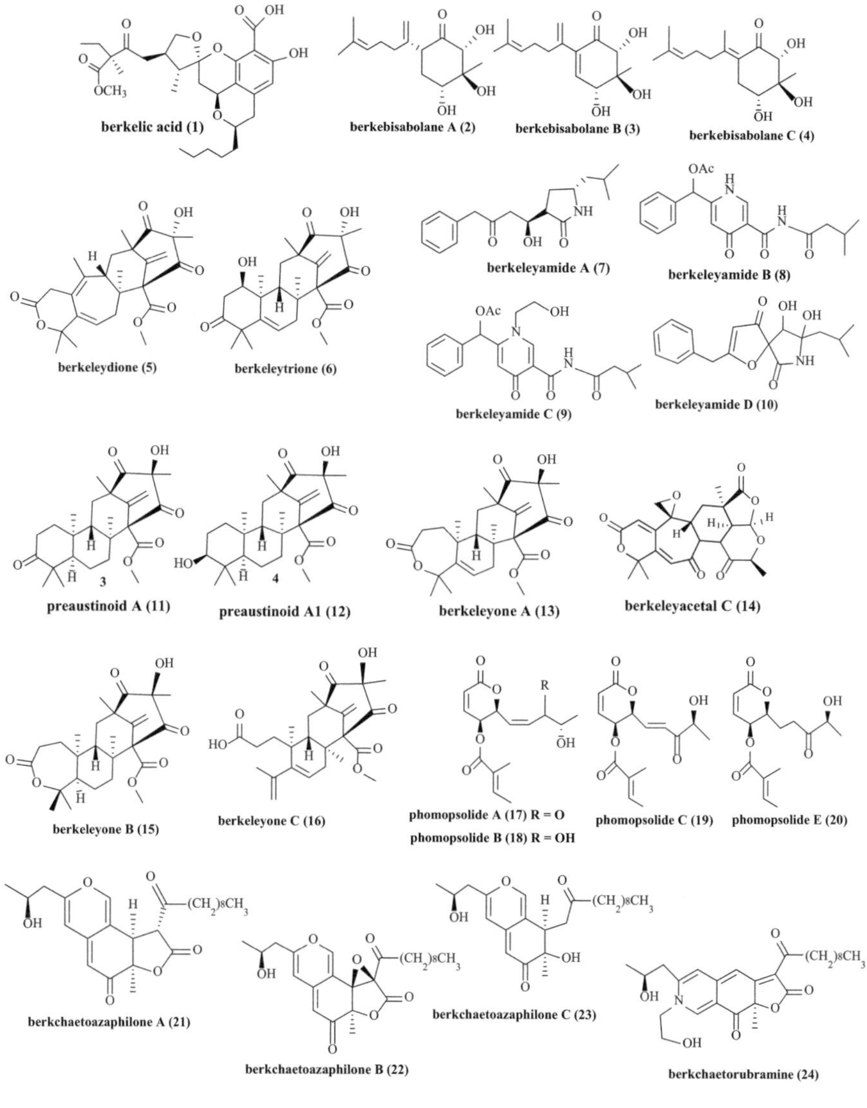

Fig. 10.7 Library of small molecule MMP-3 inhibitors from the BPL microbial collection that will be used to generate more potent analogues for the mitigation of EMT and metastatic disease

The side chain of phomopsolide E (**20**) is saturated and is an order of magnitude less enzyme inhibitory than **17** or **19**. Despite the differences in enzyme inhibitory activity all of the phomopsolides exhibited similar activity in the cancer cell line screens. Clearly their anti-proliferative/cytotoxic activities are not based solely on their abilities to inhibit MMP-3. Part of the MMP-3 inhibitor library is shown in Fig. 10.7.

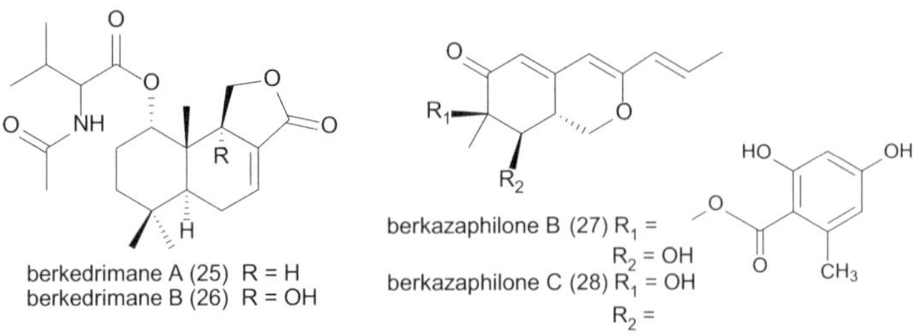

Fig. 10.8 Caspase-1 inhibitors with minimal MMP-3 activity

10.5.1.3 Caspase-1 Inhibitors

We have also developed a library of caspase-1inhibitors and have evaluated these compounds in the induced inflammasome assay. These compounds have also been sent to NCI-DTP and MSKCC for evaluation as anti-proliferative/cytotoxic agents against specific cancer cell lines. Several caspase-1 inhibitors also inhibit MMP-3 *in vitro*, although the degree of inhibition varies between the two enzymes. These included the berkedrimanes (Stierle et al. 2012a) and the berkazaphilones (Stierle et al. 2012b). The berkedrimanes were inactive against MMP-3 but were moderately active against caspase-1. They have not yet been tested against specific cancer cell lines. Both berkazaphilone B (**27**) and its isomer berkazaphilone C (**28**) were active only against specific leukemia cell lines: berkazaphilone B (**27**) exhibited a log10 GI_{50} of −5.67 against cell line RPMI-8226, and berkazaphilone C (**28**) exhibited a log10 GI_{50} of −6.42 against cell line SR (Stierle et al. 2012b). The berkeleyone family of compounds including berkeleydione (**5**), preaustinoid A (**11**), berkeleyone A (**14**), B (**15**), and C (**16**) exhibited similar activities against the two enzymes (Stierle et al. 2004, 2011). Caspase-1 inhibitors are shown in Fig. 10.8.

10.5.1.4 Results of Induced Inflammasome Assay (IIA)

Both caspase-1 and MMP-3 inhibitors were tested in the IIA, and as expected, caspase-1 inhibitors effectively mitigated the production of proinflammatory cytokines in the IIA. The most active compound tested to date was berkchaetoazaphilone B (**22**) (Stierle and Stierle 2014, 2015). At 10 μM it completely inhibited the production of interleukin 6 (IL-6) and IL-33, and mitigated production of tumor necrosis factor-alpha (TNF-α) and IL-1β by 95 %. Berkchaetoazaphilone A (**21**), which lacks the epoxide, shows similar activity at 100 μM, but no effect at 10 μM. Berkazaphilone C (**28**) also exhibited good inhibitory activity and maintained activity even at lower concentrations (Stierle et al. 2012b). Figures 10.9 and 10.10 show the results of the IIA.

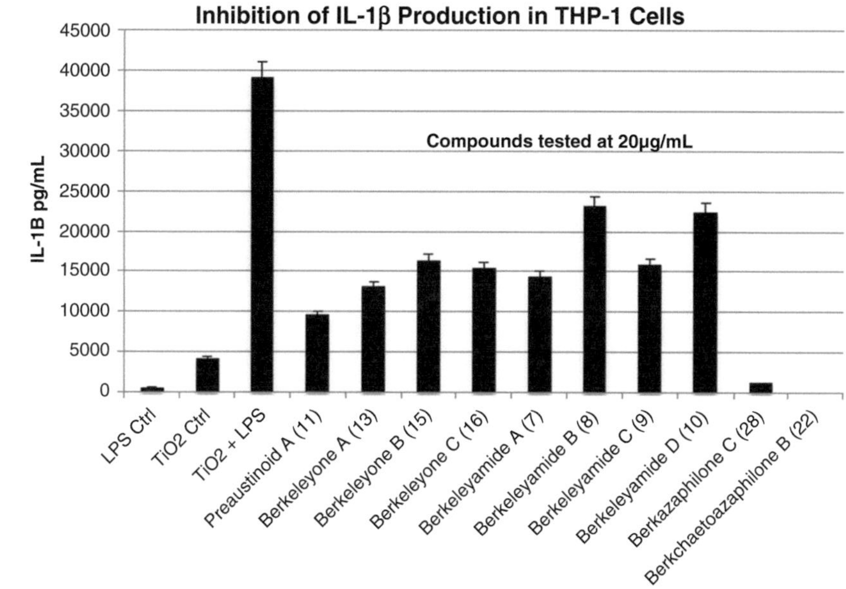

Fig. 10.9 Results of the induced inflammasome assay (IIA). Enzyme inhibitors were assessed for their ability to mitigate production of the pro-inflammatory cytokine IL-1β in THP-1 cells that were induced with titanium nanowires and bacterial LPS. The compounds included in this assay were both caspase-1 and MMP-3 inhibitors

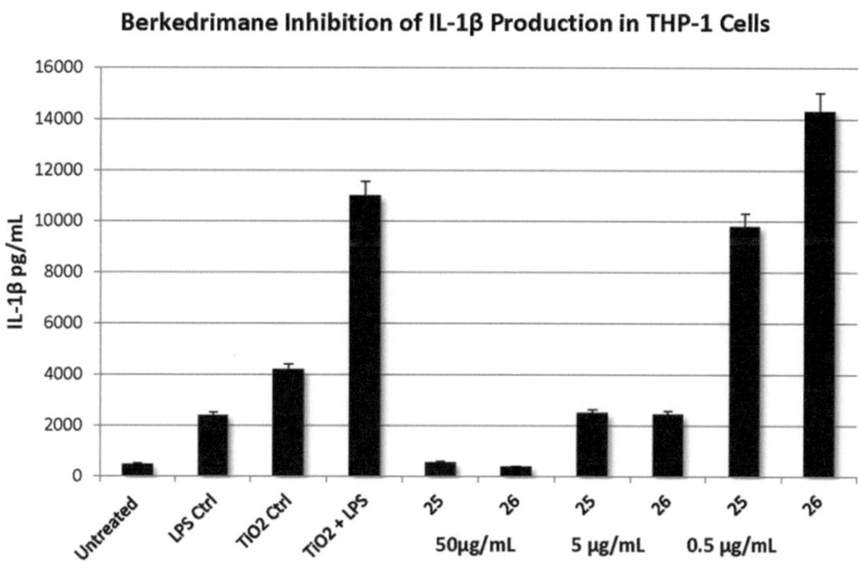

Fig. 10.10 The caspase-1 inhibitors berkedrimane A (**25**) and B (**26**) were also potent inhibitors of IL-1β production in the IIA. Compounds **25** and **26** were not cytotoxic at any level of testing, which is advantageous in anti-inflammatory agents

10.5.2 *Use of SAR Information to Direct the Synthesis of More Potent Analogues*

The production of secondary metabolites by fungi, bacteria, higher plants, or any other source, generally offers first order, "endogenous" structure/activity relationships (SAR). Most of the compounds produced by these microbes were produced as part of a series of closely related analogues. Enzyme inhibition assays were used to guide the isolation of bioactive metabolites. After the bioactive compounds were isolated and the structures were elucidated, "NMR-guided" isolation was used to purify inactive (or significantly less active) compounds with similar structural characteristics. These compounds constitute a series, and different members of that series might have dramatically different bioactivities. The SAR profiles can provide insights into what key structural features are responsible for observed activity. They can also assist in the synthesis of new analogues with enhanced biological activity.

Molecular computational methods will facilitate these studies. Using well-defined crystal structures of target enzymes, we are modeling the inhibitor collection in the *catalytic domains* of these enzymes. The (a) ligand structural features conferring high inhibition potency and/or selectivity, (b) protein-ligand interaction sites, and (c) ligand modifications which would enhance interaction within these sites will be determined. These data will help us design and synthesize analogues, assess their enzyme inhibitory activities, and determine their antitumor or antiproliferative activity on specific and established human cancer cell lines. This work is undertaken with the National Cancer Institute-Developmental Therapy Program, Eisai, Inc. and the Memorial Sloan Kettering Cancer Center. The caspase-1 and MMP-3 inhibitory activities and anticancer activities of the natural product candidates for analogue development have been determined to facilitate initial structure/activity analysis of synthetic compounds. Results from these functional assays will directly test SAR hypotheses generated by computational and molecular modeling efforts, allowing for refinement of the models and guiding further iterations of organic synthesis of novel compounds.

10.6 Conclusions and Future Directions

Although the research coupling inflammation and EMT pathways and exploring SAR is at an early stage, previous studies have already demonstrated that the microbes of the BPL are a valuable source of new and interesting secondary metabolites. The identification of drug-like molecules and therapeutic leads often begins with chemical or biological screening of compound libraries. Natural product libraries isolated from bacteria, fungi, higher plants, and assorted marine invertebrates represent a vast and diverse source of leads, often exhibiting exquisite biological activities. However, translating promising natural product leads into therapeutically relevant molecules is often impeded by their scarcity, complexity, and non-optimal *adsorption/distribution/metabolism/excretion* characteristics. This collection of BPL extremophile metabolites will provide the basic scaffolds for the synthesis of new analogues for further study and development. It is difficult to determine at this point what synthetic methods will be applied in the development of analogues until the compounds have been studied fully. Computational and molecular computational methods as well as x-ray crystallographic studies will help generate structural hypotheses that will direct modification of appropiate scaffolds to enhance selectivity and binding characteristics. It should be noted that all of the candidate molecules adhere to the medicinal chemistry guidelines for physiological relevance, *Lipinski's Rule of Five*:

- Not more than 5 hydrogen bond donors (amino or hydroxyl groups).
- Not more than 10 hydrogen bond acceptors.
- A molecular mass less than 500 amu
- An octanol-water partition coefficient log *P* not greater than 5

It is not often that scientists have the opportunity to explore such a unique environment. Based on previous studies, "bioprospecting" in the Berkeley Pit will continue to yield interesting new chemistry that will, at best, provide new insight into how two important life processes - inflammation and epithelial mesenchymal transition – interact. A collection of small molecules to probe this connection will be generated which will serve as pharmacophores for analogues with enhanced potency. These will provide (a) intriguing targets for synthesis, (b) excellent probes for the specific pathways they inhibit, and (c) pharmaceutical agents. As for the secondary metabolites and their microbial producers – they could be the richest products ever mined from "the richest hill on earth".

References

American Cancer Society. Cancer Facts & Figures (2014) Atlanta: American Cancer Society, pp 17–18

Barnes KA (2013) Extreme life. https://www.kevinabarnes.com/extreme-life. Accessed 10 Mar 2015

Boyer N, Morrison KC, Kim J, Hergenrother PJ, Movassaghi M (2013) Synthesis and anticancer activity of epipolythiodiketopiperazine alkaloids. Chem Sci 4(4):1646–1657

Brock TD, Freeze H (1969) *Thermus aquaticus* gen. n. and sp. n., a nonsporulating extreme thermophile. J Bacteriol 98(1):289–297

Carmona-Gutierrez D, Frohlich KU, Kroemer G, Madeo F (2010) Metacaspases are caspases. Doubt no more. Cell Death Differ 17:377–378

Coffelt SB, de Visser KE (2014) Cancer: inflammation lights the way to metastasis. Nature 507:48–49

Correia AL, Mori H, Chen EI, Schmitt FC, Bissell MJ (2013) The hemopexin domain of MMP3 is responsible for mammary epithelial invasion and morphogenesis through extracellular interaction with HSP90b. Genes Dev 27:805–817

Coussens LM, Werb Z (1996) Matrix metalloproteinases and the development of cancer. Chem Biol 3:895–904

Coussens LM, Fingleton B, Matrisian LM (2002) Matrix metalloproteinase inhibitors and cancer: trials and tribulations. Science 295:2387–2392
Crutzen PJ, Stoermer EF (2000) *Global Change Newsletter* 41:12–13
Davis BK, Ting JP-Y (2010) NLRP3 has a sweet tooth. Nat Immunol 11:105–106
Duaime TE (2006) Long term changes in the limnology and geochemistry of the Berkeley pit lake, Butte, Montana. Mine Water Environ 25:76–85
Dufour A, Sampson NS, Li J, Kuscu C et al (2011) Small-molecule anticancer compounds selectively target the hemopexin domain of matrix metalloproteinase-9. Cancer Res 71:4977–4988
Elinav E, Nowarski R, Thaiss CA, Jin C, Flavell RA (2013) Inflammation-induced cancer: crosstalk between tumours, immune cells and microorganisms. Nat Rev Cancer 13:759–771
Fraenkel GS (1959) The raison d'etre of secondary plant substances. Science 129(3361):1466–1470
Franchi L, Eigenbrod T, Muñoz-Planillo R, Nuñez G (2009) The inflammasome: a caspase-1-activation platform that regulates immune responses and disease pathogenesis. Nat Immunol 10(3):241–256
Garcia-Albeniz X, Chan AT (2011) Aspirin for the prevention of colorectal cancer. Best Pract Res Clin Gastroenterol 25:461–472
Granot T, Milhas D, Carpentier S, Dagan A, Segui B, Gatt S, Levade T (2006) Caspase-dependent and -independent cell death of Jurkat human leukemia cells induced by novel synthetic ceramide analogs. Leukemia 20:392–399
Harijith A, Ebenezer DL, Natarajan V (2014) Reactive oxygen species at the crossroads of inflammasome and inflammation. Front Physiol 5:352
Ho BY, YM Wu, KJ Chang, TM Pan (2011) Dimerumic acid inhibits SW620 cell invasion by attenuating H_2O_2 -mediated MMP-7 expression via JNK/c-JUN and ERK/c-FOS activation in an AP-1-dependent manner. Int J Biol Sci 7(6):869–880
Jiang CS, Guo YW (2011) Epipolythiodioxopiperazines from fungi: chemistry and bioactivities. Mini-Rev Med Chem 11(9):728–745
Jin L, Yuan RQ, Fuchs A, Yao Y, Joseph A, Schwall R, Schnitt SJ, Guida A, Hastings HM, Andres J, Turkel G, Polverini PJ, Goldberg ID, Rosen EM (1997) Expression of interleukin-1β in human breast carcinoma. Cancer 80:421–434
Kato Y, Ozawa S, Miyamoto C, Maehata Y, Baba Y (2013) Acidic extracellular microenvironment and cancer. Cancer Cell Int 13:89–96
Kelley MJ, Rose AY, Song K, Chen Y, Bradley JM, Rookhuizen D, Acott TS (2007) Synergism of TNF and IL-1 in the induction of matrix metalloproteinase-3 in trabecular meshwork. Invest Ophthalmol Vis Sci 48(6):2634–2643
Kim HG, Shrestha B, Lim SY, Yoon DH, Chang WC, Shin DJ, Han SK, Park SM, Park JH, Park HI et al (2006) Cordycepin inhibits lipopolysaccharide-induced inflammation by the suppression of NF-κB through Akt and p38 inhibition in RAW 264.7 macrophage cells. Eur J Pharmacol 545:192–199
Kolb R, Liu GH, Janowski AM, Sutterwal FS, Zhang W (2014) Inflammasomes in cancer: a double-edged sword. Protein Cell 5(1):12–20
Lee BH, Pan TM (2013) Dimerumic acid, a novel antioxidant identified from *Monascus*-fermented products exerts chemoprotective effects: mini review. J Funct Foods 5(1):2–9
Lee HJ, Lee JH, Hwang BY, Kim HS, Lee JJ (2001) Antiangiogenic activities of gliotoxin and its methylthioderivative, fungal metabolites. Arch Pharm Res 24:397–401
Liggetta JL, Zhang X, Eling TE, Baeka SJ (2014) Anti-tumor activity of non-steroidal anti-inflammatory drugs: cyclooxygenase-independent targets. Cancer Lett 346(2):217–224
Liou G-Y, Storz P (2010) Reactive oxygen species in cancer. Free Radic Res 44(5):1–20
Lu H, Ouyang W, Huang C (2006) Inflammation, a key event in cancer development. Mol Cancer Res 4(4):221–223
MacElroy RD (1974) Some comments on the evolution of extremophiles. Biosystems 6:4–75
Mannello F, Tonti G, Papa S (2005) Matrix metalloproteinase inhibitors as anticancer therapeutics. Curr Cancer Drug Targets 5:285–298

Martinon F (2010) Signaling by ROS drives inflammasome activation. Eur J Immunol 40(3):616–619

Mason JW, Kidd JG (1951) Effects of gliotoxin and other sulfur-containing compounds on tumor cells *in vitro*. J Immunol 66:99–106

Mayor A, Martinon F, De Smedt T, Petrilli V, Tschopp J (2007) A crucial function of SGT1 and HSP90 in inflammasome activity links mammalian and plant innate immune responses. Nat Immun 8:497–503

McIlwain DR, Berger T, Mak TW (2013) Caspase functions in cell death and disease. Cold Spring Harb Perspect Biol 5:a008656

Mittal M, Siddiqui MR, Tran K, Reddy SP, Malik AB (2014) Reactive oxygen species in inflammation and tissue injury. Antioxid Redox Signal 20(7):1126–1167

Mousavi SA, Robson GD (2003) Entry into the stationary phase is associated with a rapid loss of viability and an apoptotic-like phenotype in the opportunistic pathogen *Aspergillus fumigates*. Fungal Genet Biol 39:221–229

Muerkoster SS, Lust J, Arlt A, Hasler R, Witt M, Sebens T, Schreiber S, Folsch UR, Schafer H (2006) Acquired chemoresistance in pancreatic carcinoma cells: induced secretion of IL-1β and NO lead to inactivation of caspases. Oncogene 25:3973–3981

Mullis KB (1990) The unusual origin of the polymerase chain reaction. Sci Am 262(4):56–61

Nagase H (1996) In: Hooper NM (ed) Zinc metalloproteases in health and disease. Taylor and Francis, London, pp. 153–204

Noh EM, Kim JS, Hur H, Park BH, Song EK, Han MK, Kwon KB, Yoo WH, Shim IK, Lee SJ, Youn HJ, Lee YR (2009) Cordycepin inhibits IL-1b-induced MMP-1 and MMP-3 expression in rheumatoid arthritis synovial fibroblasts. Rheumatology 48:45–48

Nordstrom DK, Southam G (1997) Geomicrobiology of sulfide mineral oxidation. In: JF Banfield, KH Nealson (eds) Reviews in mineralogy: Miner Soc Am 35:361-390

Okamoto M, Liu W, Luo Y, Tanaka A, Cai X, Norris DA, Dinarello C, Fujita M (2010a) Constitutively active inflammasome in human melanoma cells mediating autoinflammation via caspase-1 processing and secretion of interleukin-1β. J Biol Chem 285:6477–6488

Okamoto M, Liu W, Luo Y, Tanaka A, Cai X, Norris DA, Dinarello C, Fujita M (2010b) Constitutively active inflammasome in human melanoma cells mediating auto-inflammation via caspase-1 processing and secretion of interleukin-1β. J Biol Chem 285:6477–6488

Orlikova B, Legrand N, Panning J, Dicato M, Diederich M (2014) Anti-inflammatory and anticancer drugs from nature. Cancer Treat Res 159:123–143

Pan XQ, Harday J (2007) Electromicroscopic observations on gliotoxin-induced apoptosis of cancer cells in culture and human cancer xenografts in transplanted SCID mice. In Vivo 21:259–266

Paugh BS, Bryan L, Paugh SW, Wilczynska KM, Alvarez SM, Singh SK, Kapitonov D, Rokita H, Wright S, Griswold-Prenner I, Milstien S, Spiegel S, Kordula T (2009) Interleukin-1 regulates the expression of sphingosine kinase 1 in glioblastoma cells. J Biol Chem 284:3408–3417

Radisky DC, Levy DD, Littlepage LE, Hong L, Nelson CM, Fata JE, Leake D, Godden EL, Albertson DG, Nieto MA, Werb Z, Bissell MJ (2005) Rac1b and reactive oxygen species mediate MMP-3-induced EMT and genomic instability. Nature 436:123–127

Saiki RK et al (1988) Primer-directed enzymatic amplification of DNA with a thermo-stable DNA polymerase. Science 239(4839):487–491

Sanders TJ, McCarthy NE, Giles EM, Davidson KL, Haltalli ML, Hazell S, Lindsay JO, Stagg AJ (2014) Increased production of retinoic acid by intestinal macrophages contributes to their inflammatory phenotype in patients with Crohn's disease. Gastroenterology 146(5):1278–1288

Schimmel TG, Coffman AD, Parsons SJ (1998) Effect of butyrolactone I on the producing fungus, *Aspergillus terreus*. Appl Environ Microbiol 64:3703–3712

Schlosser S, Gansauge F, Ramadani M, Beger H-G, Gansauge S (2001) Inhibition of caspase-1 induces cell death in pancreatic carcinoma cells and potentially modulates expression levels of bcl-2 family proteins. FEBS Lett 491:104–108

Smith DG, Guillemin GJ, Pemberton L, Kerr S, Nath A, Smythe GA, Brew BJ (2001) Quinolinic acid is produced by macrophages stimulated by platelet activating factor, Nef and Tat. J Neurovirol 7(1):56–60

Steenport M, Khan KM, Du B, Barnhard SE, Dannenberg AJ, Falcone DJ (2009) Matrix metalloproteinase (MMP)-1 and MMP-3 induce macrophage MMP-9: evidence for the role of TNF-alpha and cyclooxygenase-2. J Immunol 183(12):8119–8127

Sternlicht MD, Lochter A, Sympson CJ, Huey B, Rougier J-P, Gray JW, Pinkel D, Bissell MJ (1999) The stromal proteinase MMP3/stromelysin-1 promotes mammary carcinogenesis. Cell 98:137–146

Stetler-Steveson WG, Hewitt R, Corcoran M (1996) Matrix metalloproteinases and tumour invasion: from correlation and causality to the clinic. Semin Cancer Biol 7:147–154

Stierle AA, Stierle DB (2005) Bioprospecting in the Berkeley pit: bioactive metabolites from acid mine waste extremophiles. In: Atta-Ur-Rahman (ed) Bioactive natural products, vol 32. Elsevier Science, Amsterdam, pp 1123-1175

Stierle A, Stierle D (2013) Bioprospecting in the Berkeley pit: the use of signal transduction enzyme inhibition assays to isolate bioactive secondary metabolites from the extremophilic fungi of an acid mine waste lake. In: Atta-Ur-Rahman (ed) Bioactive natural products, vol 39. Elsevier Science, Amsterdam, pp 1-47

Stierle AA, Stierle DB (2014) Bioactive secondary metabolites of acid mine waste extremophiles. Nat Prod Commun 9(7):1037–1044

Stierle AA, Stierle DB (2015) Azaphilones from the acid mine waste extremophile *Pleurostomophora* sp. J Nat Prod. J Nat Prod 78:2917–2923

Stierle DB, Stierle AA, Hobbs JD, Stokken J, Clardy J (2004) Berkeleydione and berkeleytrione, new bioactive metabolites from an acid mine organism. Org Lett 6:1049–1052

Stierle AA, Stierle DB, Kelly K (2006) Berkelic acid, a novel spiroketal with highly specific antitumor activity from an acid-mine waste fungal extremophile. J Org Chem 71:5357–5360

Stierle DB, Stierle AA, Patacini B (2007) The berkeleyacetals, three meroterpenes from a deep water acid mine waste *Penicillium*. J Nat Prod 70:1820–1823

Stierle AA, Stierle DB, Patacini B (2008) The berkeleyamides: four new amides from *Penicillium rubrum*, a deep water acid mine waste fungus. J Nat Prod 71:856–860

Stierle D, Stierle A, Patacini B, McIntyre K, Girtsman T, Bolstad E (2011) Berkeleyones and related meroterpenes from a deep water acid mine waste fungus that inhibit the production of interleukin 1-β from induced inflammasomes. J Nat Prod 74:2273–2277

Stierle D, Stierle A, Girtsman T, McIntyre K, Nichols J (2012a) Caspase-1 and -3 inhibiting drimane sesquiterpenoids from the extremophilic fungus, *Penicillium solitum*. J Nat Prod 75:262–266

Stierle A, Stierle D, Girtsman T (2012b) Caspase-1 inhibitors from an extremophilic fungus that target specific leukemia cell lines. J Nat Prod 75:344–350

Stierle AA, Stierle DB, Mitman GG, Snyder S, Antczak SC, Djaballah H (2014) Phomopsolides and related compounds from the alga-associated fungus, *Penicillium clavigerum*. Nat Prod Commun 9(1):87–90

Sugui JA, Pardo J, Chang YC, Zarember KA, Nardone G, Galvez EM, Müllbacher A, Gallin JI, Simon MM, Kwon-Chung KJ (2007) Gliotoxin is a virulence factor of *Aspergillus fumigatus*: gliP deletion attenuates virulence in mice immunosuppressed with hydrocortisone. Eukaryot Cell 6(9):1562–1569

Thiery JP, Acioque H, Huang RYJ, Nieto MA (2009) Epithelial-mesenchymal transitions in development and disease. Cell 139(5):871–890

Toyokuni S (2009) Role of iron in carcinogenesis: cancer as a ferrotoxic disease. Cancer Sci 100(1):9–16

Tschopp J, Schroeder K (2010) NLRP3 inflammasome activation: the convergence of multiple signalling pathways on ROS production? Nat Rev Immunol 10(3):210–215

Tsiatsiani L, Breusegem F, Gallois P, Zaviolov A, Lam E, Bozhkov PV (2011) Metacaspases. Cell Death Differ 18:1279–1288

Tsuzaki M, Guyton G, Garrett W, Archambault JM, Herzog W et al (2003) IL-1β induces COX2, MMP-1, -3 and -13, ADAMTS-4, IL-1β and IL-6 in human tendon cells. J Orthop Res 21(2):256–264

Understanding Autoimmune Diseases (1998), NIH Publication Number 98-4273, May 1998

Vandooren J, Van den Steen PE, Opdenakker G (2013) Biochemistry and molecular biology of gelatinase B or matrix metalloproteinase-9 (MMP-9): the next decade. Crit Rev Biochem Mol Biol 48(3):222–272

Vigushin DM, Mirsaidi N, Brooke G, Sun C, Pace P, Luman L, Mood CJ, Coombes RC (2004) Gliotoxin is a dual inhibitor of farnesyltransferase and geranyltransferase I with antitumor activity against breast cancer *in vitro*. Med Oncol 216:21–30

Voronov E, Shouval DS, Krelin Y, Cagnano E, Benharroch D, Iwakura Y, Dinarello CA, Apte RN (2003a) IL-1 is required for tumor invasiveness and angiogenesis. Proc Natl Acad Sci U S A 100:264–2650

Voronov E, Shouval DS, Krelin Y, Cagnano E, Benharroch D, Iwakura Y, Dinarello CA, Apte RN (2003b) IL-1 is required for tumor invasiveness and angiogenesis. Proc Natl Acad Sci U S A 100:2640–2650

Warburg O, Posener K, Negelein E (1924) Über den Stoffwechsel der Tumoren (On metabolism of tumors). Biochem Z 152:319–344

Wolf JC, Mirocha CJ (1973) Regulation of sexual reproduction in *Gibberella zeae* (*Fusarium roseum* 'Graminearum') by F-2 (zearalenone). Can J Microbiol 19:725–734

Wu Y, Zhou B (2009) Inflammation: a driving force speeds cancer metastasis. Cell Cycle 8(20):3267–3273

Yabu M, Shime H, Hara H, Saito T, Matsumoto M, Seya T, Akazawa T, Inoue N (2011) IL-23-dependent and -independent enhancement pathways of IL-17A production by lactic acid. Int Immunol 23(1):29–41

Zucker S, Cao J, Chen W-T (2000) Critical appraisal of the use of matrix metallo-proteinase inhibitors in cancer treatment. Oncogene 19:6642–6650

Chapter 11
Insect Bioprospecting Especially in India

S.K. Srivastava

Abstract Bioprospecting has been an important phenomenon involving discovering new drugs and has occurred since the dawn of civilization. The Convention on Biological Diversity (CBD) Secretariat defines bioprospecting as the exploration of biodiversity for commercially valuable genetic and biochemical resources. In the early stage, bioprospecting was focused on plants and various plant-based drugs were discovered. However, insects were also explored, and are of special interest to India. In a broader sense, bioprospecting is the collecting and identifying of biological samples (plants, animals, microorganisms) and the amassing of indigenous knowledge to help in discovering genetic or biochemical resources. Bioprospecting is intended for economic purposes e.g., new drugs, crops, industrial products. However, studies suggested that the value of bioprospecting is higher if it is based on the knowledge of local people because the wisdom that indigenous people have regarding bioprospecting is embedded in their belief system and their culture. Insects as a food play an important role in the new insect focus. Ants, bees, termites, caterpillars, water bugs, beetle larvae, flies, crickets, katydids, cicadas, and dragonfly nymphs are among a long list of edible insects that provide nutrition for the people of Asia, Australia, Africa, South America, the Middle East, and the Far East. Insects represent an important food source for a wide variety of other animal species. By weight, termites, grasshoppers, caterpillars, weevils, houseflies and spiders are better sources of protein than beef, chicken, pork or lamb. Insects are also used as medicine by traditional healers. Chemicals produced by insects for self-defense can be used for antibacterial and anticancer drugs. In the present chapter, contributions that insects make to diets/food security are discussed in detail. Legal regimes and concepts of bioprospecting are described. Information on insects as (a) medicine, (b) natural dyes, (c) food, (d) animal feed, and (e) aesthetic value are assessed. A discussion on the constraints of bioprospecting is provided as are suggestions to promote insect bioprospecting. The nutritional and economic value of edible insects is of much importance and we should further encourage their collection and commercialization, given the benefits to the environment and human health. It is an interesting concept, managing pest insects by developing them into a delicacy and

S.K. Srivastava (✉)
ICAR-Central Institute for Women in Agriculture, Bhubaneswar, Odisha, India
e-mail: sksdrwaicar28@gmail.com

R. Paterson, N. Lima (eds.), *Bioprospecting*, Topics in Biodiversity and Conservation 16, DOI 10.1007/978-3-319-47935-4_11

medicine. The chapter will attract the attention of farmers, entrepreneurs, scientists, research workers, the media and the decision-makers to explore insects for the future need of society.

11.1 Introduction

People worldwide have revered insects for millennia as they have been (a) considered sacred, (b) celebrated in art and literature, (c) used as pollinators, (d) nutrient recyclers, (e) carcasses and dung decomposers and (f) human food. They have also been feared as predators and parasites. Insects are fundamental in all terrestrial and aquatic food chains and provide us with silk, honey, waxes, medicines and dyes. Exploitation and utilization of insect resources are broadly classified into four different categories, the (a) insects of edible and therapeutic purposes. This includes the utilization of various stages and forms of insects; (b) utilization of insects for industrial resources. This level includes the utilization of silk worms, honeybees, lac insects, dye insects and aesthetic insects; (c) use of insects in forensic investigation. By analyzing the stages of succession of insects, rough estimations of the post mortem intervals can be achieved and (d) insects of ecological importance. As biomass recyclers, house fly larvae are used to recycle organic wastes to produce protein and fat. Dung beetles recycle cattle dung as organic manure. Insects are also used as bio-indicators to assess the cumulative effects of environmental stressors such as pollutants. Many insect species act as potential predators and parasites of destructive insects. Despites these fascinating benefits, insect resources are often neglected due to limited documentation, expertise, and advance commercial enterprises in these fields (Zhang et al. 2008).

11.2 What Is Bioprospecting?

Bioprospecting has been an important phenomenon involving discovering new drugs and has occurred since the dawn of civilization. The Convention on Biological Diversity (CBD) Secretariat defines bioprospecting as the exploration of biodiversity for commercially valuable genetic and biochemical resources (UNEP/CBD/COP/5/INF/7 2000). However, its definition varies in different countries, with some defining bioprospecting narrowly to include only the search for valuable genetic materials, whereas others encompass the development and application of such materials. For example, the New Zealand Biodiversity Strategy defines bioprospecting as ‘the search among biological organisms for commercially valuable compounds, substances or genetic material (Slobodian et al. 2011). The South African Biodiversity Act defines bioprospecting as any research on, or development or application of, indigenous biological resources for commercial or industrial exploitation, and includes the systematic search, collection or gathering of such resources, or making extractions from such resources for purposes of such research, development or application. Hence, the commercialisation aspects of bioprospecting and potential profitability remain the critical impasse regarding its legal definition. Article 136 of the United Nations Convention on the Law of the Sea (UNCLOS) states that 'The Area and its resources are the common heritage of mankind'. The CBD includes conservation and sustainable use as well as access and benefit sharing obligations with regards to biological diversity. However, it is not clear that these obligations apply to bioprospecting given its low estimated environmental impact and lack of proprietary states, or peoples in the domains where common pooled resources are found. The Global Commons specifically refers to domains which do not fall within the jurisdiction of any one nation, thus all states have legal access. The CBD imposes a general obligation on parties to cooperate in conserving and sustainably using biodiversity in areas beyond national jurisdiction. In the early stage, bioprospecting was focussed on plants and various plant-based drugs were discovered (Taylor 2000). However, insects and algae were also explored (Kumar and Tarui 2004).

In a broader sense, bioprospecting is the collecting and identifying of biological samples (plants, animals, microorganisms) and the amassing of indigenous knowledge to help in discovering genetic or biochemical resources. Bioprospecting is intended for economic purposes e.g., new drugs, crops, industrial products (Srivastava et al. 2009). However, studies suggested that the value of bioprospecting is higher if it is based on the knowledge of local people (Martin 2001), because the wisdom that indigenous people have regarding bioprospecting is embedded in their belief system and their culture.

11.3 Legal Regimes and Concepts

1. Open-access regime: refers to the system of law in which no sovereign state controls the resource or area in question, yet they are free to exploit and profit from these resources/areas if they adhere to generally accepted principles and obligations of international law.
2. The Global Commons: also known as 'the Commons', is a term used to describe domains where common pooled resources are found. The Global Commons specifically refers to domains which do not fall within the jurisdiction of any one nation, thus all states have legal access.
3. Common Heritage of Mankind (CHM): this legal concept dictates that certain resources/areas are the communal property of all humankind. Hence, no person or state has exclusive legal rights to these resources/areas, in so far as all uses and benefits must be shared equally for the benefit of current and future generations.
4. Access and benefit-sharing (ABS): commonly describes legal regimes that seek an equitable right to developed and developing states regarding the exploitation

and any derivate benefits from scientific research and commercial development of biological resources.
5. Precautionary Principle: an approach which dictates that in the absence of adequate scientific evidence, decision-makers must err on the side of caution and adopt laws/policies that prevent suspected risks of harm to the environment or human health.

11.4 Insect as a Source of Medicine/Zootherapy/ Entomotherapy

Insects have been used in traditional medicine since time immemorial (Srivastava et al. 2009). The therapeutic application of honeybee venom (bee venom therapy) has been used to treat diseases such as arthritis, rheumatism, back pain, cancerous tumors, and skin diseases. Traditional healers use insects as medicine. Chemicals produced by insects as a result of their self-defense mechanisms can be used for antibacterial and anticancer drugs. In addition, silkworm pupae were traditionally used as medicine in countries such as China, Japan, Korea, India, and Thailand (Talukdar 2009 and Mishra et al. 2003). The waste liquor containing sericin yielded through the process of degumming silk fiber, is a raw material for the production of sericin powder which is used in various applications including medicine (Zhu 2004).

In many parts of New South Wales, Australia, people use medico-entomological drugs daily (Cambell et al. 1994). Costo neto (2003) coined the term "entomotherapy" for the use of insects for therapeutic purposes. Other accounts include Antonio (1994); Alexiades (1999); Zimian et al. (1997); Green (1998); Namba et al. (1988); Maya (2000) and Padamanbhan and Sujana (2008). Gallnuts of commercial value are produced on various species of oaks and other trees by certain *Eurasion cynipid* wasps. It is also called Allepo, Mecca, Chinese or Turkey galls and the best grades, containing more than 50 % tannic acid, come from Iran, Turkey and Syria. These are used in making dyes and medicine (Quijano and Vergara 2007). A list of insect preparations of use in Zaire is provided in Table 11.1.

One of the most commonly used insects for medicinal purposes is the blow fly (*Calliphoridae*) larvae. During World War II, military surgeons noticed that wounds which were left untreated for several days, and were infested with larvae, healed better than treated and non infested wounds. It was later discovered that the larvae secreted a chemical called allantoin which had a curative effect (Maya 2000). The use of traditional knowledge could be extended further in modern medicine systems by identifying the proactive biomolecules with pharmacological action (Anonymous 2009; Hider 1988; Werner 1970 and But et al. 1991).

The therapeutic application of honeybee products has been used in traditional medicine to treat various diseases e.g. diarrhoea, tuberculosis, impotency, asthma, exophthalmic goiter, and mouth galls (Maya 2000). The practice of using honeybee

Table 11.1 Some examples of traditional medicinal uses of insects in Zaire. These are provided only to give examples of traditional use and have no valid scientific basis. The author of the review and the editors do not recommend using these methods

Name	Local name	Used for the treatment	Method of treatment
Praying mantis	Kayakua	Epilepsy	A healer places the whole mantis in a pot with boiled aromatic leaves, and washes the entire body of the patient. The patient also drinks the preparation. Duration of treatment is 1 week.
Builder/worker caterpillar	Kenbul Mpiak	Haemorrhage during childbirth or during pregnancy	A healer crushes the caterpillar nest, the caterpillar, and the red earth, and the preparation is called largile nkol. The mixture is solubilized which the woman drinks.
Termite	N'zo Musien	Internal haemorrhage	The specialist removes the bark of the *Muton* tree (which produces red bark) and these are placed in a pot with a nest of termites. The patient drinks a little of this solution.
Glowworm	Nkwazeb	"Chasing the spirits/boogey-men from an infant having nightmares"	The healer takes several glowworms, mixes them with ash from a cooking fire near to where the infant resides. The patient drinks a small quantity in water. The same mixture is rubbed on the forehead, head, ears and nape of the infant.
Bee	Ngobo	Stuttering	The healer places several bees in a *calabasse* (gourd-bowl) of palm wine which the patient drinks.
Domestic cricket	Mpayenzo	Stuttering	The patient eats the cricket.
Butterfly	Kenguapob	Illnesses of the ears	The healer puts the nymph and cocoon in a cone made from wild leaves. The aromatic plant *Losaal Nzian (aile de Dieu;* wings of God) is added, and it is a lit on hot coal. The smoke is blown on the ears of the patient.
Tse-tse fly	Kebty	To avoid sleeping sickness after having been bitten by the tse-tse fly	Tse-tse fly is crushed and rubbed on the skin, and an incision made on the skin.
Aquatic bee	Ngundumugun and Kender Maza	Cure heavy menstruation	The healer places the bee in a pot with a piece of clothing from the woman which are burned to make ash. These are applied to the reproductive part after bathing.
Soldier termites	Mbwiidi	Revive a syncope/blackout/fainting fit	Soldier termites are placed in a container containing the sap of tobacco leaves/tobacco. The live insects are absorbed into this substance. This is placed them on the body of the patient, when the termites bite the patient, he awakens.

(continued)

Table 11.1 (continued)

Name	Local name	Used for the treatment	Method of treatment
Salivating insects	Bentiey	Stop exaggerated salivation	The insects are cooked with the meat of the Ntambien gourd which are eaten for a week.
Water bug/lion bug	Kenzi & Nziie	Used to cure in insanity	The patient eats the insects mixed with mud from the same river.
Lion-ant	Munkuuk	Used to relieve and cure high fever	The lion ant is made to bite the patient.
Trembling red ant *Solenopsis invicta*	L.nkaam	Muyeem (bronchitis)	The sticky saliva of the ants is thought to help recover normal respiration. The entire ants are used. The healer places the ants in a bowl made of forest leaves; he mixes it with pure and clear water and gives it to the patient to drink. Duration of treatment is up to 1 week. These insects also have a common name meaning "child-birth aid".
Grasshopper	Mpaylaar	Violent headaches	The healer crushes the dry grasshoppers into ash. The ash is mixed with a little organic salt. An incision is made on the nape and front of the patient and then the solution is applied.
Worker wasp	Ngankoy	Strengthens weak infant	The nest of the wasp is crushed in a glass of water, and is drunk by patient. Also, it is rubbed into the skin. The nest of the worker wasp has a acetylcholine-like substance, which is thought to cause the effect.
Cockroach	Kembaar	Scabies/mange	Cockroaches are burnt to ashes and are mixed with palm oil. It is rubbed on the body after each bath until the scabies is cured.

products for medicinal purposes is known as Apitherapy. One of the major peptides in bee venom, called melittin, is used to treat inflammation in sufferers of rheumatoid arthritis and multiple sclerosis. Melittin blocks the expression of inflammation genes, thus reducing swelling and pain (Maya 2000). The therapeutic application of bee venom therapy has been used as a traditional medicine to treat a variety of conditions, such as arthritis, rheumatism, back pain, cancerous tumors, and skin diseases (Padamanbhan and Sujana 2008). Bee venom contains at least 18 active components, including enzymes, peptides, and biogenic amines, which have a wide variety of pharmaceutical properties. It was reported that melittin inhibited the DNA binding activity of NF-kB, a critical transcriptional factor regulating inflammatory gene expression, by inhibiting IkB phosphorylation (Faulkner 1992). Bee venom also has anticancer activity: Several cancer cells including renal, lung, liver, prostrate, bladder, mammary cancer cells and leukemia cells are targets of melittin (Lazarus and Attila 1993; Park et al. 2004 and Liu et al. 2005). Pharmaceutical companies are currently funding extensive research into the potential of venom as the next generation of cancer fighting drugs (Moon et al. 2006). Honey has been used traditionally in various medicine preparations (Cao et al. 1996). The propolis of the bee hive is used in lip balms and tonics, whereas royal jelly is used with the intention to strengthen the human body, for improving appetite, preventing ageing of skin, leukemia and for the treatment of other cancers. About 80 % of all honey by weight is used directly in medicines and 10 % in traditional Ayurvedic medicine and pharmaceutical production. Thus, pharmaceuticals from bees can be cited as a spectacular example of medicinal insects. There may be many such insects having similar or superior medicinal properties.

Cantharidin is a medicine obtained from the blister beetle, *Cantharis vesicatoria*, an insect belonging to the order Coleoptera, family Meloidae. Its medical use dates back to Hippocrates (460–377 BC) (Moon et al. 2006). It was administered as a diuretic and to alleviate epilepsy, asthma, rabies, and sterility. The eggs of red ants, *Solenopsis invicta* are said to be used as a constituent of medicine for the control of malaria. An extract of cocoons of mulberry silkworm is believed to check profuse menstruation and chronic diarrhoea (Sharma et al. 2006). Pierisin, a protein from pupa of the cabbage butterfly, *Pieris rapae*, exhibit cytotoxic effects against human gastric cancer. Extracts of the body fluids of other cabbage butterflies, *P. brassicae* and *P. napi*, also contains pierisin (Davis 1918). Tang et al. (2008) proposed that antimicrobial molecules from insects may serve as significant sources of antibiotics as revealed from the Chinese traditional medicine made from edible housefly larvae of *Musca domestica.*

Traditional healers in Chhattisgarh, India have used the green coloured larva Jatropha leaf miner, *Stomphosistis thraustica*, as a medicinal insect for many decades (Srivastava et al. 2009). The larvae collected just before pupation, are considered optimal for effectiveness. The larvae are dried in shade and are converted into a dry powder. The traditional healers use this powder internally with lukewarm water in order to increase the flow of milk in lactating women. Many healers use it boiled in water and drank to obtain better effects. The traditional healers of Southern Chhattisgarh use the larvae in the treatment of fever. For the preparation of medicine, they dry the larvae in moonlight and powder it, which is taken internally in combination with other herbs, mainly Kalmegh (*Andrographis paniculata*) (http:// botanical.com 2012).

A common practice in North Gujarat is to feed animals which fail to come into heat with two or three grasshoppers (*Hieroglyphus niegrorepletus*) along with chapatti (bread) or fodder. It is believed that animal comes into estrous within 15 days after this treatment. A single treatment is enough to obtain the desired result. The grasshopper is locally called titighodo. It is generally found during the rainy season on cactus (*Euphorbia* sp) and akada (*Calotropis* sp) and it has yellow and greenish stripes on the body (Patel and Patel 1995). Natives of Sambar Village of Chhattisgarh, India use the common agricultural pest Kambal Keeda (*Diacrisia oblique*) in case of dog bite. The patients are advised to eat freshly laid eggs in order to reduce the rabies effect caused by the virus that attacks the central nervous system. It is also applied externally on affected parts. It is a promising treatment as they have used it for generations, apparently with success. Bhavri Keeda (*Gerridae*) an aquatic insect commonly known as water striders, is used in other villages of the Bagbahera region, for the treatment of dog bite and rabies (http://botanical.com 2012). However, all these practices require scientific validation.

Pieris rapae, P. brassicae and *P. napi* butterflies produce antibacterial proteins including cecropins, defensins and lysozymes. Cercopin has been reported to be cytotoxic against mammalian lymphoma and leukemia cells. Some butterflies may be a good source of other novel bioactive materials such as anti-bacterial, and anticancer drugs. In India, 1501 species of butterflies are found, hence the country has a tremendous potential in butterfly bioprospecting (http://lib.bioinfo.pl). A cockroach, *Periplaneta americana* locally known in Odisha, India as Asarpa, is used for treating asthma, coughs and colds. Local women collect the cockroach and make tea by boiling one until the water is reduced by 50 %, and use it by taking three doses three times a day to cure the ailments. Farmers of the Sonarhi and Terhi villages of Banda district of Uttar Pradesh feed insects hosted on babool tree to their cattle and buffaloes to initiate the estrous cycle. The insect appears to contain a hormone, which induces estrous in the animals (Singh 2003). Soil collected from termite hills is made into a paste with warm water and applied to the wounds of sheep and goats for 2–3 days for healing in the Makarbilli village of Nuapada district in Odisha (Behera 2003). Finally, some examples of traditional medicinal uses of insects in Zaire described by Tango (1994) are provided in Table 11.1. It requires emphasis that these treatments are non scientific and require further research: They are not recommended as treatments by the current author or the editors of the book.

11.5 Natural Dye from Insect

The demand for natural dyes is constantly growing with an increased awareness of the ecological and environmental problems associated with synthetic dyes (Sharma et al. 2006). Producing natural dye in India from insects has been suggested by

Prasad (2007) with a view to exploitation. Currently, the coccid, *Dactylopius coccus* (Hemiptera: Dactylopiidae) is the most important species due to its use for the extraction of carmine acid, a natural red dye used in food, pharmaceuticals, and cosmetics industries (Subramanian et al. 2005). The coccid lives on cladodes of prickly pears (*Opuntia ficus indica*) and dried females are a source of the dyes (Prasad 2007). *D. opuntiae* is another wild species found in Mexico and has a shorter lifespan and reproduction cycle with a larger number of generations per year (Vigueras and Porlillo 2001). The non-dye residuals from extraction can be used to enrich food for avian species or to prepare fertilizers, as they have a high content of proteins and mineral (Mendez et al. 2004).

The female of *D. coccus* is also found in central and South America. The insects are handpicked and dried and 100,000–150,000 insects yield 1 kg of raw cochineal. Total world production was reported to be 150–180 tonnes/year. Peru is the biggest producer accounting for 90 % of cochineal production. However, large scale production of cochineal has emerged in Guatemala. Cochineal was used as a dye by the Aztec and Maya peoples of Central and North America where it was a commodity of much value, comparable to gold (Mann 1969). Cochineal is used to produce scarlet, orange, and other red tints. The production and exploitation method of the dye has been studied by many workers. The insects are killed by immersion in hot water, exposure to sunlight, steam, or dry heat. Each method produces a different colour which results in the varied appearance of commercial cochineal.

Oak galls have been used commercially as a source of tannic acid. It was a principal ingredient in wool dyes and black hair colourants used during the Greek empire as early as the fifth century BC. It is still used commercially in the leather industry for tanning dying and ink manufacture. Tannic acid was obtained from the Aleppo gall found on oak trees (*Quercus infectoria* Olivier) in Asia and Persia. The trees produce gall tissues in response to polyphenolic molecules secreted by the larvae of wasps (*Cynips gallae tinctoriae* Olivier; Hymenoptera: Cynipidae) that infest the trees. Approximately, 50–75 % of the gall's dry weight is composed of tannic acid (Aldama et al. 2005). Gallnuts of commercial value for making dyes are produced on various species of oaks and other trees by certain Eurasion Cynipid wasps. They are also called Allepo, Mecca, Chinese or Turkey galls and the best grades, containing more than 50 % tannic acid, come from Iran, Turkey and Syria (Quijano and Vergara 2007). The aspect of exploring and utilizing natural dye producing insects is uncommon in the India. However, the north-eastern region is a region of high oak cultivation, and there is scope for bioprospecting in this area.

Lac insects (*Laccifer lacca*) are exploited commercially for lac resins and dyes. Lac is the hardened resin secreted by the insects. There are 87 lac insect species reported from the world, representing nine genera, of which 19 species belonging to two genera are found in India (Anonymous 2006). Of the Indian species, *Kerria lacca* is mainly exploited for commercial production of lac. *K. chinensis* in the northeastern states and *K. sharda* in coastal regions of Odisha and West Bengal are also cultivated. Lac resin being natural, biodegradable and nontoxic, finds applications in food, textiles, and pharmaceutical industries in addition to surface-coating, electrical and other fields.

11.6 Insect as a Source of Food

Some of the renowned works on edible insects from different parts of India are those of Singh et al. (2007); Alemla and Singh (2004) and Singh and Chakravorty (2008). The long history of human use of insects as foods indicates that they do not pose any significant health problem (Capinera 2004). This trend toward reducing the bias against insects as food is promising. Despites the benefits, modernization has led indigenous populations around the world away from this traditional food source, without providing nutritionally-equivalent alternatives (Fromme 2005).

Over 1500 species of edible insects have been recorded as being consumed by 300 ethnic groups from 113 countries (Grieve 2009). Human consumption of silkworm pupae has been practiced in China (Zhou et al. 1996) and India by many tribal communities (Arora and Gupta 1979). Indeed, silkworm pupae have been put in the list of "Novel food resources managed as common food" by the Ministry of Health, China (Mishra et al. 2003). Ants, bees, beetle larvae, caterpillars, cicadas, crickets, dragonfly nymphs, flies, katydids, termites and water bugs are among a long list of edible insects that provide nutrition for the people of Asia, Australia, Africa, South America, the Middle East, and the Far East (Grieve 2009). In some ethnic groups, insects provide 5–10 % of animal protein input as well as fat, calories, vitamins, and minerals (MacEvilly 2000) Studies on nutrient analysis for various insects were conducted by many authors in different countries, such as (a) Quin (1959) in South Africa; (b) Oliveira et al. (1976) in Angola; (c) Malaisse and Parent (1980) in Zaire; (d) Gope and Prasad (1983) in India; (e) Sungpuag and Puwastien (1983) in Thailand and (f) Ramos and Pino (1989) in Mexico. Insects generally have higher food conversion efficiency than higher animals. For example, the house cricket (*Acheta domesticus*) when reared at 30° C or more, and fed a diet of equal quality as that used to rear conventional livestock, showed a food conversion twice as efficient as pigs and broiler chicks, four times that of sheep, and six times higher than steer when losses in carcass trim and dressing percentage were counted (Capinera 2004). Protein production from insects for human consumption would be more effective and consume fewer resources than vertebrate protein. Interestingly, increased consumption of grasshoppers and locusts has coincided with decreased pesticides use in Asia and Oceania (Defoliart 1995, 1999).

Native Americans of western North America expended much organization and effort in harvesting insects (Srivastava et al. 2009). Ordinarily, insects are not used as emergency food to ward off starvation, but are included as a planned part of the diet throughout the year or when seasonally available (Srivastava et al. 2015). The *Yukpa* people of Colombia and Venezuela prefer certain traditional insect foods to fresh meat, as do the *Pedi* of South Africa (Ruddle 1973) at least in some cases. When *mopanie* caterpillars (*Gonimbrasia belina* Westwood) were in season, the sale of beef was seriously affected (Quin 1959). According to the Entomological Society of America, termites, grasshoppers, caterpillars, weevils, houseflies and spiders are better sources of protein than beef, chicken, pork or lamb (Srivastava et al. 2009) on a weight basis. Insects are also low in cholesterol and fat and make some food products more nutritious. According to Ramos (1997), 80 % of the world's population eats insects intentionally and 100 % eat them unintentionally! They have served as traditional foods in most cultures of non-European origin.

At least two billion people include more than 1900 species of insects in their food (FAO 2013) and they are sometimes the only source of essential proteins (amino acids), fats, vitamins and minerals for tribes living in forests. In the Central African Republic, 95 % of forest people are dependent on eating insects for their protein intake. Pupae of *Formica* are often eaten and the optimal time to collect pupae is 1 h after the rays of the sun have contacted the pupae "mound" in the morning and the pupae can be collected just under the surface of the mound at this time. Mexican "caviar" or *ahuahutle*, is composed of the eggs of several species of aquatic Hemiptera; these have formed the basis for aquatic farming in Mexico for centuries (Srivastava et al. 2009).

In parts of Africa, ants, termites, beetle grubs, caterpillars, moths, butterflies and grasshoppers are eaten. Moth larvae are collected and roasted, and may often be bought in the markets. Some insects such as termites are eaten raw soon after catching, while grasshoppers, caterpillars, and young beetles are fried. Ants are eaten raw or ground-up into a paste. Locusts are typically boiled and salted prior to eating and are a particularly important source of nutrition in Africa as they contain protein, fat, vitamins and minerals. In the Congo (Kinshaza) (formerly Zaire), more than 30 species are harvested (see Table 11.1). Some caterpillars are sold not only in the local village markets, but are shipped by the ton from one country to another and there are processing plants where caterpillars are canned in Botswana and South Africa. In the rural countryside, they are usually dried in the sun before being sold in the market. The larvae of the Mopone Emperor Moth are now a cash crop with an annual production of 2,000 tonnes in Southern Africa.

However, termites are most widely used as food in Africa. They are highly attracted to lights, even candlelight, and that is one way they are captured for use as food. The wings are broken off, and the body is fried. The queens are considered a special treat and are often reserved for children or grandparents. *Brachytrupes membranaceus*, a large, fat cricket which is destructive to root crops, is regarded as a particular delicacy and is collected by digging them up from their burrows in the ground. *Gryllotalpa africana* Palisot (mole crickets) adult is used as a food (Fladung 1924). In some cultures, bee nests are collected as much for their bee grubs as for the honey. They are considered a great delicacy. Larvae, pupae and/or adults of many beetles are used as food. The hard parts (wings, legs and head) are removed during preparation for cooking. Walking sticks and leaf insects, *Extatosoma tiaratum* are used as food in Asia and in Papua New Guinea. In Mexico, grasshoppers and other edible insects are sold in village markets and are fried before being eaten. Many are sold in cans as fried grasshoppers, chocolate covered ants, etc. Fried grasshoppers are also canned commercially and sold in supermarkets and local grocery stores. High in protein and low in fat, they may be fried or ground into meal and mixed with flour to make tortillas. Tortillas are served with red and white agave worms in many Mexico city restaurants (Sahagun 1557). *Mopane* (the caterpillars of a moth species) are a huge industry in numerous African countries. It is reported that tons of the caterpillars are harvested, processed, and sold in markets or by the truckload (Dreyer and Wehmeyer 1982). Ethiopian tribes preserved bugs in salt, as did the Algerians who sell them in their markets. There is a considerable trade in termites in some areas. Sun dried termites are found in the local markets in many East African towns and villages. They are sometimes transported long distances to markets. The Baganda community who live around the northern shore of Lake Victoria in Uganda use termites and fried grasshoppers as snacks between the main meals. In many Bantu speaking parts of the country, boiled and dried termites are on sale in the markets in some seasons of the year. One species in Asia, the giant waterbug, is reported to be exported from Thailand to the Asian food shops in the United States (Pemberton 1988).

In the United States, on the menu of some restaurants are interesting dishes such as stir-fried mealworms and caterpillar crunch (a combination of trail mix and fried caterpillars) (Triplehorn and Johnson 2005). In addition to raising your own food-insect supply or catching from the wild, there are numerous stores in various American cities that sell frozen insects from Thailand and other countries in SE Asia. There are already cricket farms in America, which raise these and other insects for the pet trade. They produce tons of insects per week (Anonymous 1991). Processed insects for sale as cocktail snacks, etc., are apparently no longer imported into the United States (Fasoranti and Ajiboye 1993).

However, several processed insects are commercially available in Japan. The most widely eaten is *inago* (the grasshopper, *Oxya velox* F.), which is preserved by boiling in soy sauce. This product appears as a luxury item in supermarkets throughout the country, including Tokyo. There is currently an effort to incorporate several insects that were important in aboriginal diets into the Australian cuisine. In Canada, attempts are under way to apply industrial methods to the production of insects as food (Fasoranti and Ajiboye 1993). Commercially grown insects available to special interest groups (from types of bait and pet foods) in the USA and Europe include the cricket, *Acheta domesticus*, the mealworm, *Tenebrio molitor* L. (a beetle grub), and the greater waxmoth larva, *Galleria mellonella* (L.). A list of insects used as food and methods of prepration in the different parts of world are presented in Table 11.2. More than 80 recipes based on these insects and honeybee pupae (*Apis mellifera*) are included in The Original Guide to Insect Cookery, Taylor and Carter (1976).

Many research teams at universities study insects as human food, using the term *Micro-livestock* to categorize the insects that can be eaten (Srivastava et al. 2009). In some parts of the world it is termed as *Entomophagy*. Some insects such as the mormon cricket, grass hoppers and pandora moth caterpillars yielded a very high energy return for the energy expended in their harvest, often much higher than return rates from seeds or other plant food resources. When dried, the insects can be stored for use as a winter food. Insects contain high value minerals such as potassium, calcium, magnesium, zinc, phosphorus and iron, and vitamins. Due to their high nutritional value, in some regions, flour made from caterpillars is mixed to prepare pulp which is given to children to counter malnutrition. In New Guinea,

Table 11.2 A list of insects used as food and method of use in the different parts of world

Name of the insect	Place	Country	Methods of preparation	Reference
Red ants and termites	Kandhamal, Koraput, Sundergarh, Keonjhar and Mayurbhanj districts of Odisha	India	Roasted. Eaten as snacks or with rice. Termites collected at the time of swarming, while red ants were collected as and when required from the plants where nests of ants were found.	Jishing (2003)
Demta eggs (a red ant on mango trees)	Pithra village, Simdega district, Jharkhand	India	Eggs are found in the curled leaf of mango. They are fried with salt, chilli, spices and mustard oil.	Verma (2003)
Pandora moth caterpillars *Coloradia pandora*	Central Africa	Africa	Caterpillars are harvested. *Piuga* is regarded as a tasty, nutritious food good for sick people. They are collected in trenches dug around the bases of Jeffery pine trees. They are eaten after roasting, by mixing them with hot sand.	Blake and Wagner (1987)
Grasshoppers, crickets, red ants, and larvae of mulberry silkworms	Phek, Dimapur and Kohima districts of Nagaland	India	Grasshoppers are available in the local markets during August and September and usually collected after the harvest of paddy, especially at night. Roasting after removing the wings and legs is the method of cooking.	Srivastava et al. (2009)
Shore flies *Hydropyrus hians* pupae	Nevada border region	California, USA	By drying in the sun and mixing with acorns, berries, grass-seeds, and other articles of food gathered up in the mountains, they make a mixture called *cuchaba*, a type of bread.	Defoliart (1994)
Mormon cricket (*Anabrus simplex*)	Indians, all over the West	India	They are dried, and ground, making a fine flour. Used to make bread, in a manner similar to sugar used in cakes.	Gottfredson (1874)
Jumiles stinkbugs	Oaxaca, Guerrero, Morelos	Mexican states	People frequently cook as a salsa. Also these bugs are eaten live with the traditional *tacos* as they have an aromatic and deep flavor like mint or cinnamon.	Srivastava et al. (2009)
Ant (*Atta cephalotes)* pupae known as escamoles *and* Grasshoppers	Mexico	Mexico	Winged female ants are consumed in the rainy season. Contains 42 % protein. *Escamoles*, are found on the menu in the finest restaurants, served fried with butter, or fried with onions and garlic. Grasshoppers are fried prior to eating.	Srivastava et al. (2009)
Oecophylla	Australia	Australia	Eaten as bush food. People prefer to fry them prior to eating.	Cherry (1991)

(continued)

Table 11.2 (continued)

Name of the insect	Place	Country	Methods of preparation	Reference
White beetles *Cyclocephala*, Cerambicid's larvae and Cicadas, Ant	Cotocollao, Quito, Amazonian region	Ecuador	People cook with pork meat and vegetables. People eat the Cerambicid's larvae and Cicadas. Lemon ant and *Hormiga Culona* ants are fried.	Dufour (1987)
Cicada	Irian Java	Indonesia	*Ekagi* people regularly eat caterpillars and grubs as source of protein.	Scholtz and Holm (1985)
Caterpillars, Migratory locust (*Locusta migratoria*); the red locust (*Cyrtacanthacris septemfasciata*); and the desert locust (*Schistocerca gregaria*). Termites *Macrotermes bellicosus, M. falciger*, and *M. subhyalinus*	Central, South and East Africa	Central, South and East Africa Africa, Korea	Collected from forests, eaten as source of protein The termites are eaten raw or lightly fried in their own fat Sometimes they are ground and added to sauces.	Scholtz and Holm (1985)
Honeybee larvae, Lake fly *Chaoborus*, winged termites	Eastern Uganda	Uganda	Larvae are collected as food. Adults of lake fly are used for making cakes as a source of protein. Termites are induced to emerge by beating the nearby ground with sticks. Drumming was observed to induce termite emergence near Namwnda in Bulmogi county of Busoga.	Osmaston (1951)
Giant water bug (*Lethocerus indicus*)	All over Asia	Asia	Gathered at night near water sources. Roasted whole and eaten as a delicacy.	Anonymous (1991)
Ants, beetles, crickets, grasshoppers, katydids, locusts and dragonfly	Philippines	Philippines	Larvae are fried or boiled prior to eating. They are also sauted with vegetables.	Srivastava et al. (2009)
Silk moth	China	China	Workers in Chinese silk factories eat the pupae.	Anonymous (1991)

Termites, palm grubs and ants	Columbia	Columbia	Ants are ground and used as a spread on breads. They are used as ingredients in recipes. It is the ant's larvae and/or pupae that are usually eaten. Roasted leafcutter ant abdomens are sold, instead of popcorn, in movie theatres.	Anonymous (1991)
Sago grubs, larvae of moths, wasps, butterflies, dragonflies, beetles, adult grasshoppers, cicadas, stick insects, moths, locusts and crickets	Papua New Guinea, Korea	Papua New Guinea, Korea	Larvae of a wood-boring beetle are considered a delicacy. The islanders boil the larvae or roast them over an open fire to serve as a main meal. Fried locusts are mostly used in Korea.	Anonymous (1991)
Pupae of the fly *Ephvdra hians*, ants, crickets, mealworms, grasshoppers, larvae of moths and yellow jackets	Western United States	United States of America	Food was called *Koo-tsabe*. Deep fry pieces are eaten.	Triplehorn and Johnson (2005)
Dryopoid beetles	Andes	South America	Tribes dry, grind and uses them as a spicy additive for food flavouring.	Pemberton and Yamasaki (1995)
Candied grasshoppers, known as *inago*, bee or wasps larvae/ pupae	Japan	Japan	As cocktail snack. Canned wasps, and wings are sold.	Pemberton and Yamasaki (1995)
Oak grubs (*Quercus velutina)*	Rome	Rome	A delicacy purposely fattened on flour.	Pemberton and Yamasaki (1995)

Table 11.3 Nutritional content of edible insects and other animals based on a 100 gm serving

Name of the insects and other animals	Energy (Kcal)	Protein (gm)	Iron (mg)	Thiamine (mg)	Riboflavin (mg)	Niacin (mg)
Termite *(Macrotermes subhyanlinus)*	613	14.2	0.75	0.13	1.15	0.95
Caterpillar *(Usata terpsichore)*	370	28.2	35.5	3.67	1.91	5.2
Weevil *(Rhynchophorus phoenicis)*	562	6.7	13.1	3.02	2.24	7.8
Beef *(Lean ground)*	219	27.4	3.5	0.09	0.23	6.0
Fish *(Broiled cod)*	170	28.5	1.0	0.08	0.11	3.0

Table 11.4 Nutritive value of different insects

Insect	Protein (gm)	Fat (gm)	Carbohydrate (gm)	Calcium (mg)	Iron (mg)
Giant water beetle	19.8	8.3	2.1	43.5	13.6
Red ant	13.9	3.5	2.9	47.8	5.7
Silk worm pupae	9.6	5.6	2.3	41.7	1.8
Meal worms	20.27	12.72	N/A	13.3	N/A
Wax worms	15.50	22.19	N/A	28.3	N/A
Super worms	17.41	17.89	N/A	12.4	N/A
Fly larvae	15.58	7.81	N/A	87.4	N/A
Dung beetle	17.2	4.3	2.0	30.9	7.7
Cricket	21.32	6.01	5.1	75.8	9.5
Small grasshopper	20.6	6.1	3.9	35.2	5.0
Large grasshopper	14.3	3.3	2.2	27.5	3.0
June beetle	13.4	1.4	2.9	22.6	6.0
Caterpillar	6.7	N/A	N/A	N/A	13.1
Termite	14.2	N/A	N/A	N/A	35.5
Weevil	6.7	N/A	N/A	N/A	13.1

N/A = Not Analyzed

many villagers are currently engaged in the innovative commercialization of the spectacular lacewing butterflies as a source of food. Taiwan exports several hundred million dollars of wild butterfly specimens annually as food. There is a significant trans-border trade in edible insects within central African countries including Sudan and Nigeria. On a smaller scale, they are exported to France and Belgium, which import about 5 and 3 tonnes, respectively of a dried Sapelli caterpillars species (*Imbrasia oyemensis*) annually from the Democratic Republic of Congo (Fasoranti and Ajiboye 1993).

The nutritional content of edible insects and other animals based on a 100 g serving is presented in Table 11.3 (William 1991) and the nutritive value of insects is presented in Table 11.4 (Dunkel 1996 and Berenbaum 1996). Research shows that

100 g of insects provide more than 100 % of the daily requirements of the required minerals and vitamins (FAO 2004). Due to their high nutritional value, in some regions of Central African Republic, flour made from caterpillars is mixed to prepare pulp given to children to counter malnutrition. As a treatment for weakness and anemia, termites are prescribed as they are particularly high in iron while red ants are rich in bone building calcium (Srivastava et al. 2009).

11.7 Insects as Animal Feed /Insectivory

Insects represent an important food source for a wide variety of other animal species. Freshwater game fish such as trout, bass and bream feed extensively on aquatic insects such as mayflies, stoneflies, or hellgrammites. Many toads, frogs, turtles, snakes, and lizards also consume insects as a major part of their diet. *Insectivory* is common among land dwelling birds. Purple martins, barn swallows, vireos, warblers, flickers, whippoorwills, and swifts, for example, survive almost exclusively on insects. Other birds such as egrets, quail, geese, plovers, snipes, and bluebirds have a more varied diet, but they still derive a large percentage of their total nutrition from insects. There are some insectivorous mammals such as shrews, moles, bats, armadillos and anteaters. When other food is scarce, even foxes, raccoons, skunks and bears also turn to insects as a source of food (McHargue 1917; Frost 1942). The vast majority of studies in the West have dealt with the nutritional value of muscoid (Diptera) larvae or pupae used to recycle nutrients from poultry manure or other organic wastes as a high-protein source for broiler production (Defoliart 2002). According to Davis (1918), there is no difference in the taste of eggs from grub-fed hens. Cotton and George (1929) also summarized the early use of the meal worm, *Tenebrio molitor* as animal feed. Insect larvae of various kinds are also commonly used as fishing bait (http://www.hobbyandlifestyle.com/fishingbait.html).

11.8 Aesthetic Value of the Insects

Although something of a subjective subject, the beauty of some insects, as well as the graceful flight of butterflies, provides a source of comfort and pleasure to many people. The aesthetic value of insects is phenomenal – many hours could be spent examining the brilliant colours of delicate butterfly wings or the hardened bodies of brightly coloured beetles. Artists, poets, songwriters, and designers have used insects as inspiration for their art, which has brought joy and satisfaction to many admirers (Fig. 11.1). In Japan, some people often keep insects as pets, because their call is believed to be soothing, reminding them of a simple, less hectic life. The diversity of beauty in the insect world is tremendous.

Fig. 11.1 Use of insects for aesthetic purpose

The body colouration, beauty, and mode of life of the insects always attract us. Coloured wing and elytra of many coleopterans are used in jewellery, embroidery, pottery, and basket makings (Cambell et al. 1994). Among the insects of aesthetic value, butterfly attains maximum attention from museums and collectors. For satiating the growing need of butterflies amongst the collectors, numerous butterfly farms have been developed in European countries (Hammond 1995). In such butterfly farms (e.g. Brinckerhoffs), all the pupae are captives, reared exclusively for sale as live insects, which yield $100,000,000 annually (VKRTEX-Tutorials 2009), indicating a source of revenue for bioprospecting. There are many reasons why butterflies and moths are important, both in their own right but also as quality of life indicators. Many countries had developed butterfly parks for conserving butterflies around the world.

Notably, butterfly and beetle species have been used for various aesthetic and decorative purposes. Designs based on insects have been employed in art, jewelry, fashion and other decorative motifs. Thousands of stamps currently display insects(http://itc2.utk.edu). Musical composition has been influenced by insects.

11.9 Constraints

The most serious legal issue facing bioprospecting is the lack of clear rules and guidelines. Various environmental, trade, and geographically-specific agreements currently offer incomplete, ambiguous, or conflicting provisions relating to bioprospecting activities. Consequently, there are no clear rules on ownership, access, benefit-sharing, and environmental responsibility for bioprospecting in the Global Commons. Lack of clarity and distinct gaps in the existing laws encourages bioprospecting by companies keen to exploit the fragmented legal frameworks and policies for their own commercial benefit. It also obstructs the participation of developing States in exploration and use of the rich biological resources in areas designated as the Global Commons.

11.10 Conclusions

According to UN projections, the global urban population will increase to 66 percent by 2050, from the present 54 % in 2014 (United Nations 2014). This change will be marked in countries like India, which would add 404 million urban dwellers by 2050. Despite this, India would still house the largest rural population. All this will lead to qualitative changes in the requirement of various food commodities. An ever increasing middle-class, elite urban populations and life style changes will stimulate demand for less explored food delicacies. The nutritional, medicinal, aesthetic and other profitable values of insects are often neglected and we should further encourage their collection and commercialization, given the benefits to the environment and human health. It is an interesting concept, managing pest insects by developing them into a sought after delicacy. The collection of edible insects is a good source of income as they require little capital input if gathered by hand. Insects have long been a significant dietary factor and remedies for illnesses in many regions of the world. Their collection, commercialization and mass-rearing needs to be explored by merging traditional knowledge and modern science for the development of rural entrepreneurs. Because of the high protein content, high digestibility, variety in food diets, high conversion efficiency, and great reproductive potential associated with a short life cycle, the useful biomass obtained would be significant when compared to other products. In developing countries, insect food could be helpful in improving livelihoods and nutritional security. To promote insect bioprospecting important suggestions are given below:

- Indigenous knowledge and practices of edible/medicinal insect should be documented.
- Edible/medicinal insect species amenable to farming should be identified including possibilities to increase their nutritional value.
- Institutional mechanisms for promotion and support including capacity building programmes are required for establishment of micro livestock production (edible insects) parks for insect rearing at the village level with greater participation of both genders.
- IT-based network of insect entrepreneurs for exchange of information and consultations should be established.

References

Aldama-Aguilera C, Llanderal-Cazares C, Hernandez MS, Castillo-Marquez LE (2005) Produccion de grana- cochinilla (Dactylopius coccus Costa) in plantas de nopal a la intemperiey in microtunnels. Agrociencia 39(2):161–171

Alemla AM, Singh HK (2004) Utilization of insect as human food in Nagaland. Indian J Entomol 66(4):308–310

Alexiades MN (1999) *Ethnobotany of Ese Eja: plants, health and changes in an Amazonian society*, Ph.D. thesis, University of New York, New York, NY, USA

Anonymous (1991) Ohio State University, *Extension Fact Sheet Entomology*, Kenny Road, Columbus, OH 43210-1000 Insects as Human *Food (Micro-livestock) HYG-2160*-96 edited by,William F, Lyon (http://ohioline.osu.edu/hyg-fact/2000/2160.html

Anonymous (2006) Honey buyer-seller meet and conference, Indian Society of Agri-business Professional, New Delhi, India, 26-27 June

Anonymous(2009)Ethnoentomology,http//en.wikipedia.wikipedia.org/wiki/ethnoentomology

Antonio TMF (1994) Insects as re*medies for illness in Zaire*. Food Insects Newslett 73(3):4–5

Arora GS, Gupta IJ (1979) Taxonomic studies of some of the Indian non-mulberry silkmoths (Lepidoptera: Saturniidae). Mem Zool Surv India 16:1–63

Behera B (2003) *Inventory of indigenous technical knowledge in agriculture*, Document 2, editors Das P, Das SK, Mishra A, Arya HPS, Bujarbaruah KM, Singh RP,Verma LR, Subba Reddy G, Geetha Rani, Gupta HS, Kavia ZD , Ray DP, (Mission Unit, Division of Agricultural Extension, Indian Council of Agricultural Research, NewDelhi, Published by Director, DIPA, ICAR, KrishiAnusandhan Bhavan, Pusa, New Delhi): 405

Berenbaum M (1996) Bugs in the system. Food Insects Newslett 9(2) Montana State University

Blake EA, Wagner MR (1987) Collection and consumption of *pandora* moth, *Coloradia pandora lindseyi* (Lepidoptera: Saturniidae) larvae by Owens Valley and Mono Lake Paiutes. Bull Entomol Soc Am 33:23–27

But PH, Tam YK, Lung LC (1991) Ethnopharmacology of rhinoceros horn II: antipyretic effects of prescriptions containing rhinoceros horn or water buffalo horn. J Ethnopharmacol 33(1-2):45–50

Cambell MH, Holtcamp RH, Cormick LHM (1994) Biological control of native shrubs *Cassinia* sp (Hemiptera: Kerriidae) in New South Wales. Plant Protect Quar 9:64–68

Cao H, Shi HF, Sun HC (1996) Exploitation of silkworm pupae food Chinese. Cuisine Res 13(1):48–51

Capinera JL (2004) *Encyclopedia of entomology*. Kluwer Academic, Boston

Cherry RH (1991) Use of insects by Australian aborigines. Am Entomol 37:13

Costo-neto EM (2003) Consideration on the man/insect relationship in the state of Bahia, Brazil. In: MF Elisabeth, MC Jacqueline (eds) Ethnocsience. pp 95–104

Cotton RT, George St RA (1929) The mealworm. USDA Technol Bull 95:3–37

Davis JJ (1918) Common white grubs. USDA Farmer's Bull 940:12–28

de Sahagun FB (1957) Original Codex, Biblioteca Nacional de Mexico, Mexico, DF Introduction

Defoliart GR (1994) Some insect foods of the American Indians: and how the early whites reacted to them, Food Insects Newslett 7(3) 1–2 &10–11

Defoliart GR (1995) Edible insects as minilivestock. Biodivers Conserv 4(3):306–321

Defoliart GR (1999) Insects as food: why the western attitude is important. Annu Rev Entomol 44:21–50

Defoliart G R (2002) The human use of insects as food resource: a bibliographic account in progress, University of Wisconsin- Madison, Madison,Wis, USA

Dreyer JJ, Wehmeyer AS (1982) On the nutritive value of mopanie worms. South Afr J Sci 78:33–35

Dufour DL (1987) Insects as food, a case study from the northwest Amazon. Am Anthropol 89:383–397

Dunkel FV(1996) Food Insects Newslett 9(2) Montana State University

FAO (2004) Edible insects, important source of protein in central Africa, FAO Newsroom. 8 November 2004, Rome.
FAO (2013) Forest products critical to fight hunger – including insects, Rome, 13 May 2013, www.fao.org/news/story/en/item/175922/icode/
Fasoranti JO, Ajiboye D O (1993) Some edible insects of Kwara State,Nigeria, Am Entomol, 39 (2): 116.
Faulkner DJ (1992) Biomedical uses for natural marine products. Oceanus 35:29–35
Fladung EB (1924) Insects as food. Maryland Acad Sci Bull 8
Fromme A (2005) Edible insects. ZooGoer 34(4)
Frost SW (1942) *General entomology*. McGraw–Hill, New York
Gope B, Prasad B (1983) Preliminary observations on the nutritional value of some edible insects of Manipur. J Adv Zool 4:55–61
Gottfredson P (1874) From the Gottfredson family history Ms on file, Utah State Hist Soc, Salt Lake city. J Perter Gottfredson 15–16
Green SV (1998) The bushman as an entomologist. Antenna 22(1):4–8
Grieve M (2009) A modern herbal, http://www.botanical.com
Hammond PM (1995) The currentmagnitude of biodiversity. In: Heywood VH (ed) *Global biodiversity assessment*. Cambridge University Press, Cambridge, pp. 113–138
Hider RC (1988) Honeybee venom: a rich source of pharmacologically active peptides. Endeavour 12(2):60–65
http://www. hobbyandlifestyle.com/fishin*gbait.html.*
http://botanical.com (2012)
http://itc2.utk.edu/web/jgrant/InsectIntro/HumImpactPosAesthetic.html
http://lib.bioinfo.pl/meid:186381
Jishing A (2003) *Inventory of indigenous technical knowledge in agriculture*, Document 2, edited by Das P, Das SK, Mishra A, Arya HPS, Bujarbaruah KM, Singh RP, Verma LR, Subba Reddy G, Geetha Rani, Gupta HS, Kavia ZD & Ray DP, (Mission Unit, Division of Agricultural Extension, Indian Council of Agricultural Research, New Delhi, Published by Director, DIPA, ICAR, Krishi Anusandhan Bhavan, Pusa, New Delhi): 623
Kumar P, Tarui N (2004) Identifying the contribution of indigenous knowledge in bioprospecting for effective conservation strategy. In Bridging scales and epistemelgioes conference, Alexandria, Ezypt March 17–20, 2004
Lazarus LH, Attila M (1993) The toad, ugly and venomous, wears yet a precious jewel in his skin. Prog Neurobiol 41(4):473–507
Liu XY, Wu SX, Wang YY, Wang W, Zhou L, Li YQ (2005) Changes of 5-HT receptor subt*ype mRNAs in rat dorsal* root ganglion by bee venom-induced inflammatory pain. Neurosci Lett 375(1):42–46
MacEvilly C (2000) Bugs in the system. Nutr Bull 25:267–268
Malaisse F, Parent G (1980) Les chenilles comestibles du Sh*aba meridionnale (Zai*re). Les Natual Belges 62:2–24
Mann J (1969) Cactus—feeding insects and mites. US Natl Mus Bull 256:1–158
Martin GJ (2001) Mutual impact: the relationship between etnobiological research and the convention on biological diversity, Building bridges with traditional knowledge II summit ,May 28 – June 2 , Honolulu, Hawaii.
Maya EMA (2000) Ethnoentomologia de la Comunidad. Hnahuunu EL-Dexthi-San Juanico, Hidalgo
McHargue JS (1917) A study of proteins of certain insects with reference to their value as food four poultry. J Agric Res 10:633–637
Mendez J, Gonzalez M, Lobo MG, Carnero A (2004) Color quality of pigments in cochineals (*Dactylopius coccus* Costa). Geographical origin characterization using multivariate statistical analysis. J Agric Food Chem 52(5):1331–1337
Mishra N, Hazarika NC, Narain K, Mahanta J (2003) Nutritive value of non-mulberry and mulberry silkworm pupae and consumption pattern in Assam, India. Nutr Res 23(10):1303–1311

Moon DO, Park SY, Heo MS, Kim KC, Park C, Ko WS (2006) Key regulators in bee venom—induced apoptosis are Bcl-2 and caspase-3 in human leukemia U937 cells through down regulation of ERK and Akt. Int Immunopharmacol 6(12):1796–1807

Namba T, Ma YH, Inagaki K (1988) Insect derived crude drugs in the chinese song dynasty. J Ethnopharmacol 24(2-3):247–285

Oliveira JF, Carvalho SJ, Passos de Sousa RFX, Bruno de Simao M, Sinao MM (1976) The nutritional value of four species of insects consumed in Angola. Ecol Food Nutr 5(2):91–97

Osmaston HA (1951) The termite and its uses for food. Uganda J (Kampala) 15:83

Padamanbhan P, Sujana KA (2008) Animal products in traditional medicine from Allapady hills of Western Ghats. Indian J Tradit Knowl 7(2):326–329

Park HJ, Leu SH, Sow DJ, Oh KW, Kim KH, Song HS (2004) Anti-arthritic effect of bee venom: inhibition of inflammation mediator general by suppression of NF-kapp B through interaction with the p50 subunit. Arthritis Rheum 50(11):3054–3515

Patel SN, Patel IK (1995) Inducing animal into heat feeding grasshoppers. Honey bee 6(3):14

Pemberton RW (1988) The use of the Thai giant waterbug, *Lethocerus indicus* (Hemiptera: Belostomatidae) as human food in California. Pan-Pacif Entomol 64:81–82

Pemberton RW, Yamasaki T (1995) Insects : old food in new Japan. Am Entomol 41:229

Prasad B (2007) Insect diversity and endemic bioresources. In: Endemic bio-resources of India-conservation & sustainable development with special reference to North East India, Bishen Singh Mahendra Pal Singh, Dehra Dun, India

Quijano PW, Vergara V (2007) Determination productive in rumplaza de harina de pescado en dietes para pollos de carne, http://www.engormix.com/MA-avicultura/nutricion/articulos/determinacion-energia-metabolizable-subproducto-t1554/141-p0.htm

Quin PJ (1959) Foods and feeding habits of the pedi. Witwatersrand University Press, Johannesburg, p 428

Ramos EJ (1997) Insects: a sustainable source of food?, edited by MG Paoletti, SG F Bukkens, Special Issue: Mini livestock. Ecol Food Nutr 36(2-4):247–276

Ramos EJ, Pino JM (1989).*Los Insectos Comestiblesen el M´exico Antiguo Estudio Etno-Entomologico.* Edited by Mexico DF, Sahag un FB, 1st edn

Ruddle K (1973) The human use of insects: examples from the Yukpa. Biotropica 5:94–101

Scholtz CH, Holm E (1985) Insects of Southern Africa, (Butterworths, Durban): 534

Sharma KK, Jaiswal AK, Kumar KK (2006) Role of lac culture in biodiversity conservation. Curr Sci 91(7):894–898

Singh SK (2003) Inventory of indigenous technical knowledge in agriculture, Document 2. (Das P, DasSK, Mishra A, Arya HPS, Bujarbaruah KM, Singh RP,Verma LR, Subba Reddy G, Geetha Rani, Gupta HS, Kavia ZD, Ray DP, eds) (Mission Unit, Division of Agricultural Extension, Indian Council of Agricultural Research, New Delhi, Published by Director, DIPA, ICAR, Krishi Anusandhan Bhavan, Pusa, New Delhi), 324

Singh OT, Chakravorty J (2008) Diversity and occurrence of edible orthopterans in Arunachal Pradesh with a comparative note on edible orthopterans of Manipur and Nagaland. J Nat Conserv 19:169–176

Singh OT, Chakravorty J, Nabom S, Kato D (2007) Species diversity and occurrence of edible insects with special reference to coleopterans of Arunachal Pradesh. J Nat Conserv 19:169–176

Slobodian L; Kinna R; Kambu A, Ognibene L (2011) Bioprospecting in the global comm*ons: legal issues br*ief United Nations Environment Programme- Division of *Environmental Law a*nd Conventions Environmental Law and Governance. http://www.unep.org/delc/Portals/119/Biosprecting-Issuepaper.pdf

Srivastava SK, Naresh B, Hema P (2009) Traditional insect bioprospecting – as human food and medicine. Indian J Tradit Knowl 8(4):485–494

Srivastava SK; Naresh B, Hemrom A (2015) Insect as a source of food -women's perspective, *Oral paper* in Souvenir and Book of Abstracts National Entomologists' Meet ,organised by Society for Advancement of Natural Resins and Gums; ICAR-Indian Institute of Natural resins and

Gums and Network project on Conservation of Lac Insect Genetic Resources, Namkum , Ranchi, Jharkhand held from February 5-7, 2015 at ICAR-Indian Institute of Natural resins and Gums, Namkum, Ranchi: 27-28
Subramanian M, SS Kannan, R Geethamalini (2005) Natural dyes—a glimpse, *Pakistan Textile Journal*, vol. 54, no. 4, p. 53.
Sungpuag P, Puwastien P (1983) Nutritive value of unconventional protein source: insect. Pochanagan Sarn 1:5–12
Talukdar RB (2009) A fillip for the silk industry, in North–east India, http://southasia.oneworld.net/fromthegrassroots
Tang YL, Shi YH, Zhao W, Hao G, Le GW (2008) Insertion mode of a novel anionic antimicrobial peptide MDpep5 (Val-Glu-Ser-Trp-Val) from Chinese traditional edible larvae of housefly. *J Pharm Biomed Anal* 48(4):1187–1194
Tango M (1994) *Les Insectes CommeAliments de LHomme.* Published in 1981, reviewed in Food Insects Newslett 7(3): 3–4
Taylor L (2000) Plant based drug and medicines. From the Raintree Internet Nutrition http://www.rain.tree.com
Taylor RL, Carter BJ (1976) Entertaining with insects, or: the original guide to insect cookery. Woodbridge Press, Santa Barbara 160pp
Triplehorn CA, Johnson NF (2005) Introduction to the study of insects, 7th edn. Thomas Brooks, Cole. ISBN 0030968356, pp 643
UNEP/CBD/COP/5/INF/7 (2000) Conference of the parties to the convention on biological diversity, Fifth meeting, Nairobi, 15-26 May 2000: pp 16
United Nations (2014) Department of Economic and Social Affairs, Population Division, World Urbanization Prospects: The 2014 Revision, Highlights (ST/ESA/SER.A/352)
Verma LR (2003) Inventory of indigenous technical knowledge in agriculture, Document *2* (Das P, Das SK, Mishra A, Arya HPS, Bujarbaruah KM, Singh RP, Verma LR, Subba Reddy G, Geetha Rani, Gupta HS, Kavia ZD, Ray DP, eds) (Mission Unit, Division of Agricultural Extension, Indian Council of Agricultural Research, New Delhi, Published by Director, DIPA, ICAR, Krishi Anusandhan Bhavan, Pusa, New Delhi): 542
Vigueras AL, Porlillo LM (2001) Usos del pigmento de la grana Cochinilla. In: Llanderal C, Neito Hernandez R (eds) *Produccion de Grana Cochinilla.* Colegio de Postgraduados, Texcoco, pp. 93–103
VKRTEX-Tutorials (2009) Manufacture of all kinds of auto loom fabrics and natural dye fabrics, http://www.vkrtex.com
Werner D (1970) Healing in the sierra madre. Nat Hist 79(9):61–66
William FL (1991) Ohio State University *Extension Fact Sheet Entomology* Kenny Road, Columbus, OH 43210-1000 Insects as Human Food (Micro-livestock) HYG-2160-96 (http://ohioline.osu.edu/hyg-fact/2000/2160.html)
Zhang CX, Tang XD, Cheng JA (2008) The utilization and industrialization of insect resources in China. Entomol Res 38(1):38–47
Zhou JH, Sun XB, Xu HB, Li YZ, Zhao JH (1996) Research progress on silkworm pupae. Spec Wild Econ Anim Plant Res 18(1):38–40
Zhu LS (2004) Exploitation and utilization of the silkworm *Antheraea pernyi.* N Sericult 25(101):32–33
Zimian D, Yonghua Z, Xiwu G (1997) Medicinal insects in China. Ecol Food Nutr 36(2–4):209–220

Chapter 12
Phages Against Infectious Diseases

Luís D.R. Melo, Hugo Oliveira, Sílvio B. Santos, Sanna Sillankorva, and Joana Azeredo

Abstract (Bacterio)phages (viruses that infect bacteria) are the most abundant entities on earth. Phages are the natural predators of bacteria and therefore have a great antibacterial potential. Immediately after their discovery, in the early twentieth century and before the antibiotic era, they were extensively used to treat infectious diseases. Now, on account of the spread of antibiotic resistance, the interest in phage therapy has been reborn and the recent advances on phage biology and host interactions have reinforced their therapeutic potential. This chapter describes the most important features of phages that make them valuable alternatives to antibiotics in controlling infectious diseases and the challenges that phage therapy is facing for them to be brought into clinical practice.

12.1 Phage Discovery

The observation that an "invisible microbe" present in bacteria-free filtrates of stool samples from dysentery patients was responsible for bacterial lysis in liquid cultures, and patches of clear zones in the bacterial lawns grown in the surface of agar seeded with the dysentery bacillus, led the French-Canadian microbiologist d'Herelle to the discovery of bacteriophages in 1917 (d'Herelle 1917). However, in 1896 the UK bacteriologist Ernest Hankin observed that the waters of the Ganges and Jumna rivers in India, after being passed through fine porcelain filters, presented strong antibacterial activity against *Vibrio cholerae* and was responsible for stopping cholera epidemics from being spread by ingestion of the water (Hankin 1896). Similar observations were made by the Russian bacteriologist Nikolay Gamaleya with *Bacillus subtilis* in 1898, and subsequent researchers (van Helvoort 1992). It was only in 1915 that the English medical bacteriologist Frederick Twort, observing a similar phenomenon while working with *Micrococcus*, suggested for the first time that a virus could explain these observations, although he was not

L.D.R. Melo • H. Oliveira • S.B. Santos • S. Sillankorva • J. Azeredo (✉)
Laboratório de Investigação em Biofilmes Rosário Oliveira, Centre of Biological Engineering, University of Minho, Braga, Portugal
e-mail: jazeredo@deb.uminho.pt

R. Paterson, N. Lima (eds.), *Bioprospecting*, Topics in Biodiversity and Conservation 16, DOI 10.1007/978-3-319-47935-4_12

convinced this was the most likely explanation. Like the former research, these findings were not further explored (Summers 2006).

Félix d'Herelle was the first to clearly describe that this antibacterial particle was a virus capable of multiplying indefinitely at the expense of living bacteria, leading to cell lysis in order to accomplish the multiplication process (d'Herelle 1917). Based on this ability, he used the Greek suffix *phage* (eat) and called them *bacteriophages*. Unlike other researchers, the author systematically investigated the nature of these viruses and found that the ability of the phage to form plaques could be used to enumerate them and showed that phage multiplication was made in waves (or steps), each one representing a new phage cycle of infection, multiplication and release (d'Herelle 1917).

12.2 Phage Structure and Classification

For more than 20 years phages were explored without knowledge of their morphologic characteristics due to the lack of tools that enabled study of such small particles. It was only in the 1940s, with the advent of the electron microscope (EM), that the phage size, capsid symmetry and the length of the tail fibers (when they existed) were disclosed (Luria et al. 1943). Moreover, the EM enabled observation of the interactions between phages and bacteria. Discrimination of phages according to their morphology was possible which, coupled with the newer molecular biology methods, permitted classification of phages based on their morphologies and nucleic acid sequences (Nelson 2004).

A phage is composed of a nucleic acid genome encapsulated in a capsid, and may contain lipids in the particle wall or in the envelope (when present) (Ackermann 2006). The capsids may present different morphologies ranging from small hexagonal structures, to filaments, or highly complex structures consisting individually of a head and a tail. The phage genomes can vary from as few as 5 kb (ex. phage phiX174) to as many as 500 kb such as in *Bacillus* phage G, the phage presenting the biggest known genome. Despite their genome size, phages do not possess the necessary machinery for energy production nor ribosomes for proteins synthesis and thus are metabolically inert (Guttman et al. 2005). Consequently, phages depend on the host to produce progeny and their genome is devoted to direct the host for that purpose. In most cases, phage genetic information is carried in double-stranded DNA (dsDNA) but can also be as single stranded DNA (ssDNA), single-stranded RNA (ssRNA) or rarely as double-stranded RNA (dsRNA).

The polythetic species concept adopted by the International Committee on Taxonomy of Viruses (ICTV) grouped phages in (a) one order, the *Caudovirales*, composed of tailed phages with binary symmetry distributed in three very large and phylogenetically related families, and (b) 17 families comprising the polyhedral or cubic (due to their cubic symmetry and icosahedral shape), filamentous and pleomorphic phages (PFP or CFP). The latter are extremely diversified in their basic properties suggesting they constitute many different lines of descent (Ackermann 2006).

Approximately 96 % of phages are tailed and belong to the order *Caudovirales*. The dsDNA phages, which are not limited to the *Caudovirales*, represent by far the largest group of phages. Within the *Caudovirales*, almost 66 % have long and non-contractile tails and represent the *Siphoviridae* which are the most representative family, not only among *Caudovirales*, but among all classified phages. Approximately, 25 % of tailed phages have contractile tails (*Myoviridae*) and the remainder have short tails (*Podoviridae*) (Ackermann 2007). Due to their abundance and predominance, *Caudovirales* are the best studied phages and represent the most diversified of all virus groups. The PFP phages that belong to families not grouped inside an order, represent less than 4 % of the observed phages (Ackermann 2007).

Phages are an extremely diversified group and it has been estimated that ten phage particles exist for each bacterial cell. This fact accounts for the estimated size of the global phage population to be approximately 10^{31} particles, making phages the most abundant living entities on earth (Ackermann 2007). They are able to infect at least ten Archaeal and 144 Eubacterial genera, with the *Enterobacteria* family being the most infected. The origin of phages is situated in a time before the phylogenetic divergence of Bacteria and Archaea, since there are phages able to infect these representatives of these two domains (Ackermann 2007).

12.3 Phage Life Cycles

Phages can behave differently upon host infection and two main life cycles are established: the lytic and the lysogenic. The lytic cycle always leads to phage replication and progeny release resulting in death of the host cell, while the lysogenic cycle enables bacterial lysogenization, *i.e.*, the phage genome assumes a quiescent state called "prophage", coexisting in a stable form with the host (Fig. 12.1). Phages are consequently classified as "virulent" when they can only follow the lytic pathway and as "temperate" when they are able to follow the lysogenic pathway. Temperate phages are also able, in certain conditions, to follow the lytic cycle (Skurnik and Strauch 2006).

The beginning of a phage life cycle is the adsorption of the phage particle to its host as a consequence of a random encounter. While phage specificity (or phage host range) depends on the highly variable adhesion regions of the tail fibers or spikes (in *Caudovirales*) and the corresponding receptors on the cell surface (adsorption), the phage-bacteria affinity (or phage adsorption rate) depends on the number of receptors present in the host surface. Examples of receptors in the case of Gram-negative hosts include capsules, different parts of lipopolysaccharides (LPS), flagella, fimbriae and many other surface proteins (Guttman et al. 2005). The number and type of receptors depend to a large extent on the host physiological state that in turn is influenced by environmental conditions. Any variation in these factors will thus have consequences in phage-bacterium affinity (Guttman et al. 2005).

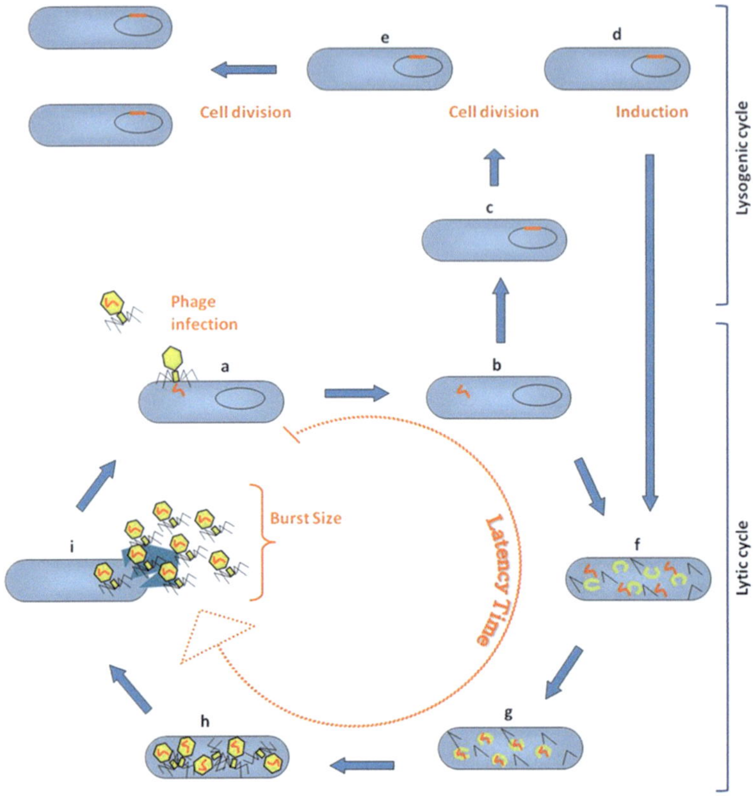

Fig. 12.1 Lytic and lysogenic life cycles of phages. (**a**) Phage adsorption and DNA injection; (**b**) Host infection, lytic or lysogenic cycle can be followed; (**c**) Prophage establishment, phage DNA integrates into the host chromosome; (**d**) Prophage induction; (**e**) Prophage replication along with the host chromosome; (**f**) Phage DNA takes control of the host transcriptional machinery to direct the expression of its own genes; (**g**) Packaging of phage DNA into the procapsids; (**h**) Assembly of the phage particles components to form the mature viruses; (**i**) Cell burst and progeny release

After successful adsorption, the phage tail tip is inserted through the outer membrane and host cell wall in order to transfer the DNA that was stored in the capsid through the tail and into the host cell (Fig. 12.1a, b) (Kutter et al. 2005). In the lytic pathway, once the phage DNA is inside the host it will direct all the host machinery to produce its progeny by making use of the phage's strong promoters. These regulate the early genes in order to direct the host RNA polymerase to transcribe the phage genes. Transcription of the early genes will often shut down the host bacterium system, inactivating host proteases and blocking restriction enzymes, or destroying some host proteins and enabling the phage to take control of the cell. The RNA polymerase will transcribe the next set of genes, the middle genes, encoding all the necessary proteins for phage DNA replication. This is usually accomplished by degradation of the host genome and reutilization of the obtained nucleotides. Finally, the late genes, encoding for all the protein components that compose the mature phage particles (capsid, tail and fibers) and others necessary for their assembly, are transcribed (Fig. 12.1f) (Kutter et al. 2005).

The phage DNA is then packed into the capsid and the phage particles components are assembled to form the mature viruses. In the case of tailed phages, two proteins, endolysin and holin, are also produced (Fig. 12.1h). This holin-endolysin two-component cell lysis system is known as the lambda paradigm, thought to be present in almost all dsDNA phages (Kakikawa et al. 2002; Sao-Jose et al. 2000). The holin will produce pores in the inner membrane to enable the endolysin to reach and digest the peptidoglycan. The consequent osmotic pressure induced by the cell wall digestion leads to cell lysis with the consequent release of the phage progeny that will be available to find a host and begin a new infection cycle (Fig. 12.1i) (Kutter et al. 2005).

On the other hand, DNA injection of a temperate phage will not stop normal bacterial metabolism and induce progeny production, but the phage will coexist with the host by integrating into the host genome as a prophage (Fig. 12.1c) or, in fewer cases, remain separated as a plasmid. The phage will be replicated as part of its host every time the cell reproduces. This dormant state may remain indefinitely until specific environmental conditions (usually adverse to the host cell) induce the prophage to follow the lytic cycle (Fig. 12.1d) and the phage assures its continuity even after the cell dies. Prophage induction leads to phage DNA excision from the host genome, which in some cases can lead to incorporation of bacterial DNA into the phage genome. This new transducing phage will be able to lysogenize new hosts, providing potential modifications to the infected cells (Guttman et al. 2005).

12.4 Early Phage Therapy Studies

Understanding the nature of phages as described above and observing that phage titres increased in recovering patients with infectious diseases, led d'Herelle to realize there was potential to use them as therapeutic agents (Kutter et al. 2015). He conducted experiments using phages to control *Salmonella gallinarum* in chickens and dysentery in rabbits *in vitro (*d'Herelle 1926*)*. The use of phages reduced the number of deaths and the duration of infection and were able to prevent reinfection by the pathogenic bacteria (Skurnik et al. 2007). These encouraging results led d'Herelle to perform the first studies on the therapeutic use of phages in humans. He developed an anti-dysentery phage product that was able to completely cure a 12 years old boy and three other patients of severe dysentery within a few days through a single dose (d'Herelle 1926). Recovery signs were seen after 24 h, confirming the therapeutic efficacy. Phages were also tested to treat a human skin disease caused by *Staphylococcus* and the recovery signs were observed within 24–48 h (Skurnik et al. 2007). Many other studies on the use of phages to treat diseases were conducted and the success was such that in the early 1930s the pharmaceutical companies Eli Lilly, E. R. Squibb & Sons and Swan-Myers, a division of Abbott Laboratories, all manufactured "phage" preparations (Häusler 2006).

The optimism concerning the use of phages in therapy led to a growing number of studies on their application. However, some inconsistencies and contradictory results implied the treatments were unreliable. This was a consequence of the lack of knowledge on the nature of phages, bad general practices and poor quality control when preparing and storing the phage stocks, sometimes leading to phage preparations deprived of active viable particles. The major problem with these studies was the absence of suitable control groups making it impossible to assign good results to phage activity. The eagerness to sell phage products led to an emergence in the 1930s of preparations marketed with wildly exaggerated claims, which also stretched credibility (Kutter et al. 2015).

Furthermore, the discovery of antibiotics (i.e. penicillin and streptomycin) in the 1940s revolutionized medicine and led to the near eradication of bacterial diseases. These compounds also pushed phage therapy research to the background, exacerbated by their debilitated reputation. The Scot, Sir Alexander Fleming predicted that bacteria would develop resistance to antibiotics when giving his Nobel lecture which has now occurred. However, there was no realisation that it would happen so soon. The use, abuse and misuse of antibiotics in animal food and human medicine have led to the emergence of antibiotic resistant bacteria, making treatment of the affected individual difficult and sometimes impossible. There has been an exponentially growing number of cases of antibiotic resistance (Skurnik et al. 2007). The introduction and wide use of antibiotics have shown that it is likely to always select for resistant strains, leading inevitably to their emergence (Johnson 2015; Melnyk et al. 2015).

12.5 Phages as Alternatives to Chemotherapy

The viral nature confers to phage therapy many potential advantages over antibiotics. The first, and probably the most important, is specificity. Phages are highly specific to their hosts, a specificity that is usually at the species level but can be towards strains. This characteristic enables phage therapeutic application without disturbing the normal flora, in contrast to chemical antibacterials, consequently reducing the likelihood of super-infection and other complications of normal-flora reduction. On the other hand, phage specificity means that diagnosis of the bacteria involved in the infection is required before therapy can be applied (Barrow and Soothill 1997).

The self-replicating nature of phages causes an increase of their concentration in the bacterial host and thus phages will be present and persist at a higher concentration in the place of infection, which is where they are more needed. This will lead to a reduction in the need for multiple doses to treat infectious diseases and to a higher efficacy of the treatment. The reverse is also true, they are self-limiting since that when the target organism is not present the phages will not replicate and will be removed from the system (Dabrowska et al. 2005). Moreover, some authors claim that the ability of phages to rapidly distribute throughout the body and reach organs that are usually not readily accessible to drugs (*e.g.* prostate gland, bones and brain) and then multiply in the presence of their hosts, constitutes a valuable advantage over antibiotics, enabling to be treated otherwise untreatable infections (Dabrowska et al. 2005). The mechanisms developed by bacteria to resist antibiotics do not also interfere with phage efficacy, therefore phages are able to target and kill antibiotic resistant bacteria (Matsuzaki et al. 2005).

However, as with antibiotics, bacteria are able to develop resistance against phages usually through losing or modifying molecules that the phage uses as receptor. Since bacterial receptors for phage adsorption are often virulent determinants or crucial molecules to the bacterial cell, the development of resistance will decrease the bacterium virulence or produce loss of important functions and competitiveness, which will also contribute to infection control (Levin and Bull 2004). Moreover, phages are able to rapidly modify themselves in response to the appearance of phage-resistant mutants, making them efficient in combating the emergence of newly arising bacterial threats (Matsuzaki et al. 2005).

It is common to find different phages for the same bacterium recognizing different cell surface receptors (i.e. lipopolysaccharides, proteins, teichoic acids, etc.) on the same bacterium. This means that when a bacterium develops resistance to a particular phage (by changing or hiding the receptors to that phage), it will still be sensitive to the other phages that target different receptors. Besides the availability of new phages able to infect phage resistant bacteria, it is much cheaper, faster and easier to develop a new phage system than a new antibiotic, which is a long and expensive process, which is another advantage of phages over antibiotics (Kutter et al. 2005, 2015).

12.6 Developing a Phage Product

Step 1: Choosing the Phages. Phages are ubiquitous in nature. Due to their high abundance and presence in almost every ecological niche, it is straightforward to isolate new phages. Consequently, every environmental source that contains a given pathogen is likely to contain specific phages for that microorganism. Phages can be isolated from various environments (e.g. soil, water, sewage effluent, hospital effluent, hot springs, and fecal material). Furthermore, they are commensal in humans and animals and especially abundant in the gastrointestinal tract (Bachrach et al. 2003; Brussow et al. 1998; Hitch et al. 2004; Sozzi et al. 1973). In fact, human body metagenome studies have shown that phages are key components of human microbiomes (Minot et al. 2011; Modi et al. 2013; Waller et al. 2014). The daily contact between humans and phages might explain why no adverse side effects have been detected in human studies involving phages (Bruttin and Brussow 2005; Rhoads et al. 2009).

The advances in the last decades in diagnostic tools and technologies have greatly improved the selection of appropriate phages for therapy (Keen 2012). The selection is the crucial first stage and the knowledge of its properties is fundamental to characterizing a phage as a good/bad candidate for therapy. This can be done taking into account as the following:

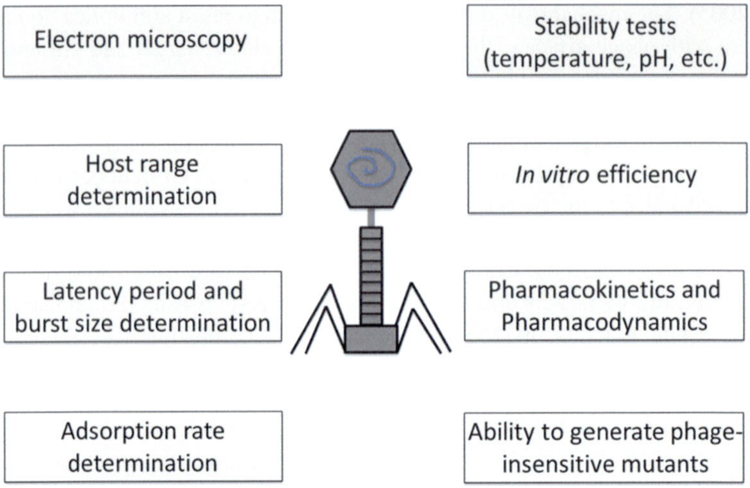

Fig. 12.2 Characterization of phages according to different parameters, in order to determine their suitability for therapeutic applications

(i) **The use of lab-adapted phages should be avoided**. "Lab-domesticated" phages tend to have a narrow host range (Weinbauer 2004). These phages should be avoided, for industrial and commercial purposes, as their potentiality to be patentable is limited.

(ii) **The bacterial host choice for phage isolation**. This is of great importance. Generally, a panel of host strains should be used, to increase the diversity of phages and to broaden the lytic coverage of the isolated phages. However, if assays on specific bacterial strains are performed, then only these should be used for isolation (Gill and Hyman 2010).

(iii) **Phages should be characterized in detail**. After the isolation, selected phages should be further characterized in detail (Fig. 12.2) to determine their suitability for therapeutic applications. Phages can be characterized in terms of their lytic efficiency against a broad range of bacterial targets. In this scenario, phages that display poor killing potential against their target hosts should be excluded (Loc-Carrillo and Abedon 2011). After this stage, life cycle parameters should be identified. Phage-host adsorption rate, latent period and burst size can provide very useful data on phages-host interaction properties and help the selection of the best bacterial candidate for phage production. Ideally, any phage cocktail should include phages with high host adsorption rates, quick replication cycles and large burst sizes (Payne and Jansen 2001). For pharmacological uses, phages should be very active against their targets, have low potential to harm patients, and good ability to reach their target and survive where the infection is present (Abedon and Thomas-Abedon 2010).

Phages should also be observed by EM and their genomes should be fully sequenced in order to avoid the possibility of selecting phages with a temperate trait (see above) or bacterial origin virulence-factors (i.e. toxins). In the pre-genomic era, phage observation by EM was considered crucial for a full phage characterization. Nowadays, sequencing a phage genome is fast and cheap, and researchers are recommended to sequence the entire genome of isolated phages. In a scenario where a phage genome has low homology with publicly available phage genomes deposited in databases, it is recommended to undergo microscopic observation of the phage particles, in order to complement phage characterization (Loc-Carrillo and Abedon 2011). Finally, the frequency of the target bacterium to develop resistance towards the phage should be determine in order to select for phages that induce low resistant rates (Skurnik and Strauch 2006).

(iv) **Use of temperate phages should be avoided**. For therapy, the use of temperate phages is not recommended for several reasons. The genomes of these phages can integrate into the host cell and alter their physiology after lysogeny establishment. Furthermore, several temperate phages can mediate transduction of antibiotic resistance or virulent encoding genes. Temperate phages can display superinfection immunity that is the phenomenon by which phage-sensitive bacteria are converted in insensitive ones after lysogeny establishment. In this sense, the lysogeny is immune to superinfection by the same phage or similar ones. Although some virulent phages can perform generalized transduction, the proportion of temperate phages with this ability is higher (Volkova et al. 2014).

Summarizing, a phage based product for therapeutic proposes should include strictly lytic and highly virulent (i.e. high adsorption rate and burst size and short latent period) phages, in order to quickly lyse bacteria, persist and migrate to the remaining infection sites by releasing virion progeny.

Step 2: Creating Phage Cocktails. The concept of conventional phage therapy relies on the use of carefully selected and specific phages to treat bacterial infections. Although monophage approaches report significant bacterial reductions, the use of phage cocktails has shown far more potential (Abdulamir et al. 2014; Fu et al. 2010; Gu et al. 2012; Mendes et al. 2014; Zhang et al. 2010). Constructing the proper cocktail in order to obtain the maximum effectiveness can, nevertheless, be challenging. The best way is to select phages that recognize different types of bacterial receptors and that demonstrate synergetic antibacterial effects, to cover a broader range of species and prevent or delay resistances. Moreover, phage cocktails should have complementary host ranges in order to guarantee a broader host range (Goodridge 2010). The success of phage cocktails is reported to be five- to six-fold higher than single phage usage (Sulakvelidze et al. 2001).

Generally, phages for cocktails are produced and purified separately and mixed prior to use. One of the challenges of developing phage cocktails is interference between phages. While in the majority of the studies interference has not been reported, researchers have noted that different phages could interfere with each other by competition or because their infection probability was dependent on bacterial density (Hall et al. 2012). To improve the efficacy of phage cocktails, it has been suggested to use phages without antagonistic interactions between each other or to increase the concentration of each phage in the formulation.

A step-by step approach can be used to formulate a cocktail exhibiting a low probability to induce bacteria resistance. Phage resistant mutants are induced and used to isolate and select new phages in a series of steps. In this way, phages with lytic activity to the wild-type strain and phage-resistant variants can be selected. Gu et al. (2012) used this method to establish a cocktail consisting of three phages selected successively, with different but complementary host ranges. The three phages were more effective in reducing bacterial mutation frequency, compared to a single phage and exhibited an excellent *in vivo* activity.

Step 3: Quality and Safety of the Product. In a recent opinion paper, a group of researchers were asked by a non-profit organization to establish realistic quality and safety requirements for phage-based product development (Pirnay et al. 2015). Possible quality and safety risks and the tests to decrease the risks were:

(i) Phage products should be handled in a clean environment with high air quality in order to avoid contamination. The biosafety level required will be determined by the host bacteria used during production procedures. All the equipment and material should minimize hazards to staff and recipients. Good Manufacturing Practice (GMP) should be applied during the entire process. All equipment must be maintained, cleaned and disinfected regularly with proper records kept. Furthermore, all devices and material should be identified and frequently inspected. Precise measuring equipment should be well calibrated. All production procedures must specify all materials with a detailed description of the protocols being available and including specifications for all critical materials and reagents. For example, equipment and culture conditions need to be validated. According to the current version of the Note for Guidance on Minimizing the Risk of Transmitting Animal Spongiform Encephalopathy Agents via Human and Veterinary Medicinal Products (EMEA/410/01), animal component free culture media should be used. In the case of it being impossible to avoid the usage of animal-products, a certification should be obtained for all components of animal origin. To ensure the identity, the activity of the phages and purity of the organisms used in the production, phenotypic and genotypic markers should be designed for bacteria and phages. Moreover, both should be deposited in certified Biological Research Centers that work as repositories.

(ii) The host bacterial strain used for phage production should be the least pathogenic as possible. The strain should have a known origin and its history should be traceable. Non-lysogenic bacteria should be used and strains with mutator phenotypes should be avoided. Finally, the preservation and storage conditions must be optimized.

(iii) Some of the phage requirements that will be part of a product were described above, nonetheless, they will be reemphasized here as they are particularly important. Firstly, their origin should be well-known and documented. The family should be determined using morphological and molecular methods. The phage should be a non-transducing one, and its genome should be fully sequenced. Each phage must not codify in its genome, any gene that confers toxicity, virulence, antibiotic resistance or lysogeny. Each selected phage must have a broad host range and should (a) be stable in broth culture until 48h and (b) cause a low frequency of resistance from bacteria. The preservation and storage conditions must be optimized.

(iv) The phage working solution should be free of contamination by microorganisms and bacterial-origin endotoxins and DNA. The concentration of each phage in the final product should be well-known and the pH well defined.

(v) Prior to clinical use and depending on the usage, phages should be diluted and can be added to carriers which are substances that improve the delivery and the effectiveness of drugs, in this case specifically phages. The carrier choice must ensure phage activity during the application period. Before commercialization, relevant pharmacological data must be added to the product along with storage conditions to provide information for the user.

(vi) The shelf life of the phages, and the sterility and pH of product and working solutions under storage conditions should be determined periodically.

(vii) All phage-based therapy products must be surveyed and described in a centralized and publicly available pharmacological system to verify the safety of medicinal products.

Naturally, the detailed tests that need to be performed will depend on the product application, route of administration and legislation.

12.7 Testing Phages in Laboratory Animals *In Vivo*

Phages have been widely tested *in vitro* against pathogenic bacteria involved in many human diseases, particularly those relating to chronic bacterial infections that are resistant to other treatments. Although *in vitro* studies provide extremely useful data, testing phages *in vivo* in animal models provides a better understanding of the overall effects of phages. Many animal models are available for reliable studies (e.g. hamster, guinea pig, rabbit, and dog). Due to the rapid increase in phage *in vivo* publications in the last decades and the great diversity of animal models, we will focus in this chapter only in experiments carried out in mice published within the last 5 years (Table 12.1).

The mice studies described (Table 12.1) have been performed with single or cocktail phage preparations. Most of the *in vivo* tests were undertaken against bacteria involved in, or commonly isolated from: (i) cystic fibrosis (*B. cenocepacia*, *P. aeruginosa*); (ii) urinary tract infections (*C. turicensis*, *E. coli*); (iii) respiratory tract infections (*K. pneumoniae*, *P. aeruginosa*, *S. aureus*); and wound infections (*P. aeruginosa*, *S. aureus*, *V. vulnificus*). Phages have been applied by oral, topical, intraperitoneal, intramuscular, intragastric and intranasal administration and some

Table 12.1 *In vivo* mice animal models tested with phage(s)[a]

Phage host	Methodology and treatment outcome	References
Burkholderia cenocepacia	Respiratory infection model and treatment with one of five phages (aerosol or intraperitoneal administration)	Semler et al. (2014)
	Mice receiving aerosolized phage demonstrated significant decreases in bacterial loads. Mice receiving phage treatment by intraperitoneal injection did not demonstrate significantly reduced bacterial loads	
Cronobacter turicensis	Urinary tract infection model and treatment with phages (intraperitoneal administration)	Tothova et al. (2011)
	Phages reduced *Cronobacter* levels by 70 % in the kidney. The expression of pro-inflammatory cytokines tumor necrosis factor-alpha and monocyte chemoattractant protein-1 increased due to *Cronobacter* infection and were attenuated by phage therapy	
Escherichia coli	Mouse intestinal model and treatment with a cocktail of three virulent phages (added to drinking water)	Bull et al. (2012), Dufour et al. (2015), and Maura et al. (2012)
	Bacterial levels in ileal were greatly decreased and weakly decreased in fecal samples	
	Two classes of phages resulted in different levels of recovery in experimentally infected mice	
	Phages requiring the K1 capsule for infection rescued virtually all infected mice, phages not requiring the capsule rescued only ~30 %	
Klebsiella pneumoniae	Mouse lethal pneumonia model and intranasal administration of phage	Gu et al. (2012), Cao et al. (2015a), Hung et al. (2011), Kumar and Chhibber (2011), and Kumari et al. (2011)
	Phages were able to protect mice against lethal pneumonia. In a sublethal pneumonia model, phage-treated mice exhibited a lower level of bacteria in the lungs as compared to the untreated control. Lung lesion conditions were obviously improved by phage therapy	
	Murine bacteremia model and treatment with monophage and phage cocktail	
	Phage cocktail significantly reduced the mutation frequency of the bacteria and effectively rescued mice. The minimal protective dose (MPD) of the phage cocktail was significantly smaller than that of single monophage. A delayed administration of the phage cocktail was still effective in protection against the bacteria	
	Intragastric model of infection and treatment with phage (intraperitoneal and intragastric administration)	

(continued)

Table 12.1 (continued)

Phage host	Methodology and treatment outcome	References
	A single dose of phage (after 30 m of bacterial infection), administered by the two means, was able to protect mice from death in a dose-dependent manner. Phage administered after 24 h of inoculation of the bacteria was still protective. Intraperitoneal treatment with phage was more efficient than intragastric treatment as a result of the dissemination of bacteria into the circulation at 24 h postinfection	
	Murine burn wound model and daily topical treatment with silver nitrate, gentamicin, and phage in hydrogel	
	Efficacy assessed on the basis of percentage of survival of infected mice following treatment. The level of protection given by these two agents was lower than that given by the phage therapy	
Pseudomonas aeruginosa	Bacteremic mice model and treatment with phage (intraperitoneal administration)	Alemayehu et al. (2012), Cao et al. (2015b), Fukuda et al. (2012), Henry et al. (2013), and Shivshetty et al. (2014)
	A single injection of phage protected both diabetic (90 %) and nondiabetic (100 %) bacteremic mice. The protection rate was reduced in diabetic mice when phage was administered after 4 h and 6 h of lethal bacterial challenge. Nondiabetic bacteremic mice were rescued even when treatment was delayed up to 20 h after lethal bacterial challenge	
	Murine lung infection model and treatment with a 2-phage mix (intranasal administration)	
	The phage mix was effective in clearing the bacteria from murine lungs in 6 h	
	Mouse lung infection model and treatment with phage (a set of nine phages given administered intramuscularly were tested and compared)	
	For seven phages, a good correlation was found between *in vitro* and *in vivo* activity. While the remaining two phages were active in vitro, they were not sufficiently active in vivo under similar conditions to rescue infected animals	
	Mouse model of keratitis and phage application to the corneal surface	
	Single-dose administration of phage eye-drops significantly improved disease outcome, and preserved the structural integrity and transparency of the infected cornea. Phage treatment resulted in the suppression of neutrophil infiltration and greatly enhanced bacterial clearance in the infected cornea	

(continued)

Table 12.1 (continued)

Phage host	Methodology and treatment outcome	References
Staphylococcus aureus (including Methicillin-resistant *S. aureus* (MRSA)	Mouse model of lung-derived septicemia and treatment with phage (Intraperitoneal administration)	Chhibber et al. (2014), Chhibber et al. (2013), Gupta and Prasad (2011),; Paul et al. (2011), Sunagar et al. (2010), and Takemura-Uchiyama et al. (2014)
	Phage administration 6 h postinfection reduced the severity of infection and rescued infected mice	
	Mice neutropenic model and treatment with recombinant endolysin-deficient phage	
	Treatment with the endolysin-deficient phage rescued mice from the fatal infection	
	Lethal bacteremia in streptozotocin (STZ) induced-diabetic and non-diabetic mice and treatment with phage	
	A single administration of phage significantly protected diabetic and non-diabetic mice from lethal bacteremia (survival rate 90 % and 100 % for diabetic and non-diabetic bacteremic groups versus 0 % for saline-treated groups). Protection efficiency of phage was attained even when the treatment was delayed up to 4 h in both diabetic and non-diabetic bacteremic mice	
	Bacteremia in mice treated with a single subcutaneous injection of phage	
	Phage protect mice from bacteremia and subsequent death. A considerable decline of more than 6 logs (99.9 %) of bacteria in splene was noted at the 3 days of phage treatment	
	Nasal carriage model of colonization treated with phage (intranasal administration) and/or mupirocin	
	Phage given along with mupirocin showed an additive effect and the combination was able to eradicate the bacterial population from the nares of mice by day 5	
	Acute hindpaw infection model and treatment with phage and linezolid	
	A single administration of phage exhibited efficacy similar to linezolid in resolving the course of hindpaw infection in diabetic animals. Combination therapy was more effective in arresting the entire infection process and hastening the process of tissue healing	
Vibrio vulnificus	Mouse survival model and treatment with phage (intraperitoneal administration)	(Lee et al. 2014)
	Phage significantly protected mice from infection for up to 2 months	

[a]Due to the large number of reports available in laboratory animal models, only some experiments done in mice and which were published after 2010 were included.

of the studies summarized in Table 12.1 have compared two administration routes. This is the case, for instance, of the studies performed with *B. cenocepacia* (aerosol and intraperitoneal) and *K. pneumoniae* (intraperitoneal and intragastric). In the *B. cenocepacia* studies only the aerosol route showed significant decreases in bacterial loads (Semler et al. 2014). While in *K. pneumoniae* both treatments protected the infected mice, the intraperitoneal administration was more efficient than the intragastric administration (Hung et al. 2011).

Phage therapy has also been compared *in vivo* against different antimicrobial agents. For instance, topical administration of three agents (silver nitrate, gentamicin, and phage in hydrogel) demonstrated that all agents resulted in a level of protection of infected mice. However, the protection granted by phages was highest (Kumari et al. 2011). Also, phage and linezolid have been administered individually and combined to target MRSA. While single administration of both agents exhibited similar efficacies in resolving the course of hindpaw infection in the diabetic animals, combination therapy was more effective. The combination of both agents was not only far more effective in arresting the entire infection process (bacterial load, lesion score, foot myeloperoxidase activity and histopathological analysis) but also accelerated the process of tissue healing.

The in vivo studies against bacterial pathogens, including antibiotic-resistant bacteria, provide evidence of the safe nature of phages. Nevertheless, the effectiveness of the phage treatment can be affected by the (a) administration route, (b) concentration of bacteria on the infection site, (c) phage preparation (monophage or cocktail) and dose applied, (d) the mode and time of treatment, and (e) environmental conditions among others (Ly-Chatain 2014).

12.8 Testing Phage Products – Human Clinical Trials

As viruses of prokaryotes, phages are innocuous to eukaryotic cells, making phage-associated side effects an unlikely event. This has been proven by their extensive clinical use in Eastern Europe and the former Soviet Union (Gorski et al. 2009; Sulakvelidze and Morris 2001; Miedzybrodzki et al. 2012; Pirnay et al. 2011). In fact, much of the knowledge about the practice of phage therapy comes from the Republic of Georgia and Poland and particularly from two institutes – Eliava Institute of Bacteriophages, Microbiology and Virology, Georgia and the Ludwik Hirszfeld Institute of Immunology and Experimental Therapy (HIIEP), Poland. In Georgia, phage therapy remains as a component of standard medical practice being routinely applied in hospitals and clinics for prophylactic and treatment purposes, with the major focus being on cocktail formulations. Nonetheless, many of the studies have not been properly documented, are not accessible in English and have not been properly conducted under controlled conditions (Kutter et al. 2010). On the other hand, numerous studies have been completed in patients at the HIIEP and the results are available in various respected, international publications, and show the great potential of phage therapy in patients (reviewed thoroughly in (Gorski et al. 2009; Miedzybrodzki et al. 2012)).

Despite the several reports of phages used in humans, especially from Polish studies, data from clinical trials conducted under carefully controlled conditions and following western health agency jurisdictions, is limited to some products which have started the initial phases of clinical trials. According to FDA, like any new drug, a phage product developed in a laboratory will take many years before approval for use in medicine by this regulatory agency. The path of a phage product is the same as antibiotics undertake and involves stages, starting with the submission of the Investigational New Drug (IND) application, which already has to include the preclinical testing results carried out in laboratory animals. After, the product goes through three phases of testing in humans, it is followed by formal steps for new drug approval until finally a decision is issued whether the application can be filed to be reviewed. Each of the clinical trial phases involves humans. Only healthy volunteers, typically 20 to 80, are accepted for phase I testing while phases II and III involve ill patients and the main difference is in the number of patients of phase II (typically < 300) and how large-scale studies are necessary to complete phase III which can range from several hundreds to 3,000 people.

This whole process is long and expensive. Moreover, after the completion of all phases, the regulatory agency can still refuse to file the application due to incomplete and even missing data. Hence, only a few studies have reached some level of testing in humans in different countries and under the standards of National Health Authorities, Ethics Committees and other regulatory agencies (Table 12.2).

None of the phage clinical products has reached the approval phase in Table 12.2 under the standards of the European Medicines Agency (EMA) or US FDA jurisdictions. Nevertheless, some have gone through preclinical and phase I and I/II stages. The results from the few clinical trials are worthy of being described. For instance, the venous leg ulcer phase I clinical trial carried out by the Southwest Regional Wound Care Center with (a) two phages active against *S. aureus*, (b) five phages of *P. aeruginosa* and (c) one phage against *E. coli* showed no significant differences in the frequency of adverse events, rate of healing, or frequency of healing between test and control groups, evidencing that the phage product is safe for use in patients (Rhoads et al. 2009). Safety was also assessed orally by giving an *E. coli* phage dose in drinking water. Once again, none of the 15 healthy volunteers showed adverse events related to phage application. The serum transamine levels remained in the normal range and antibodies specific for T4 phage of T4-specific antibodies were not reported (Bruttin and Brussow 2005). Biocontrol Ltd (now Ampliphi Biosciences Corporation) published their results from the phase I/II clinical trial in chronic otitis patients using their phage preparation Biophage-PA (*P. aeruginosa*) and reported statistically significant combined data (erythema/inflammation, ulceration/granulation/polyps, discharge quantity, discharge type and odour using a Visual Analogue Scale (VAS)) and significantly lower *P. aeruginosa* counts for the phage treated group compared to the placebo groups (Wright et al. 2009).

Safety results (phase I) coupled with efficacy testing with a small number of patients (phase II) and in the future with a significantly larger number of patients (phase III) will surely encourage (a) phage therapy in Western medicine and (b) research applied to the control of infections caused by other bacteria.

12.9 Challenges of Phage Therapy

Despite the phage clinical trials in humans (Table 12.2), the effective use of phages face various other challenges:

Table 12.2 Human phage clinical trials

Clinical trial*	Condition/ disease	Country, Institution/ company & Health Authority	Phage(s)	Host(s)
Experimental phage therapy of drug-resistant bacterial infections, including MRSA infections	Bacterial infections that prove incurable by antibiotics	**Poland** Institute of Immunology and Experimental Therapy of the Polish Academy of Sciences Polish Ethics Committee	Patient specific phage lysate or purified phage formulations	*Staphylococcus, Enterococcus, Pseudomonas, Escherichia, Klebsiella, Proteus, Citrobacter, Acinetobacter, Serratia, Morganella, Shigella, Salmonella, Enterobacter, Stenotrophomonas*, or *Burkholderia* strains
Evaluation of phage therapy for the treatment of *E. coli* and *P. aeruginosa* wound infections in burned patients (PHAGOBURN)	Treatment of wound infections	**France, Belgium, Switzerland** Pherecydes Pharma Agence Nationale de Securité du Médicament et des produits de santé (France); Agence Fédérale des Médicaments et des Produits de Santé (Belgium); Suissmedic, Institut Suisse des produits thérapeutiques (Switzerland)	phage cocktails	*E. coli, P. aeruginosa*
Phage effects on *P. aeruginosa* (MUCOPHAGES)	Cystic fibrosis	**France** University Hospital, Montpellier Committee for the Protection of Personnes (France)	Ten phages cocktail	*P. aeruginosa*

(continued)

Table 12.2 (continued)

Clinical trial*	Condition/ disease	Country, Institution/ company & Health Authority	Phage(s)	Host(s)
Antibacterial treatment against diarrhea in oral rehydration solution	Childhood diarrhea	**Bangladesh** Nestlé Bangladesh Medical Research Council; Ethical Review Committee	T4 phage cocktail	*E. coli*
A prospective, randomized, double-blind controlled study of WPP-201 for the safety and efficacy of treatment of venous leg ulcers	Venous leg ulcer	**USA** Southwest Regional Wound Care Center US FDA	Eight phage cocktail	*P. aeruginosa, S. aureus, E. coli*
A controlled clinical trial of a therapeutic phage preparation in chronic otitis due to antibiotic-resistant *Pseudomonas aeruginosa*	Chronic otitis	**UK** Biocontrol Ltd (UK) UK Medicines and Healthcare products Regulatory Agency; Central Office for Research Ethics Committees	Six phages cocktail	Antibiotic-resistant *P. aeruginosa*
Human volunteers receiving *E. coli* phage T4 orally	Safety test	**Belgium** Nestlé Research Center Local ethical committee	Phage T4	*E. coli*

*Institutes/companies that have received FDA clearance but have not yet recruited volunteers for the human clinical trial are not included.

(i) **Large diversity of phages**. Phages with several characteristics exist for a given target species. Although relatively cheap and easy to isolate and propagate (this is more challenging for phages targeting fastidious hosts where their availability and likelihood in the environment are relatively low), the exhaustive characterization needed to select appropriate phage candidates for therapy, may cause huge costs. Furthermore, phage activity should not be influenced by biological and physico-chemical (pH, temperature, ionic strength) factors. For example, although phages are in general regarded as active at neutral pH, if oral administration is to be considered, phages would have to withstand the stomach acid pH values, where most phages cannot survive (Watanabe et al. 2007). All these tests however, may not guarantee in vivo efficacy, and each phage requires individual tests as the in vivo data for one cannot be transferred to other phages. At a genomic level, although considered inherently non-toxic, due to their nucleic acid and protein nature, a careful design of the downstream processes is required to avoid the occurrence of bacterial resistant and toxins in the phage product (Merabishvili et al. 2009).

(ii) **Antibacterial efficiency (specificity and resistance)**. The high phage specificity represents the strength and the weakness of phage therapy. On one hand, it is possible to make targeted therapy, leaving untouched the remaining microbiota, a property that favors phages over traditional antimicrobials that can cause microbiota collateral damage. On the other hand, there is a large genetic variability of pathogenic bacteria, making almost impossible the coverage of all existence bacterial infections with phage therapy. Even the broad-spectrum phage is more selective than the narrow-spectrum antibiotic. This can hinder the commercial potential of phage products, if ineffectiveness is demonstrated against particular bacterial strains that predominate hospitals or clinical environments.

Bacterial resistances are another important threat for the efficacy of phage therapy in humans (Dennehy 2012). In the case of phages, the rate of resistances is approximately 10-fold lower than resistance to antibiotics (Carlton 1999). Possible ways to circumvent this problem, would be to simply switch to other highly virulent phages (e.g. fast adsorption and high burst size), in order to significantly reduce bacterial populations to levels which the immune system could control, or, as mentioned, with the use of a cocktail of phages with broad cross-strain lytic activity. Importantly, contrary to antibiotics, as bacteria develop resistant mechanisms, phages will also continuously adapt to altered host systems. This has been observed in all bacterial defense mechanisms previously reported towards phages, at a level of the host receptors and anti-phage DNA-restriction enzymes, including clustered regularly interspaced short palindromic repeats (CRISPR) systems (Jiang et al. 2013; Seed et al. 2013). It is reassuring that there will probably be effective phages to virtually all pathogenic bacteria, that could be rapidly isolated (usually, several months), comparing to the slow selection and screening process of novel antibiotics (usually, several years (if at all)) (Sulakvelidze et al. 2001).

(iii) **Pharmacology**. Studies demonstrating the ability of phages to diffuse in the body to reach the infection (pharmacokinetics) and their interaction with the human body (pharmacodynamics) are few. The specific pharmacokinetics-pharmacodynamics properties of individual phages must be investigated to assess their effectiveness and safety. Understanding the pharmacokinetics can be a complex issue that is dependent on the size, concentration and virulence of phage administrated, which are unique to each virus. Compared to antibiotics, phages are much larger particles and this limits their diffusion. This is of particular importance against intracellular bacterial pathogens (e.g. *M. leprae*, *Chlamydia*) that reproduce and thrive in human cells, shielded from many antimicrobials. Their larger size, also means that concentrated doses cannot be used, as they become too viscous ($\approx$1013 phages/ml). Also, owing to the self-replicating and self-limiting nature of phages, a low or single dosages will multiply if there remains a host threshold present, a detailed examination of the pharmacology indicators are needed during the course of treatment. The level of complexity increases when a cocktail of phages with different virulences (adsorption rates, latent period and burst sizes) is administrated.

Concerning pharmacodynamics, some studies have shown that phages administrated intravenously are removed by the reticuloendothelial system and inactivated by neutralizing antibodies similar to most pharmaceuticals that interact with the body's immune system (Lusiak-Szelachowska et al. 2014; Westwater et al. 2003). However, the risks are minimal because phages' activity kinetics are faster compared to the production of neutralizing antibodies by the immune system (Lusiak-Szelachowska et al. 2014). In case of when phage-neutralizing antibodies are still present at the time of a second phage post-treatment, a repetition of phage administration with similar or higher concentration, or the use of different phages might be enough to overcome these antibodies. In addition, the delivery of less-immunogenic phages with suitable nanocarriers (e.g. liposomes), or by engineering them to have non-immunogenic and biocompatible peptides on their surface (e.g. polyethylene glycol molecules) should be considered. Another important issue is the systemic administration of phages that might release prohibitory endotoxins (i.e. lipopolysaccharides) levels after bursting bacterial cells. This may limit phage treatment of Gram-negative systemic infections, as endotoxins are very allergenic substances that increase the production of cytokines (Diomede et al. 2001). The same challenges exist for some antibiotics (Holzheimer 2001) and more detailed studies will be needed for each specific phage and bacterial host.

(iv) **Absence of regulatory framework**. There is a current miss conceptualisation and classification of phage therapy. It cannot be considered a medicinal product, a biological medical product, nor an advance therapy medicinal product, where specific frameworks exist in Europe (Directive 2001/83/EC) (Westwater et al. 2003). This makes phage products difficult to commercialize, unless a dedicated framework is created in the future, either for a uniform distribution of phage products or for more specific treatments.

(v) **Consumer's acceptance**. Commercialization of phages will finally be dependent on consumer acceptance. Phages are viruses and this may raise public concern. A large part of medical society in the West is not aware of the potential of phage therapy. Therefore, a tremendous effort is needed to educate the general public and medics about the nature of phages, and their safety and effectiveness to manage bacterial infection. Equally important, will be the costs that need to be attractive from the consumer perspective, in order to bring phages from the bench to the patient.

(vi) **Profitability**. An additional factor will be if pharmaceutical companies consider there is enough profit for commercialisation and whether intellectual property rights can be gained.

These hurdles and knowledge gaps need to be solved before commercialization of phage-based products is gained.

12.10 Modification of Phages – The Future of Phage Therapy?

A possible way to overcome some of the above-mentioned constraints of phage therapy could be with the use of modified phages. Recent developments have shown that genetic and chemical modification of phages are promising strategies to enhance the phage antibacterial impact and decrease its cytotoxicity/immunogenicity properties (Moradpour and Ghasemian 2011). Genetic manipulation of phage genomes is the latest trend of phage research. With current powerful tools to manipulate viral DNA (bacteriophage recombineering of electroporated DNA (BRED), and bacteria and yeast-artificial chromosomes (BAC/YAC)), a new era of modified phages with better performance is being created (Nobrega et al. 2015).

To resolve insufficient lytic host range, chimerical phages harboring foreign proteins have been created. For instance, the recognition tail fiber proteins of the narrow spectrum *E. coli* phage T2 were replaced with those of the phage IP008, creating a mutant phage with the high bactericidal properties of T2 and the wide-spectrum of activity of IP008 (Mahichi et al. 2009). Extended host range of the *E. coli* T7 was accomplished by overexpressing endosialidase responsible for phage attachment and degradation of capsular polysaccharides (Scholl et al. 2005).

Phages have been genetically altered to express different active substances during their lytic cycle to enhance antibacterial activity. As most phages are inactive against biofilms, *i.e.*, microbial communities attached to a variety of biotic/abiotic surfaces characterized by highly dense and protective extracellular polymeric substances (also known as the matrix), the *E. coli* phage T7 was engineered by Lu and Collins (2007) to express a matrix-degrading enzyme, called dispersin B. This allowed the mutant phage to reduce nearly 100 % of the biofilm biomass, greatly enhancing the anti-biofilm activity by 100 000-fold, comparing to the wild-type phage. This result is of particular importance, as biofilms represent the most prevalent bacterial life-style, and are greatly associated with several bacterial-associated infections. The same researchers have created phages to specifically target gene networks serving as adjuvants for antibiotic therapy (Lu and Collins 2009). Hence, by using synthetic biology, proteins were expressed during the phage lytic cycle to suppress the SOS network of *E. coli*, being able to kill antibiotic-resistant bacteria, biofilm cells, and also demonstrating a strong synergistic effect with antibiotics (e.g. aminoglycosides and β-lactams).

Finally, to deal with phage immunological detection and cytotoxic issues, several other studies were conducted. To mask the phage immunogenic properties and increase plasma half-life, several inert, nonimmunogenic and biocompatible peptides (e.g. polyester and poly-D, L-amino acids), were chemically attached to the phage surface (Kim et al. 2008). An alternative method made use of an artificial selection of phages, through a serial-passage technique in mice, in order to select mutants that remained 13,000–16,000-fold longer in the circulatory system (Merril et al. 1996). The single mutation of the phage major capsid protein responsible for this "long-circulating" phenotype, was later introduced into other phages with similar results (Vitiello et al. 2005). Several other studies have also attenuated the inflamatory response by either creating lysis-deficient phages or non-replicative lethal phages, in order to avoid bacterial burst and subsequent release of large amounts of endotoxins (Paul et al. 2011; Hagens et al. 2004; Matsuda et al. 2005).

Taken together, the data reported show that is possible to transform naturally occurring phages into viruses with better therapeutic traits. However, by solving previous limitations of phage therapy with modified phages, other issues may arise, from the early preparation of these more-complex phages, to their safe use inherent to all genetically/chemically modified organisms.

12.11 Conclusions

The increasing emergence of multiresistant pathogenic bacteria coupled with a better understanding of phage biology, has led to a widespread renewed research interest in the possibilities of phage therapy. It is interesting, and even ironical, that phage therapy was put aside for so many years due to antibiotics and rediscovered due to the emergence of antibiotic resistant bacteria (Kutter et al. 2005; Barrow and Soothill 1997). Classically, phage therapy uses a bacteriophage, or cocktail of several phages, to specifically lyse target pathogenic bacteria isolated from the patients. This "tailor-made" mode of phage therapy should ideally be performed in hospitals or clinics, with fast detection and phage screening methods to guide phage therapy specifically for the target bacterium. There is, however, no regulatory framework where this therapeutic modality fits. Conversely, a standard phage product composed of a cocktail of phages for multiple usages could fall into one of the three categories established by the current legal framework for the use of medicines in Europe. This implies the performance of costly and time consuming clinical trials and the submission of a full product dossier compliant with directive 2001/83/EC. It is a fact that under this scenario the development costs will be similar to any other drug, but the profits are not as high, and pharmaceutical companies will not invest.

In summary, the lack of a regulatory framework for phage therapy, the overlong and expensive process to get the approval of a phage product by regulatory agencies and even uncertainty regarding phage efficacy and safety, have hindered the broadening of this therapeutic option. As such, while people are dying from infectious diseases and antibiotic resistance is spreading and increasing enormously, it is urgent that the scientific community, regulatory agencies and clinical practitioners join efforts to determine the most effective way to implement phage therapy.

References

Abdulamir AS, Jassim SA, Abu Bakar F (2014) Novel approach of using a cocktail of designed bacteriophages against gut pathogenic *E. coli* for bacterial load biocontrol. Ann Clin Microbiol Antimicrob 13:39

Abedon ST, Thomas-Abedon C (2010) Phage therapy pharmacology. Curr Pharm Biotechnol 11:28–47

Ackermann H-W (2006) Classification of bacteriophages. In: Calendar R (ed) The bacteriophages. Oxford University Press, New York, pp. 8–16

Ackermann HW (2007) 5500 Phages examined in the electron microscope. Arch Virol 152:227–243

Alemayehu D, Casey PG, McAuliffe O, Guinane CM, Martin JG et al (2012) Bacteriophages phiMR299-2 and phiNH-4 can eliminate *Pseudomonas aeruginosa* in the murine lung and on cystic fibrosis lung airway cells. MBio 3:e00029–e00012

Bachrach G, Leizerovici-Zigmond M, Zlotkin A, Naor R, Steinberg D (2003) Bacteriophage isolation from human saliva. Lett Appl Microbiol 36:50–53

Barrow PA, Soothill JS (1997) Bacteriophage therapy and prophylaxis: rediscovery and renewed assessment of potential. Trends Microbiol 5:268–271

Brussow H, Bruttin A, Desiere F, Lucchini S, Foley S (1998) Molecular ecology and evolution of Streptococcus thermophilus bacteriophages – a review. Virus Genes 16:95–109

Bruttin A, Brussow H (2005) Human volunteers receiving *Escherichia coli* phage T4 orally: a safety test of phage therapy. Antimicrob Agents Chemother 49:2874–2878

Bull JJ, Otto G, Molineux IJ (2012) *In vivo* growth rates are poorly correlated with phage therapy success in a mouse infection model. Antimicrob Agents Chemother 56:949–954

Cao F, Wang X, Wang L, Li Z, Che J et al (2015a) Evaluation of the efficacy of a bacteriophage in the treatment of pneumonia induced by multidrug resistance *Klebsiella pneumoniae* in mice. Biomed Res Int 2015:752930

Cao Z, Zhang J, Niu YD, Cui N, Ma Y, et al. (2015b) Isolation and characterization of a "phiKMV-like" bacteriophage and its therapeutic effect on mink hemorrhagic pneumonia. PLoS One 10: e0116571.

Carlton RM (1999) Phage therapy: past history and future prospects. Arch Immunol Ther Exp (Warsz) 47:267–274

Chhibber S, Kaur T, Sandeep K (2013) Co-therapy using lytic bacteriophage and linezolid: effective treatment in eliminating methicillin resistant *Staphylococcus aureus* (MRSA) from diabetic foot infections. PLoS One 8:e56022

Chhibber S, Gupta P, Kaur S (2014) Bacteriophage as effective decolonising agent for elimination of MRSA from anterior nares of BALB/c mice. BMC Microbiol 14:212

d'Herelle F (1917) Sur un microbe invisible antagoniste des bacilles dysentériques. CR Acad Sci Paris 165:373–375

d'Herelle F (1926) The bacteriophage and its behaviour. In: d'Herelle F (ed). The Williams & Wilkins Company, Baltimore

Dabrowska K, Switala-Jelen K, Opolski A, Weber-Dabrowska B, Gorski A (2005) Bacteriophage penetration in vertebrates. J Appl Microbiol 98:7–13

Dennehy JJ (2012) What can phages tell us about host-pathogen coevolution? Int J Evol Biol 2012:396165

Diomede L, Albani D, Bianchi M, Salmona M (2001) Endotoxin regulates the maturation of sterol regulatory element binding protein-1 through the induction of cytokines. Eur Cytokine Netw 12:625–630

Dufour N, Debarbieux L, Fromentin M, Ricard JD (2015) Treatment of highly virulent extraintestinal pathogenic *Escherichia coli* pneumonia with bacteriophages. Crit Care Med 43:e190–e198

Fu W, Forster T, Mayer O, Curtin JJ, Lehman SM et al (2010) Bacteriophage cocktail for the prevention of biofilm formation by *Pseudomonas aeruginosa* on catheters in an *in vitro* model system. Antimicrob Agents Chemother 54:397–404

Fukuda K, Ishida W, Uchiyama J, Rashel M, Kato S et al (2012) *Pseudomonas aeruginosa* keratitis in mice: effects of topical bacteriophage KPP12 administration. PLoS One 7:e47742

Gill JJ, Hyman P (2010) Phage choice, isolation, and preparation for phage therapy. Curr Pharm Biotechnol 11:2–14

Goodridge LD (2010) Designing phage therapeutics. Curr Pharm Biotechnol 11:15–27
Gorski A, Miedzybrodzki R, Borysowski J, Weber-Dabrowska B, Lobocka M et al (2009) Bacteriophage therapy for the treatment of infections. Curr Opin Investig Drugs 10:766–774
Gu J, Liu X, Li Y, Han W, Lei L et al (2012) A method for generation phage cocktail with great therapeutic potential. PLoS One 7:e31698
Gupta R, Prasad Y (2011) Efficacy of polyvalent bacteriophage P-27/HP to control multidrug resistant *Staphylococcus aureus* associated with human infections. Curr Microbiol 62:255–260
Guttman B, Raya R, Kutter E (2005) Basic phage biology. In: Kutter E, Sulakvelidze A (eds) Bacteriophages – biology and applications. CRC Press, Boca Raton
Hagens S, Habel A, von Ahsen U, von Gabain A, Blasi U (2004) Therapy of experimental pseudomonas infections with a nonreplicating genetically modified phage. Antimicrob Agents Chemother 48:3817–3822
Hall AR, De Vos D, Friman VP, Pirnay JP, Buckling A (2012) Effects of sequential and simultaneous applications of bacteriophages on populations of *Pseudomonas aeruginosa in vitro* and in wax moth larvae. Appl Environ Microbiol 78:5646–5652
Hankin EH (1896) L'action bactericide des eaux de la Jumna et du Gange sur le vibrion du cholera. Ann Inst Pasteur 10:511
Häusler T (2006) Viruses vs. superbugs – a solution to the antibiotic crisis? Macmillan, New York
Henry M, Lavigne R, Debarbieux L (2013) Predicting *in vivo* efficacy of therapeutic bacteriophages used to treat pulmonary infections. Antimicrob Agents Chemother 57:5961–5968
Hitch G, Pratten J, Taylor PW (2004) Isolation of bacteriophages from the oral cavity. Lett Appl Microbiol 39:215–219
Holzheimer RG (2001) Antibiotic induced endotoxin release and clinical sepsis: a review. J Chemother 13 Spec No 1: 159–172
Hung CH, Kuo CF, Wang CH, Wu CM, Tsao N (2011) Experimental phage therapy in treating *Klebsiella pneumoniae*-mediated liver abscesses and bacteremia in mice. Antimicrob Agents Chemother 55:1358–1365
Jiang W, Maniv I, Arain F, Wang Y, Levin BR et al (2013) Dealing with the evolutionary downside of CRISPR immunity: bacteria and beneficial plasmids. PLoS Genet 9:e1003844
Johnson AP (2015) Surveillance of antibiotic resistance. Philos Trans Royal Soc B-Biol Sci 370
Kakikawa M, Yokoi K, Kimoto H, Nakano M, Kawasaki KI et al (2002) Molecular analysis of the lysis protein Lys encoded by *Lactobacillus plantarum* phage theta g1e. Gene 299:227–234
Keen EC (2012) Phage therapy: concept to cure. Front Microbiol 3:238
Kim KP, Cha JD, Jang EH, Klumpp J, Hagens S et al (2008) PEGylation of bacteriophages increases blood circulation time and reduces T-helper type 1 immune response. Microb Biotechnol 1:247–257
Kumar V, Chhibber S (2011) Acute lung inflammation in *Klebsiella pneumoniae* B5055-induced pneumonia and sepsis in BALB/c mice: a comparative study. Inflammation 34:452–462
Kumari S, Harjai K, Chhibber S (2011) Bacteriophage versus antimicrobial agents for the treatment of murine burn wound infection caused by *Klebsiella pneumoniae* B5055. J Med Microbiol 60:205–210
Kutter E, Raya R, Carlson K (2005) Molecular mechanisms of phage infection. In: Kutter E, Sulakvelidze A, (eds). CRC Press
Kutter E, De Vos D, Gvasalia G, Alavidze Z, Gogokhia L et al (2010) Phage therapy in clinical practice: treatment of human infections. Curr Pharm Biotechnol 11:69–86
Kutter EM, Kuhl SJ, Abedon ST (2015) Re-establishing a place for phage therapy in western medicine. Future Microbiol 10:685–688
Lee HS, Choi S, Shin H, Lee JH, Choi SH (2014) *Vibrio vulnificus* bacteriophage SSP002 as a possible biocontrol agent. Appl Environ Microbiol 80:515–524
Levin BR, Bull JJ (2004) Population and evolutionary dynamics of phage therapy. Nat Rev Microbiol 2:166–173
Loc-Carrillo C, Abedon ST (2011) Pros and cons of phage therapy. Bacteriophage 1:111–114

Lu TK, Collins JJ (2007) Dispersing biofilms with engineered enzymatic bacteriophage. Proc Natl Acad Sci U S A 104:11197–11202

Lu TK, Collins JJ (2009) Engineered bacteriophage targeting gene networks as adjuvants for antibiotic therapy. Proc Natl Acad Sci U S A 106:4629–4634

Luria SE, Delbruck M, Anderson TF (1943) Electron microscope studies of bacterial viruses. J Bacteriol 46:57–77

Lusiak-Szelachowska M, Zaczek M, Weber-Dabrowska B, Miedzybrodzki R, Klak M et al (2014) Phage neutralization by sera of patients receiving phage therapy. Viral Immunol 27:295–304

Ly-Chatain MH (2014) The factors affecting effectiveness of treatment in phages therapy. Front Microbiol 5:51

Mahichi F, Synnott AJ, Yamamichi K, Osada T, Tanji Y (2009) Site-specific recombination of T2 phage using IP008 long tail fiber genes provides a targeted method for expanding host range while retaining lytic activity. FEMS Microbiol Lett 295:211–217

Matsuda T, Freeman TA, Hilbert DW, Duff M, Fuortes M et al (2005) Lysis-deficient bacteriophage therapy decreases endotoxin and inflammatory mediator release and improves survival in a murine peritonitis model. Surgery 137:639–646

Matsuzaki S, Rashel M, Uchiyama J, Sakurai S, Ujihara T et al (2005) Bacteriophage therapy: a revitalized therapy against bacterial infectious diseases. J Infect Chemother 11:211–219

Maura D, Galtier M, Le Bouguenec C, Debarbieux L (2012) Virulent bacteriophages can target O104:H4 enteroaggregative *Escherichia coli* in the mouse intestine. Antimicrob Agents Chemother 56:6235–6242

Melnyk AH, Wong A, Kassen R (2015) The fitness costs of antibiotic resistance mutations. Evol Appl 8:273–283

Mendes JJ, Leandro C, Mottola C, Barbosa R, Silva FA et al (2014) *In vitro* design of a novel lytic bacteriophage cocktail with therapeutic potential against organisms causing diabetic foot infections. J Med Microbiol 63:1055–1065

Merabishvili M, Pirnay JP, Verbeken G, Chanishvili N, Tediashvili M et al (2009) Quality-controlled small-scale production of a well-defined bacteriophage cocktail for use in human clinical trials. PLoS One 4:e4944

Merril CR, Biswas B, Carlton R, Jensen NC, Creed GJ et al (1996) Long-circulating bacteriophage as antibacterial agents. Proc Natl Acad Sci U S A 93:3188–3192

Miedzybrodzki R, Borysowski J, Weber-Dabrowska B, Fortuna W, Letkiewicz S, et al. (2012) Clinical aspects of phage therapy. Adv Virus Res, Vol 83: Bacteriophages B 83: 73–121

Minot S, Sinha R, Chen J, Li HZ, Keilbaugh SA et al (2011) The human gut virome: inter-individual variation and dynamic response to diet. Genome Res 21:1616–1625

Modi SR, Lee HH, Spina CS, Collins JJ (2013) Antibiotic treatment expands the resistance reservoir and ecological network of the phage metagenome. Nature 499:219

Moradpour Z, Ghasemian A (2011) Modified phages: novel antimicrobial agents to combat infectious diseases. Biotechnol Adv 29:732–738

Nelson D (2004) Phage taxonomy: we agree to disagree. J Bacteriol 186:7029–7031

Nobrega FL, Costa AR, Kluskens LD, Azeredo J (2015) Revisiting phage therapy: new applications for old resources. Trends Microbiol 23:185–191

Paul VD, Sundarrajan S, Rajagopalan SS, Hariharan S, Kempashanaiah N et al (2011) Lysis-deficient phages as novel therapeutic agents for controlling bacterial infection. BMC Microbiol 11:195

Payne RJ, Jansen VA (2001) Understanding bacteriophage therapy as a density-dependent kinetic process. J Theor Biol 208:37–48

Pirnay JP, De Vos D, Verbeken G, Merabishvili M, Chanishvili N et al (2011) The phage therapy paradigm: pret-a-porter or sur-mesure? Pharm Res 28:934–937

Pirnay JP, Blasdel BG, Bretaudeau L, Buckling A, Chanishvili N et al (2015) Quality and safety requirements for sustainable phage therapy products. Pharm Res 32:2173–2179

Rhoads DD, Wolcott RD, Kuskowski MA, Wolcott BM, Ward LS et al (2009) Bacteriophage therapy of venous leg ulcers in humans: results of a phase I safety trial. J Wound Care 18(237-238):240–233

Sao-Jose C, Parreira R, Vieira G, Santos MA (2000) The N-terminal region of the Oenococcus oeni bacteriophage fOg44 lysin behaves as a bona fide signal peptide in Escherichia coli and as a cis-inhibitory element, preventing lytic activity on oenococcal cells. J Bacteriol 182:5823–5831

Scholl D, Adhya S, Merril C (2005) *Escherichia coli* K1's capsule is a barrier to bacteriophage T7. Appl Environ Microbiol 71:4872–4874

Seed KD, Lazinski DW, Calderwood SB, Camilli A (2013) A bacteriophage encodes its own CRISPR/Cas adaptive response to evade host innate immunity. Nature 494:489–491

Semler DD, Goudie AD, Finlay WH, Dennis JJ (2014) Aerosol phage therapy efficacy in Burkholderia cepacia complex respiratory infections. Antimicrob Agents Chemother 58:4005–4013

Shivshetty N, Hosamani R, Ahmed L, Oli AK, Sannauallah S, et al. (2014) Experimental protection of diabetic mice against Lethal P. aeruginosa infection by bacteriophage. Biomed Res Int 2014: 793242

Skurnik M, Strauch E (2006) Phage therapy: facts and fiction. Int J Med Microbiol 296:5–14

Skurnik M, Pajunen M, Kiljunen S (2007) Biotechnological challenges of phage therapy. Biotechnol Lett 29:995–1003

Sozzi T, Maret R, Cerise L (1973) Isolation and some characteristics of 2 *Micrococcus* phages from Italian salami, type-Varzi. Arch Mikrobiol 92:313–320

Sulakvelidze A, Morris JG Jr (2001) Bacteriophages as therapeutic agents. Ann Med 33:507–509

Sulakvelidze A, Alavidze Z, Morris JG Jr (2001) Bacteriophage therapy. Antimicrob Agents Chemother 45:649–659

Summers WC (2006) Phage and the early development of molecular biology. In: Calendar R (ed) The bacteriophages. Oxford University Press, New York

Sunagar R, Patil SA, Chandrakanth RK (2010) Bacteriophage therapy for *Staphylococcus aureus* bacteremia in streptozotocin-induced diabetic mice. Res Microbiol 161:854–860

Takemura-Uchiyama I, Uchiyama J, Osanai M, Morimoto N, Asagiri T et al (2014) Experimental phage therapy against lethal lung-derived septicemia caused by *Staphylococcus aureus* in mice. Microbes Infect 16:512–517

Tothova L, Celec P, Babickova J, Gajdosova J, Al-Alami H et al (2011) Phage therapy of *Cronobacter*-induced urinary tract infection in mice. Med Sci Monit 17:BR173–BR178

van Helvoort T (1992) Bacteriological and physiological research styles in the early controversy on the nature of the bacteriophage phenomenon. Med Hist 36:243–270

Vitiello CL, Merril CR, Adhya S (2005) An amino acid substitution in a capsid protein enhances phage survival in mouse circulatory system more than a 1000-fold. Virus Res 114:101–103

Volkova VV, Lu Z, Besser T, Grohn YT (2014) Modeling the infection dynamics of bacteriophages in enteric *Escherichia coli*: estimating the contribution of transduction to antimicrobial gene spread. Appl Environ Microbiol 80:4350–4362

Waller AS, Yamada T, Kristensen DM, Kultima JR, Sunagawa S et al (2014) Classification and quantification of bacteriophage taxa in human gut metagenomes. ISME J 8:1391–1402

Watanabe R, Matsumoto T, Sano G, Ishii Y, Tateda K et al (2007) Efficacy of bacteriophage therapy against gut-derived sepsis caused by *Pseudomonas aeruginosa* in mice. Antimicrob Agents Chemother 51:446–452

Weinbauer MG (2004) Ecology of prokaryotic viruses. FEMS Microbiol Rev 28:127–181

Westwater C, Kasman LM, Schofield DA, Werner PA, Dolan JW et al (2003) Use of genetically engineered phage to deliver antimicrobial agents to bacteria: an alternative therapy for treatment of bacterial infections. Antimicrob Agents Chemother 47:1301–1307

Wright A, Hawkins CH, Anggard EE, Harper DR (2009) A controlled clinical trial of a therapeutic bacteriophage preparation in chronic otitis due to antibiotic-resistant Pseudomonas aeruginosa; a preliminary report of efficacy. Clin Otolaryngol 34:349–357

Zhang J, Kraft BL, Pan Y, Wall SK, Saez AC et al (2010) Development of an anti-*Salmonella* phage cocktail with increased host range. Foodborne Pathog Dis 7:1415–1419

Chapter 13
The Role of Biodiversity Centres in Bioprospecting: A Case Study from Sarawak

Julia Shaw

Abstract Sarawak is a mega-biodiverse region of Malaysia, with huge potential for bioprospecting activities. The Sarawak Biodiversity Centre, has been involved in research and development of natural products from the island, whilst paying full attention to Access and Benefit Sharing issues. The development of LitSara^{R} is one of their model successes.

13.1 Introduction

Sarawak, the large island part of Malaysia, has a stunningly unique, eclectic essence of natural and cultural diversity, with traditional wisdom and awareness. A fact appreciated far afield, despite the growing global pressures and momentum of globalization and environmental change. This realization has led to the establishment of the Sarawak Biodiversity Centre (SBC), Sarawak, Malaysia (http://www.sbc.org.my/) whose mission is: "*To decode biodiversity for the benefit of mankind*".

The Sarawak State Government enacted the Sarawak Biodiversity Centre Ordinance in 1997 and established SBC the following year to initiate programmes for the conservation, utilization, protection and sustainable development of biodiversity in the State. The State Government also enacted the Sarawak Biodiversity Regulations in 1998. Initially, the Centre's role was primarily inventory and regulatory – often known as the gatekeeper to Sarawak's rich biodiversity to those who wanted access to, and collection of, biological resources in the State for research or commercial purposes.

The State took cognizance of the potential for intensive biotechnology-based research and product development from this vault of diverse biological resources until 2003. However, in December 2003, the State Legislative Assembly passed the SBC (Amendment) Ordinance (2003) and passed the revision of the Sarawak Biodiversity Regulations in 2004. The amendment relieved the Centre of its role of

J. Shaw (✉)
MANTA, 108 Huynh Thuc Khang, Mui Ne, Binh Thuan, Vietnam
e-mail: juliashaw2@gmail.com

R. Paterson, N. Lima (eds.), *Bioprospecting*, Topics in Biodiversity and Conservation 16, DOI 10.1007/978-3-319-47935-4_13

conducting general biodiversity inventory and regulating general biodiversity research. SBC is now entrusted to initiate intensive biotech-based research and development on the State's biological resources – particularly those that have been utilized by indigenous communities and to facilitate the documentation of the fast disappearing traditional knowledge of indigenous communities on the utilization of biological resources. In 2014, the State Legislative Assembly passed the Sarawak Biodiversity Ordinance (Amendment) 2014 to strengthen rules and procedures on Access and Benefit Sharing (ABS).

The benefits of conserving Sarawak's bioheritage are clear: The richer the diversity of life, the greater the opportunity for biodiscovery, economic development and adaptive (survival) responses to challenges such as climate change (including the capacity for carbon dioxide extraction to help counteract temperature and sea levels rising). Also, the forests could be sources of new pharmaceuticals providing treatments for the rising resistance to infectious diseases, a situation potentially as alarming as climate change.

13.2 The Work of the Sarawak Biodiversity Centre

Three core functions of SBC are relevant:

- Traditional knowledge from a wealth of sources, including five local communities in the highlands, the Bidayuh of Kiding, LunBawang of Long Semadoh and Kelabit of Bario;
- Biodiscovery Research and Development, leading eventually to commercialization and the market;
- Implementing ABS in line with the 2010 Nagoya Protocol, discussed frequently elsewhere in this current book.

The SBC Ordinance falls in line with the final stages of Malaysia's draft law on ABS to address all three objectives of the 1992 Rio Convention on Biodiversity for conservation and sustainable use of biodiversity, including the Nagoya Protocol. These objectives are: (i) the conservation of biological diversity, (ii) the sustainable use of its components and (iii) the fair and equitable sharing of the benefits arising out of the utilization of genetic resources, including by appropriate access to genetic resources and by the appropriate transfer of relevant technologies, taking into account all rights over those resources and to technologies, and by appropriate funding.

It is a joy to see glimpses of the SBC team's long standing ethical diligence reflected in constructive outcomes throughout their core programs including:

An ongoing programme to facilitate documentation of traditional knowledge among Sarawak's diverse indigenous communities. Since 2001, in addition to literature reviews, SBC has been working with 16 indigenous communities at 90 locations across the state. Of these, 62 locations are now carrying out documentation work and activities to conserve the knowledge involved about how locals use plants.

Fig. 13.1 Sarawak's national orchid, *Phalaenopsis bellina,* growing in SBC's propagation shed

Some 5500 plants have been documented to date and more than 1200 species identified. In addition, a natural product library consisting of extracts from plants (including algae), and other microbes. Essential oils, pure compounds, enzymes, and other derivatives are studied.

- An ethnobotanical garden growing a selection of specific plants used by Sarawak's different indigenous cultures.
- Isolation and propagation of specific fragrance and essential oil producing plants, including the scent capture, analysis and reproduction of Sarawak's national orchid, *Phalaenopsis bellina*, an endemic moth orchid of Borneo (Fig. 13.1).
- Photos of local communities, after SBC training, helping to isolate essential oils, in the highlands, in particular LitSaraR, a showcase endeavor, ensuring indigenous communities play a fundamental role in adding value to their traditional knowledge.
- Awareness and appreciation-raising tours for schools, institutions of higher learning, governmental and non-governmental organisations, associations and other specific interest groups.
- Awe inspiring talks by researchers, most recently by the Godfather of Biodiversity, Dr Thomas E. Lovejoy (who introduced the term Biological Diversity to scientists in 1980), on: "A Wild Solution for Climate Change", a public talk given in Kuching.

13.3 The Story of LitSara^{R}

Relevant is the story of Litsara^{R}, a lemony, citral-eucalyptol-based essential oil, trademarked for Sarawak, and launched earlier this year in a range of hand-made personal care products without additives: The uses include, from antibacterial wet wipes to natural insect repellent, body washes, cleansing and uplifting oils, soaps and "sniff jars". Already SBC are undertaking research into the antibacterial and anti-insecticidal activity and attempting to isolate the active components for potential commercialization.

In 2005, the distinct scent of *Litsea cubeba*, in hilly forests, inspired sharing of traditional knowledge, between a team of highland communities working closely with SBC. Ethnobotanical conservation techniques were engaged: From documenting the uses of the tree, standing 5–12 m tall, to sustainable harvesting of the fruits and leaves. Although *L. cubeba* has been studied in other laboratories, laboratory extraction and isolation identified the oil to have a unique composition, distinct from the same tree species in any other location, which helped the tree's protection as a Geographical Indicator. Complex and robust laboratory testing showed the essential oil, derived from the leaves and fruits of the tree, to have both antimicrobial and insect repellant properties, trademarked now as Litsara[R], in abbreviation of its origins: *Litsea cubeba* of Sarawak (see: www.borneotalk.com/download.php).

Importantly, the Sarawak government has incorporated provisions for ABS in the SBC's Ordinance to ensure indigenous communities play a pivotal role in adding value to their traditional knowledge and local biodiversity and are duly acknowledged through recognition and fair and equitable sharing of benefits arising from product commercialisation.

The deep cleansing body wash and oils work well, as reported by the current author! Immediately the washing water darkens, but without stripping natural body oils, and means a shower before dusk, resists mosquito arrival. Finally, restoring deep respect and connection to the natural biodiversity of Sarawak with gratitude to centuries of tried and tested traditional knowledge, enhances the Quality of Life for all.

13.4 Biography of the Author

Julia Shaw's father, Datuk George Anthony Theodore Shaw Esq. CBE, was the First State Secretary of Sarawak, in the 1960s, who distilled a keen interest of the island region and its people in Julia. She has a BSc Hons (Botany) from Exeter University, UK and an MPhil on the Cyperaceae: Taxonomy, Geography, Ecology from the University of Hong Kong (HK). Julia has worked with governments, NGOs, universities and other institutions, regionally and internationally, including the Royal Botanic Gardens, Kew and consultancies for the HK government on wetlands and wetland flora. Julia established her own environment consultancy in HK (1998–2002), then Wildlife at Risk in Vietnam for Dominic Scriven Esq. OBE (2003–2009), including WAR-marine (2006–2009). Marine conservation focused on sail-training fishermen, by establishing Vietnam's first (and only) sailing school at the MANTA Sail Training Centre – a not for profit organisation (sailing skills were passed to Julia, aged 4, by her mother). The work of SBC is close to Julia's heart, hence she was inspired to write this chapter.

Acknowledgement Julia Shaw is grateful for the cooperation of the Sarawak Biodiversity Centre, and particularly Asha Kaushal, for reviewing this chapter and agreeing to its publication.

Chapter 14
Bioprospecting Insights

Russell Paterson and Nelson Lima

Abstract Insights into bioprospecting are provided by the editors of the current book based on personal experience and the reasons why bioprospecting potential has not been fulfilled are considered. The requirement to change current paradigms are emphasized. Some interesting developments on using investigators' immediate environments for sampling are discussed. A new UK funded government report covering resistant bacterial infections is an important development.

Bioprospecting promises but does not fulfil. We attempt to explain this statement in the current, prescient book, while giving examples of areas that show great promise. We emphasise that the situation has changed radically because of resistant bacterial infections tipping the balance towards bioprospecting and novel antibiotics.

The field is defined as the systematic search for, and development of, novel sources of chemical compounds, genes, microorganisms, macro-organisms, and other valuable products from nature: Bioprospecting means looking for ways to commercialize biodiversity (WHO 2001). The concept has been discussed for decades and has a mixed history, inexorably linked to natural product screening as undertaken by pharmaceutical accompanies, but can be related to agriculture, food or cosmetics. There is almost always an element of collecting "genetic resources" often from biodiverse areas of the world. Normally there are more than one organization or entity involved e.g. (a) industry, (b) academic organization and governments of countries and (c) representatives of indigenous people from where the genetic resources are taken. These often have very different interests. Industry is concerned predominately with profits, agencies by increasing knowledge from science, governments by political objectives and indigenous people by monetary reward - financially or in kind (e.g. better living conditions). These objectives can be mutually exclusive.

We need to differentiate between a biodiversity inventory and an inventory for bioprospecting: these are different but not irrevocably. Much effort and time may be expended in collecting and identifying samples taxonomically that are representa-

R. Paterson (✉) • N. Lima
Centre for Biological Engineering, University of Minho, Braga, Portugal
e-mail: russell.paterson@deb.uminho.pt

R. Paterson, N. Lima (eds.), *Bioprospecting*, Topics in Biodiversity and Conservation 16, DOI 10.1007/978-3-319-47935-4_14

tive of a particular region which may result in little in terms of exploitable properties, in an exercise where the chances of producing a successful product are already low. Hence, some sort of pre-screening is a better approach: chromatographic analysis or a rapid bioassay for bioprospecting may be more useful than identifying a sample to species. Searching in the tropics where organisms face a great deal of competition to survive, is almost certainly better that collecting from the Arctic, where competition may be low. Organisms might produce toxic secondary metabolites to control the competitors especially in the tropics and these are a source of pharmaceuticals for controlling human diseases. This form of competition would not exist to the same extent in the Arctic, although low temperature enzymes could be produced, which may be marketable. There are fascinating organism present that survive these conditions, but they may not be useful for bioprospecting.

The internet can be a source of false optimism and there are examples of the bizarre situation where promising compounds are not developed to market because they are too cheap to be profitable. A so-called miraculous treatment for cancer is reported which is based on an old report resurfacing on the blogosphere (Coghlan 2011). There was a description of a new treatment that showed promise of tackling most types of human cancer but in animal experiments. The drug was dichloroacetate (DCA), which blocked glycolysis in cancer cells and is a cheap, simple molecule that no-one has patented. Within weeks, patients were trying to get their own supplies, and entrepreneurs set up websites to sell it, that were subsequently declared illegal and closed down by the US Food and Drug Administration. It was finally tested in five patients with aggressive brain cancer by Prof Michelakis, University of Alberta in Edmonton, Canada, who had conducted the original animal experiments. The published results revealed that it probably extended the lives of four of the patients. Importantly, it was said that it would be impossible to tell whether DCA works until it is tested in a placebo-controlled trial (Coghlan 2011). This example is provided to indicate some of the potential problems that could be encountered towards the end of a bioprospecting project, even when active compounds may be discovered. In addition, it is unlikely that such a chemical would be marketed as it is simple, readily available and cheap.

Stiglitz (2006) mentions, "The establishment of the World Trade Organization … imposed US style intellectual property rights around the world. These rights were intended to reduce access to generic medicines and they succeeded. Developing countries paid a high price for this agreement. But what have they received in return? Drug companies spend more on advertising and marketing than on research, more on research on lifestyle drugs than on life saving drugs, and almost nothing on diseases that affect developing countries only. This is not surprising. Poor people cannot afford drugs, and drug companies make investments that yield the highest returns. The chief executive of Novartis, a drug company with a history of social responsibility said, "We have no model which would [meet] the need for new drugs in a sustainable way … You can't expect for-profit organizations to do this on a large scale." "

It should be stated that the pharmaceutical success in bringing new drugs to market has been dismal. Many of the companies have retreated from natural products in favour of combinatorial chemistry to produce chemical leads. But this has been without ultimate success. Predictably, the lack of investment has become critical with the development of drug resistant bacteria and where there is nothing to replace the existing and last-resort antibiotics.

The causes of antibiotic resistance are complex and include human behaviour at many levels of society: the consequences affect everybody. Similarities with the scope of climate change are evident. The popular media are beginning to realise the extent of the problem and the reasons why more has not been done to solve it in terms of investing in natural product discovery. The British Broadcasting Cooperation (2015) programme, "Antibiotic Apocalypse" was particularly revealing: The reason why drug companies do not invest is because the new drug would be used sparingly from potential resistance as mentioned by Patrick Vallance, President R and D, Glaxo Smith Kline. For example, it could be placed on the shelf for years by which time the patent would run out and so there is no incentive. The resistance problem was highlighted when it was revealed that GSK were close to having a new drug, but resistance immediately developed at the final stages of development. A solution may be for governments to pay drug companies to develop drugs as the problem is too big to fail (see the much anticipated O'Neill (2016) report).

Other approaches were described in the Guardian (2016) newspaper, which states that a possible solution to one of the world's most pressing problems might be lurking in the back garden of Dr Adam Roberts, University College of London! It is a crowd-sourced search for the next antibiotics, where Dr Roberts asks the public to sample their environment and post the swabs back to his laboratory. A "heroic" PhD student analysed hundreds of swabs posted from "all four corners of the world" – a very tedious and repetitive job. However, the student is finding a large biodiversity, although one wonders about the IPR arrangements of this project in terms of the Nagoya protocols. Isolates were challenged against a bacterium and activity was assessed as a zone of inhibition. For any novel microorganism they find, the team attempts to identify the antibiotic and if it is undiscovered, becomes an addition to the dwindling arsenal of compounds. How this would be commercialized remains to be seen.

As the article states, the vast majority of microbes growing in the environment are still unknown to science because they do not grow well under laboratory conditions (and presumably that few environments have actually been sampled given the diversity of niches that exists). The article mentions the compound aspergillomarasmine from *Aspergillus versicolor,* which was recently reported to overcome resistances of bacteria to conventional antibiotics and yet was previously little-known as a toxin involved in barley disease caused by the fungus *Pyrenophora teres.* This indicates various alternative approaches to novel drug discovery in that innovative ways of overcoming resistance are provided rather that a new drug that killed the pathogen *per se*, and the possibility of discovering new activities from chemicals that have already been discovered and have known activities. However, whether pharmaceutical companies would be interesting to develop already-discovered compounds is doubtful.

The British Broadcasting Cooperation (2015) Panorama programme mentioned previously was revealing when it discussed Prof Slava Epstein's, (North-eastern University, Boston) work of digging up soil samples in Boston and testing the activity using a so-called "Ichip" (Nichols et al. 2010). Teixobactin was discovered from this work. These innovative approaches are welcome, although getting these compounds to market may require some additionally innovative thinking. Finally, why would a pharmaceutical company develop these compounds further if they are only going to be used as drugs of last resort and not realising their full economic potential? A new paradigm is required given the current emergency of resistance, where governments should subsidise these and similar projects (see O'Neill (2016).

Nevertheless, there are numerous scientific papers highlighting activities of high pharmaceutical relevance from a wide range of organisms. Hence, this type of work is ongoing, but the results may have few financial implications in terms of marketable products, because the activity is already revealed and so would be difficult to patent. Also the organism or techniques used are known. The industry needs something over which they can have the full commercial rights, although again, O'Neill (2016) has presented a new paradigm to solve the impasse.

Ideologically, many drug companies support the position of less government involvement, yet in the developing world in particular, diseases and illnesses affect the poorest who cannot afford expensive treatments. In the past decade or so, pharmaceutical companies have therefore also been criticized for ignoring this "market" because they cannot pay (Shah 2010). Shah's website is an interesting critique of the pharmaceutical industry's attitude to funding research and development in the area. However this position may no longer be tenable as everyone is susceptible to resistant infectious diseases, rich and poor, and government support is essential. It is too important to fail as mentioned. In the O'Neill (2016) report, which was funded by the UK government, a most welcome recommendation is that a billions-of-dollars fund be established to (a) improve the numbers, pay and recognition of people working on infectious disease, (b) establish a Global Innovation Fund for early-stage and non-commercial research and (c) provide better incentives for discovering new drugs and improving existing ones. Funds should also be directed towards improving the successful implementation of bioprospecting projects, in the present authors'opinion.

One of the current authors (RRMP), has been involved in two bioprospecting projects which provide insights into the issues involved: First, the Iwokrama project was an EU funding project to isolate fungi from the Iwokrama rainforest, Guyana. This project had bioinventary and bioprospecting components. However, the funding was not renewed after the initial 3 years and no marketable products were obtained; this and surrounding issues are fully described in a chapter in the current book (Pingle 2016). Second, he undertook screening work during his employment at CABI Bioscience, UK (a not-for-profit, non-governmental organisation), involving bioassays of insect pests with extracts of fungi, some of which were entomopathogens (insect pathogenic fungi), and which had been isolated from biodiverse regions of the world. Much time was spent attempting to attract agrochemical companies to fund the work. The best option achieved was that a company paid slightly more than the catalogue price for strains which had activity in the tests used in his work. A moderate increase in payment was agreed if the activities were confirmed in the company's tests. Finally, a small percentage royalty payment would be made if a product was ever marketed, although this was not achieved. An internal report was produced which detailed the biochemical analyses undertaken and with the screening results against insect pests. Another issue was that when an academic organization becomes involved in a bioprospecting project with a pharmaceutical company there are low prospects of the academic scientists being able to publish their results, as they will almost certainly be under confidentiality agreements. This could be detrimental to scientists' careers and some other reward is desirable, such as a higher salary, or perhaps greater job security, as suggested by O'Neill (2016). This information is provided to illustrate that the good intentions of bioprospecting projects may not be rewarded with large amounts of funds and that a great deal of effort and expense is often required with only a remote chance of a good return.

Finally, the path towards successful natural products development is paved with many good intentions. However, one of the most encouraging innovations is the report by O'Neill (2016), the recommendations of which should be implemented and applied towards bioprospecting.

References

British Broadcasting Cooperation, 2015 Antibiotic apocalypse, Panorama (18th May, 2015). http://www.bbc.com/news/health-32722411. Accessed 23 June 2016

Coghlan, A., 2011. Cancer drug resurfaces and threatens false optimism. New Scientist, 16 May

Guardian, 2016 "Fighting bugs with bugs: this time it's personal" 15 May 2016. https://www.theguardian.com/science/occams-corner/gallery/2015/may/15/fighting-bugs-with-bugs-this-time-its-personal. Accessed 10 June 2016

Nichols D, Cahoon N, Trakhtenberg EM, Pham L, Mehta A, Belanger A, Kanigan T, Lewis K, Epstein SS (2010) Use of Ichip for high-throughput in situ cultivation of "uncultivable" microbial species. Appl Environ Microbiol 76:2445–2450

O'Neill, J., 2016. Tackling drug-resistant infections globally: Final report and recommendations. Welcome Trust and HM Government. pp 81. http://amr-review.org/sites/default/files/160518_Final%20paper_with%20cover.pdf. Accessed 23 June 2016

Pingle, R., 2016 In press. Iwokrama fungal/plant biopropsecting project 2000-2003 – a model for the future? In Paterson, R.R.M., Lima, N. (Eds.) Bioprospecting. Success, potential and constraints. Springer.

Shah, A., 2010. Pharmaceutical corporations and medical research. http://www.globalissues.org/article/52/pharmaceutical-corporations-and-medical-research. Accessed 22 June 2016

Stiglitz J (2006) Scrooge and intellectual property rights. Br Med J 333:1279–1280

WHO 2001. Essential medicines and health products information portal. A World Health Organisation resource. http://apps.who.int/medicinedocs/en/d/Jh2996e/6.3.html. Accessed 23 June 2016

MIX
Papier aus verantwortungsvollen Quellen
Paper from responsible sources
FSC® C105338

FSC
www.fsc.org

If you have any concerns about our products,
you can contact us on
ProductSafety@springernature.com

In case Publisher is established outside the EU,
the EU authorized representative is:
Springer Nature Customer Service Center GmbH
Europaplatz 3, 69115 Heidelberg, Germany

Printed by Libri Plureos GmbH
in Hamburg, Germany